Organic Synthesis, Reactions and Mechanisms

Of Propellanes — and of Spirans

D. Ginsburg

A New Treatment of Chemical Reactivity: Development of EROS, an Expert System for Reaction Prediction and Synthesis Design

J. Gasteiger, M. G. Hutchings, B. Christoph, L. Gann, Ch. Hiller, P. Löw, M. Marsili, H. Saller and K. Yuki

Transition-metal Catalyzed Decomposition of Aliphatic Diazo Compounds — New Results and Applications in Organic Synthesis

G. Maas

THEORIA CUM PRAXI

AKADEMIE-VERLAG BERLIN

Organic Synthesis, Reactions and Mechanisms

Organic Synthesis, Reactions and Mechanisms

With 33 Figures and 26 Tables

AKADEMIE-VERLAG BERLIN
1987

Die Originalausgabe erscheint im Springer-Verlag
Berlin · Heidelberg · New York · London · Paris · Tokyo
in der Schriftenreihe "Topics in Current Chemistry",
Volume 137

Vertriebsrechte für die sozialistischen Länder:
Akademie-Verlag Berlin
Vertriebsrechte für alle Staaten mit Ausnahme der sozialistischen Länder:
Springer-Verlag Berlin · Heidelberg · New York · London · Paris · Tokyo

ISBN 3-540-16904-0 Springer-Verlag Berlin · Heidelberg · New York
ISBN 0-387-16904-0 Springer-Verlag New York · Heidelberg · Berlin

ISBN 3-05-500343-8 Akademie-Verlag Berlin

Erschienen im Akademie-Verlag Berlin, DDR-1086 Berlin, Leipziger Straße 3—4
Lizenznummer: 202 · 100/507/87
Printed in the German Democratic Republic
Gesamtherstellung: VEB Druckerei „Thomas Müntzer", 5820 Bad Langensalza
LSV 1275
Bestellnummer: 763 717 2 (3072/137)

13800

Editorial Board

Table of Contents

Of Propellanes — and Of Spirans[1]

David Ginsburg

Department of Chemistry, Technion — Israel Institute of Technology, Haifa, Israel

'The time has come,' the Walrus said,
'To talk of many things;
Of shoes — and ships — and sealing wax —
Of cabbages — and kings —'
Lewis Carroll, "The Walrus and the Carpenter"

Table of Contents

The synthesis of propellanes and spirans is reviewed, attempting to explain why a starting material sometimes yields a member of one class or the other, but apparently not a mixture of both.

1 Introduction

I should like to attempt to explain why a starting material that may apparently afford a propellane and/or a dispiran sometimes gives one or another, apparently not a mixture of both. There does not appear to be a denominator common for all the cases to be discussed but perhaps discussion of cases pertaining both to carbocyclic and heterocyclic compounds may cast some light on the problem; there need not be a unique reason for the behavior in the two series.

To emphasize this statement I should like to begin with a heterocyclic case which provides an outlet for the molecule's behavior that leads to neither propellane or spiran albeit, on paper, both of these types might be expected to form. In Mülheim/Ruhr the following reaction was studied:

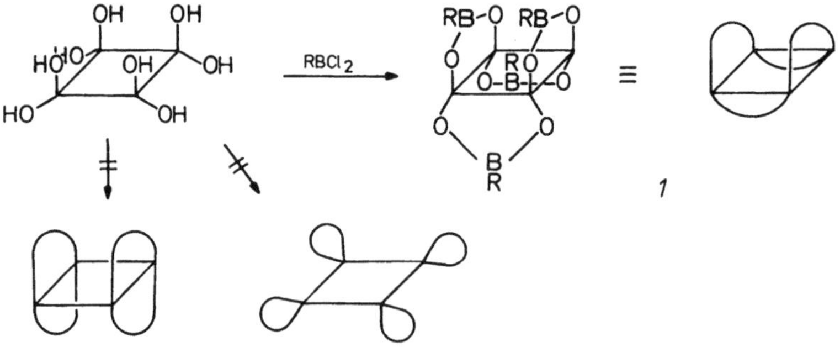

1

2 3

The product *1* is formed exclusively [2]. No propellane is formed. An explanation has been given by a group interested more in propellanes than in spirans and therefore considered only the relative stability between *1* and *2* [3].

MNDO calculations indicate that *1*, rather than *2*, forms because of repulsion between lone-pairs on proximate oxygens which would occur in the propellanes but not in compounds of type *1* [3]. Compunds *1* and *3* were not compared. In the two systems which were compared, the hetero-rings are five-membered whilst in *3* they would be four-membered, perhaps sufficient reason without further ado to ignore the importance of *3* being potentially formed. The simplistic argument with respect to relatively greater strain in 4-membered rings may not be the only consideration, however. Here too electrostatic interactions between oxygens in the spiro-rings might be of even greater weight.

Often there are two (or more) courses that a reaction may take, say, cyclization of a common starting material (under whatever conditions that are being used) leading to several possible products. In this chapter, I want to discuss just such a case where cyclization of a generalized 1,1,2,2-tetrasubstituted ring *4* may lead to a propellane *5* and/or to a dispiran *6*.

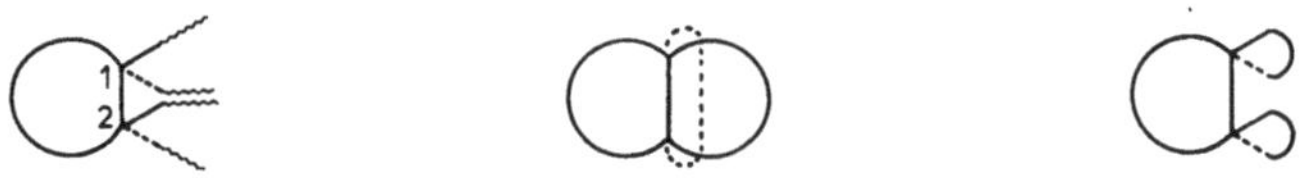

4 5 6

It may be useful either at the outset or post factum to use molecular mechanics to calculate which of these products may be the more stable. A priori there is no way to tell whether either of the two products is the generally preferred one. In the present case such calculations do not appear to have been carried out at the outset; not surprisingly. This sort of thing was not done at the time the work was conducted. I know of only one very recent paper in which such calculations appear, comparing intermediates between a dispiran with its isomeric propellane (see below [40]).

Although we are dealing with work described in the literature by means of a posteriori molecular calculation it is useful to see the relative calculated heats of formation of the isomeric propellanes and dispirans and note particularly the right-most columns so as to be in a position to gauge these against the experimental results or vice versa.

Amnon Stanger [3b] has kindly calculated the heats of formation H_f and strain energies E, of the two sets of isomers shown in the Table.

Table 1. (Program MM2, QCPE No. 395)

Compound	ΔH_f (kcal/mol)	Strain Energy, E	$\Delta\Delta H_f$	ΔE
	chair 9.0	50.7		
	boat 13.2	54.9		
			7.5	9.7
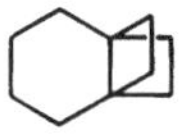	chair 16.5	60.4		
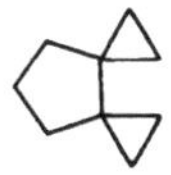	20.4	56.3		
			3.1	4.4
	23.5	60.7		

Thus, the experimental results follow in the path of the calculations.

Incidentally, PE spectra of these dispirans in addition to several others, have been reported [3c].

2 Carbocyclic Propellanes and Dispirans

Let us now turn to simpler systems in which there are two, not three, structural alternatives for the potential product. Buchta and his collaborators published many papers on the preparation of spiro-compounds [4]. His work stemmed primarily from his greater interest in these [4] rather than in the propellane by-products obtained in certain cases [4d–g]. The case was reversed for our group. Propellanes were paramount but sometimes spiro-compounds were obtained [5]. First we shall discuss a number of carbocyclic examples.

At the time our work was done (1968) very few "small-ring" propellanes were known [5]. Nevertheless it was clear that these molecules would be strained and that the smaller the rings the more care must be exercised in choosing the reaction conditions for the last step(s) in their synthesis; the limiting cases certainly wouldn't be formed under relatively stringent reaction conditions.

A propellane containing, say, a six-membered ring and two four-membered ones (a [4.2.2]propellane) would presumably be easier to prepare than one with three four-membered rings. This assumption was proved amply true when the time came (1971) and a [4.2.2]propellane derivative was used to prepare compounds successively containing the [3.2.2] and the [2.2.2]propellane skeleton [6].

We tried in 1968, as it turned out, unsuccessfully, to prepare a propellane having the [4.2.2] nucleus (actually it was a [4.2.2]propellene). But we were most successful in preparing the isomeric dispiran [7]. None of the desired propellane was formed (see reaction scheme):

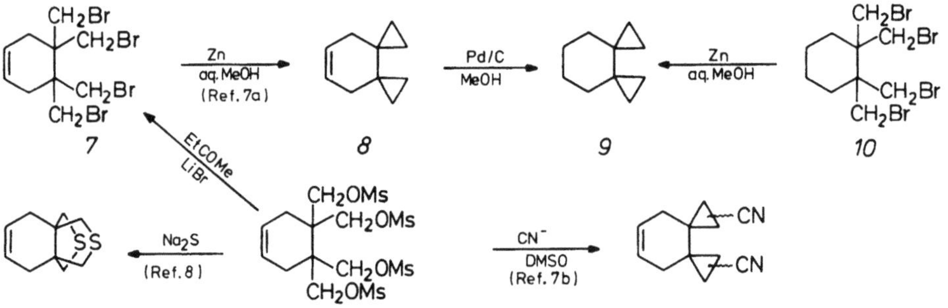

12 11 13: 4 of 6 possible isomers

The structures of the 4 isomeric nitriles listed in the above reaction scheme were determined by NMR spectroscopy and dipole moment measurements [7b].

The tetramesylate *11* used had been reported previously[4e,8]. We shall see below that it nonetheless is a useful intermediate in synthesis of propellanes [8].

The dispiro[2.4.2.0]dec-5-ene was later used to prepare the compound with a conjugated diene in the six-membered ring [9]. There is no sign of rearrangement under the conditions used.

8 15

Buchta and his coworkers have contributed to the problem we are discussing. The tetraester of the [3.3.1]propellane shown, *17*, is formed "überraschenderweise" from 1,1,2,2-tetrakis-hydroxymethylcyclopropane tetramesylate *16* and sodio-diethyl malonate, in 53% yield [4f].

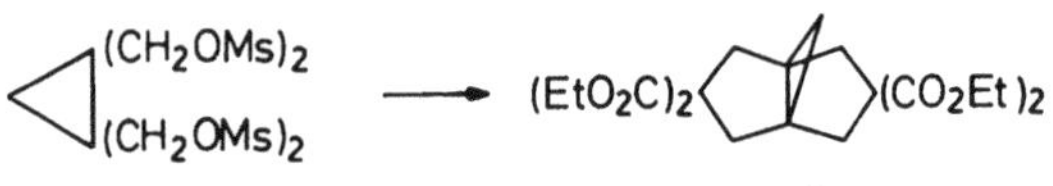

16 17

Wherefore "überraschenderweise"? For the corresponding cyclobutane-1,1,2,2-tetramesylate *18* gave the homologous [3.3.2]propellane tetramethyl ester *19* in only 0.5% yield, the major product being 7-oxa[3.3.2] propellane-3,3-diethyl ester *20* [4d].

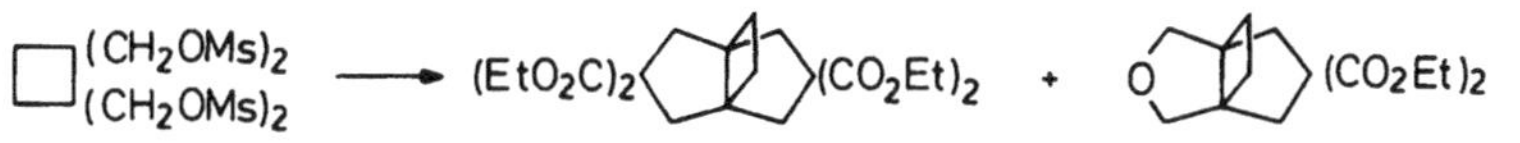

18 *19*, 0.5% *20*, main product

Dispiro[2.2.2.0]octane *22* was obtained in good yield along with a fragmentation product *23* by treating 1,1,2,2-tetrakis-bromomethyl-cyclobutane *21* with zinc dust in aqueous ethanol [4g]. The cyclobutene analog of *22* has also been reported [4h].

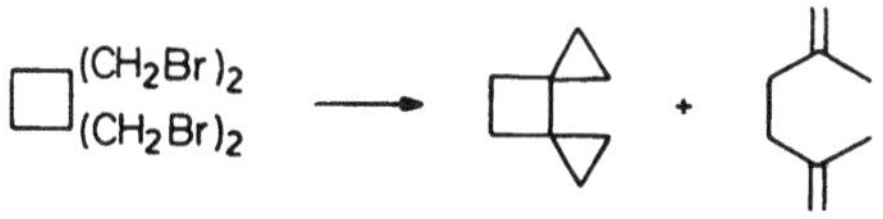

5.5 parts 1 part

21 *22* *23*

A Wurtz reaction, using sodium, of the same tetrabromide gave a wealth of fragmentation products whose formation may be reasonably explained mechanistically [4g].

When the cyclopropane homolog *24*, with the same 1,1,2,2-tetrabromide array, was treated with zinc, the analogous fragmentation reaction occurred, leading in this case to an 84% yield of 2,4-dimethyl-penta-1,4-diene, *25* [4g]. This is in contradistinction to the reaction of the corresponding tetramesylate *16* with sodio-diethyl malonate [4f], (vide supra).

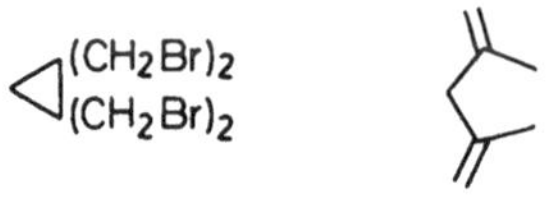

24 *25*

In the abovementioned cases preparation of the propellanes was direct. A very nice instance exists, however, of rearrangement of a dispiran, *26*, to Dewar benzenes *27* which happen to be [n.2.2]propelladienes. Silver ion (silver perchlorate at —20 °C) promotes the isomerization, as shown [10].

26 *27* *26b* *27b*

a, n = 4
b, n = 3
c, n = 5

Various Dewar benzenes of type *27* are formed depending upon the size of the alicyclic ring.

To be quite formal about the connection between spirans and propellanes an interesting pathway may be cited, although admittedly, it is farfetched.

The spiroketone *28* was converted into its tosylhydrazone whose sodium salt *29* was heated without solvent in a high vacuum. The bicyclic olefin *30* was formed and being a cyclobutene, Woodward and Hoffmann allowed it to ring-open to afford *31* and these two products were collected in a trap cooled by liquid nitrogen. When the mixture was permitted to warm up, an exothermic reaction (again allowed by W & H) set in and the cyclobuteno[4.2.2]propellane *32* was formed [11].

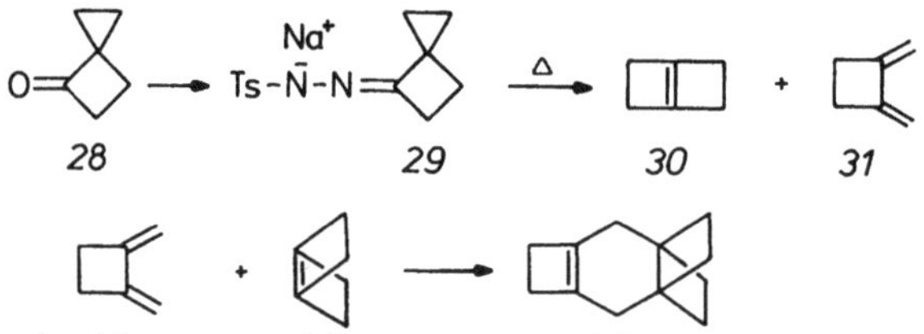

28 29 30 31

31 30 32

3 Heterocyclic Propellanes and Dispirans

Let us now discuss heterocyclic propellanes and dispirans. 8,11-Dioxa[4.3.3]propell-3-ene *34* was prepared (in 73% yield) by heating the tetrol *33* with $KHSO_4$ at 190–200 °C [4e]. This was accompanied by the bicyclic ether *35* (10% yield) but no

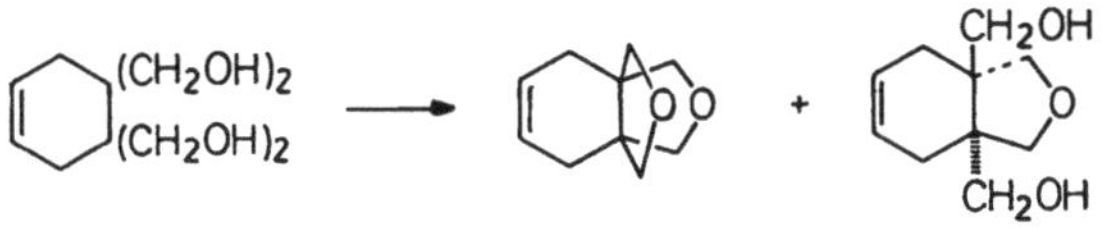

33 34 35

mention was made of any accompanying olefinic dispiran containing two oxetan rings. A saturated isomeric dispiran analog was prepared by another route [4d]. Thus, rather than form a dispiran a *trans*-fused bicyclic product is preferred.

Again, only oxa-propellanes, not dispirans, were formed when 1,1,2,2-cyclobutane derivatives were used as starting materials [4d]. The tetratosylate *37* was formed by esterification of the corresponding tetrol *36* with *p*-TsOH accompanied by the bicyclic

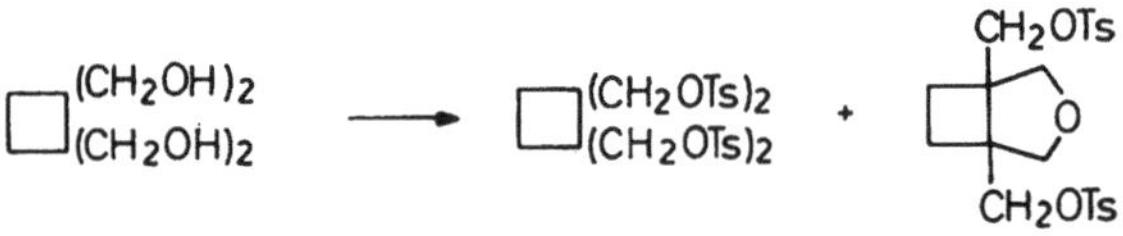

36 37 38

ether ditosylate *38*. The tetratosylate obtained, when treated with sodio-malonic ester, did not give a [3.3.2]propellane tetraester (vide supra [4d]) but rather the propellane ether diester *39*. This product was also obtained when the bicyclic ether ditosylate *38* was treated with sodio-malonic ester [4d]. Treatment of the tetrabromide *21* with

39

sodio-malonic ester gave the bicyclic compound *40*. It gave after heating in methanol with KOH a salt which upon acidification gave the dicarboxylic acid *41*. At first

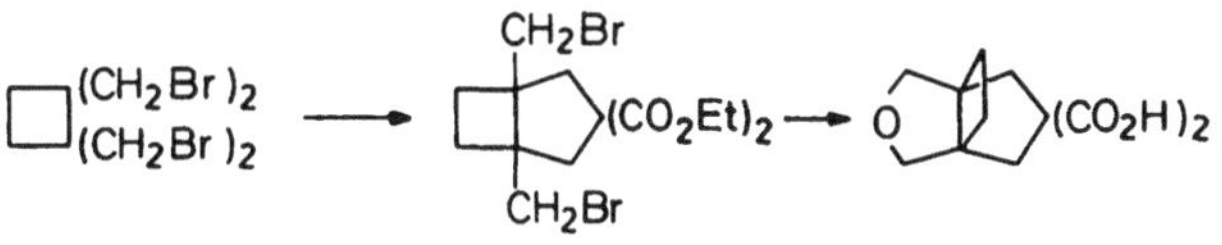

21 *40* *41*

glance the conditions leading to the oxa[3.3.2]propellane dicarboxylic acid may appear strange but though the reaction course is not explained one can rationalize it by stepwise nucleophilic attacks, involving one set of cis-disposed bromomethyl groups in each case:

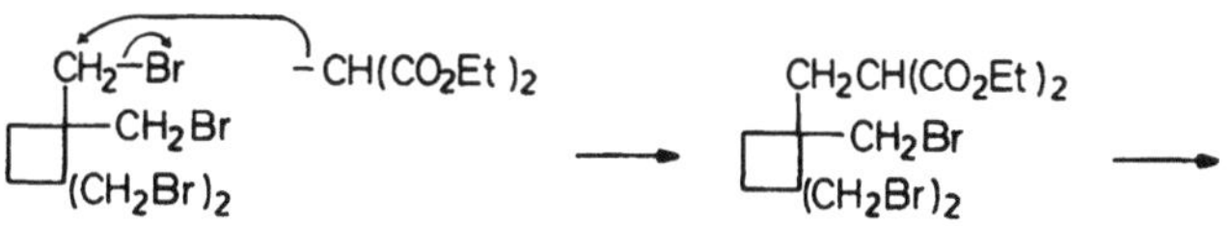

21

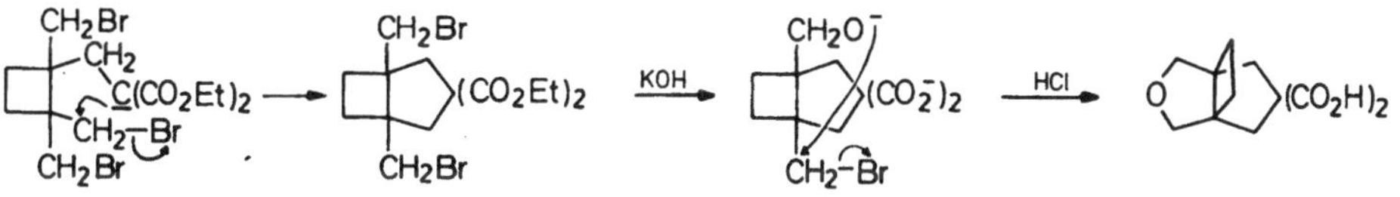

40 *41*

Both successive nucleophilic cyclizations to a ring are kosher because exo-nucleophilic attack is involved in each [12]. In any event one can just as easily write on paper, for example:

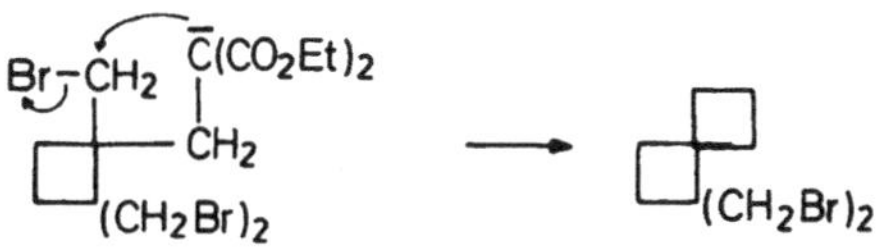

This clearly need not necessarily occur under nucleophilic conditions. But we note that in the above case, mesylate, a better leaving group is involved whilst in the present case bromide ion is the leaving group. Does one always get a propellane with the worse leaving group and a dispiran with the better one? No, we shall see plenty of examples of propellane formation with a mesylate or a tosylate leaving group (vide infra). But here too we note that the nucleophiles being compared in the various cases differ as do the substrates, as do the strain energies of the potential propellane or dispiran products. Many parameters differ and we must recall one of the teachings of a great man when he gave the famous course on natural products: "Never compare apples with pears". This advice is useful far beyond the field of chemistry and was in fact more generally intended (as I was later told explicitly over "tea").

There are many more syntheses of heterocyclic propellanes from 1,1,2,2-substituted carbocyclic starting materials. The tetrol discussed above, when treated with $KHSO_4$ at 170–190 °C affords the dioxa[3.3.2]propellane shown; no isomeric spiran is mentioned. Although the yield is only 50%; perhaps some dispiran is hiding in the "brauner Rückstand" from which the propellane diether is either crystallized at low

36 42

37 43

temperature or sublimed?[4d] Heating of the corresponding tetratosylate *37* with $Na_2S \cdot 9\,H_2O$ in ethanol affords the oxathia[3.3.2]propellane *43*[4d]. (It should be noted that in other cases dithioethers are obtained using analogous starting materials with the same sulfide.) Using the tetramesylate *16* instead of the tetratosylate *37* in dioxan/ethanol, albeit in the lower homolog (two parameters change: substrate and solvent) the dithia[3.3.1]propellane *44* is obtained in 66% yield. When DMSO is the solvent the yield of *44* rises to 72%[4f].

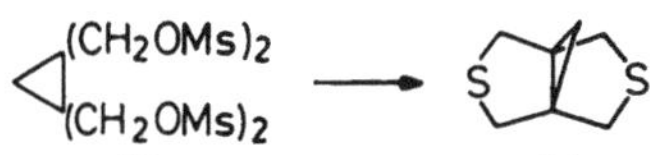

16 44

Returning to the higher member of the homologous series but nevertheless changing the sulfonate type and solvent as compared to the above case of the cyclobutane-1,1,2,2-tetratosylate *37* the corresponding tetramesylate *18* affords the dithioether *45* instead of the ether-thioether *43* (dioxan/ethanol; 77%).

18 45

The result common to all of these cases is that apparently no dispiran is formed under these conditions, both acidic ($KHSO_4$) and basic (Na_2S).

Starting from 1,1,2,2-substituted derivatives of tetrahydrofuran *46*, *47* and of thiophan, *48*, *49*, trioxa[3.3.3]propellane *50*, oxadithia[3.3.3]propellane *51*, dioxathia-[3.3.3]propellane *52* and trithia[3.3.3]propellane *53*, respectively, were obtained [8b,c]. No dispirans were detected!

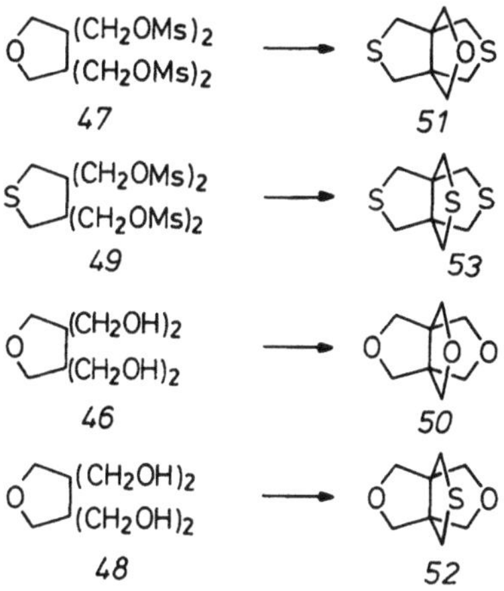

To the Buchta heterocycles the higher homologs must also be added. The cyclopentane-1,1,2,2-substituted tetrol *54* was cyclized, in this case heated rapidly with H_2SO_4 at 160–170°, to give the dioxa[3.3.3]propellane *55* in 74% yield, no dispiran by-product being mentioned here either [13].

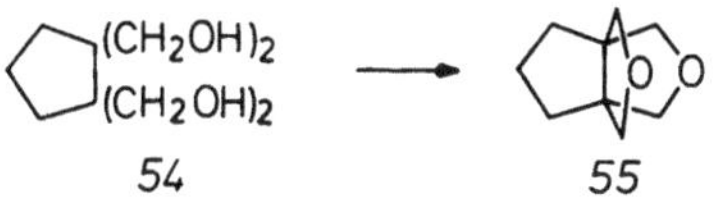

A somewhat different method than those described above led to dithia[3.3.3]propellane. When the same tetrol *54* was treated with *p*-TsCl in pyridine, the ditosylate *56* was formed at room temperature. At −5 °C the tetratosylate *57* was formed without formation of the five-membered ether ring [13].

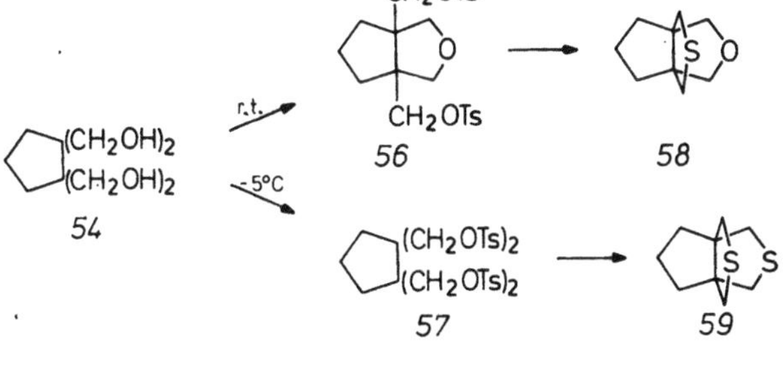

Sodium sulfide then gives products *58* and *59* of the hetero[3.3.3]propellane series; no dispirans are reported. The same approach was used for preparation of the corresponding oxa-thia and dithia[3.3.1] and [3.3.2]propellanes, some reported earlier by Buchta. Somewhat different reaction conditions have been published [14].

We turn now to the work of Jamrozik who has published several papers pertaining to our theme. A new and different parameter is involved. If until now we have suspected (but not proved) that a dispiran may form in lieu of a propellane with relatively small rings because of higher strain in the latter, now we must bring a set in which medium rings are involved. The relative difficulty in the formation of such rings as compared to 5- and 6-membered rings on the one hand, and the so-called large rings on the other, is too well known to require documentation.

When two equivalents of *trans*-1,2-dimercaptocyclohexane are heated with the tetrabromide *24* (a substrate used earlier by Buchta [4g]), again a priori there is a possibility that a propellane *60* will form, a dispiran *61* or a mixture of both [15].

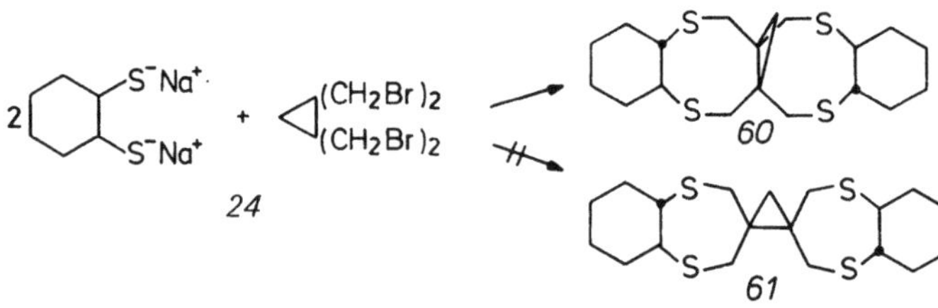

Jamrozik decided between the two on the basis of what he himself regards as a tenuous argument involving UV spectroscopy. The arguments based on NMR spectroscopy support the propellane structure shown, *60*, rather than the one with a cis-arrangement *62* of the abovementioned hydrogens:

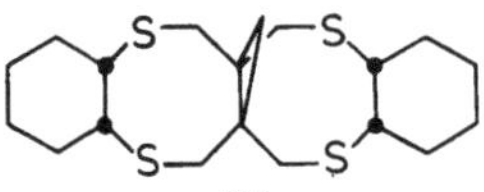

62

A mixture of the two is not obtained. Further mass spectral fragmentation did not show fragments based upon cleavage of the cyclopropane ring, a cleavage which is common for spirans containing such a ring, including that of the monospiran *63* (containing two sulfur atoms) shown [15].

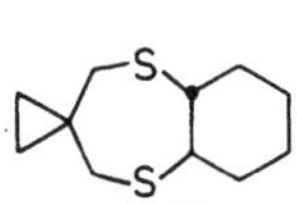

63

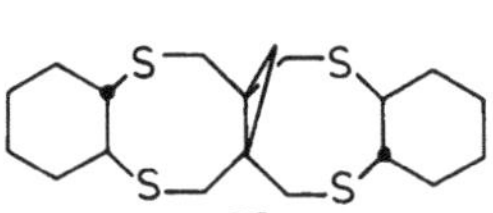

60

The yields of the tetrathia[6.6.1]propellane *60* and of the spirodithiabicyclo[5.4.0]-undecane *63* are only 25% and 18%, respectively. Perhaps this is the manifestation of the relative difficulty in formation of medium rings. It is not otherwise obvious why the yields should be so low. Of course propellanes with larger rings exist but the medium ring isn't cyclized at the bicyclic stage. It is already there [16].

In an analogous study similar arguments were used to show that a tetraoxa[7.7.1]-

propellane *64*, and again not a dispiran *65*, is formed from the same 1,1,2,2-tetrabromomethylcyclopropane and from 1,8-dihydroxynaphthalene [17].

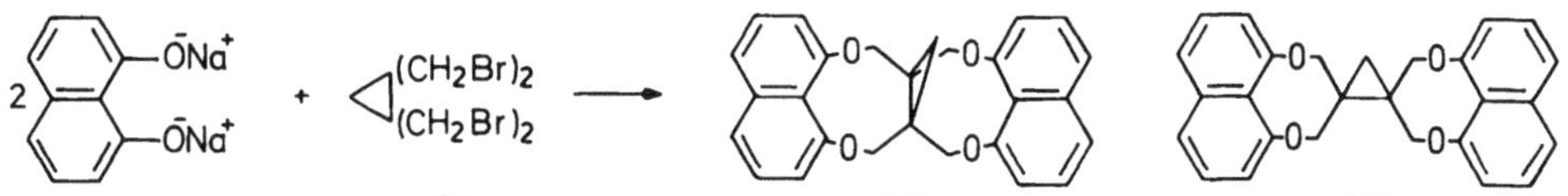

24 64 65

With catechol or with 2,3-dihydroxynaphthalene, dispirans of the [6.0.6.1]type are formed [18]:

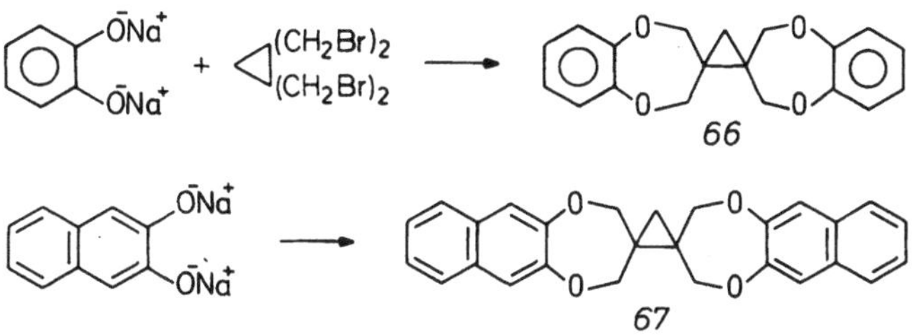

66

67

The author [17] explains the difference by involvong the "peri-effect" in 1,8-dihydroxynaphthalene.

Another similar case has been discussed. 2,2,3,3-Tetrahydroxymethyltetralin *68* was dehydrated by the method liked by Buchta, $KHSO_4$ at 170–190°, followed by sublimation. An ether *69* and a propellane diether *70* are formed in about equal amounts (4:5, respectively). Neither monospiran diol nor dispiran are formed [19].

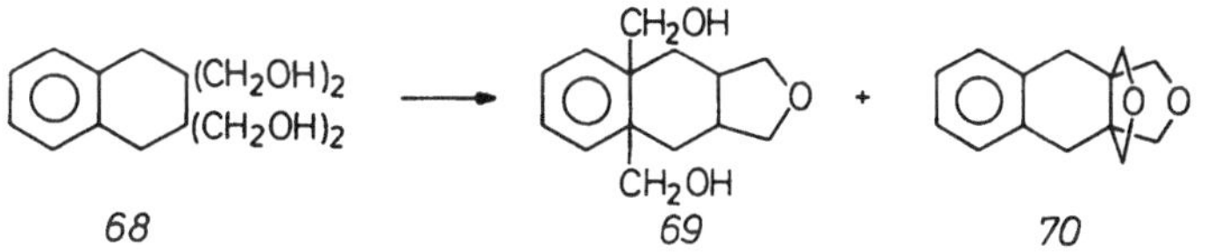

68 69 70

The twice aromatic analog *71* of the propellane *60* containing cyclohexane rings [15] (in which there are, of course, no bridgehead hydrogens), has been obtained from the tetrabromide and 1,2-benzenedithiol. Again no dispiran is formed [20].

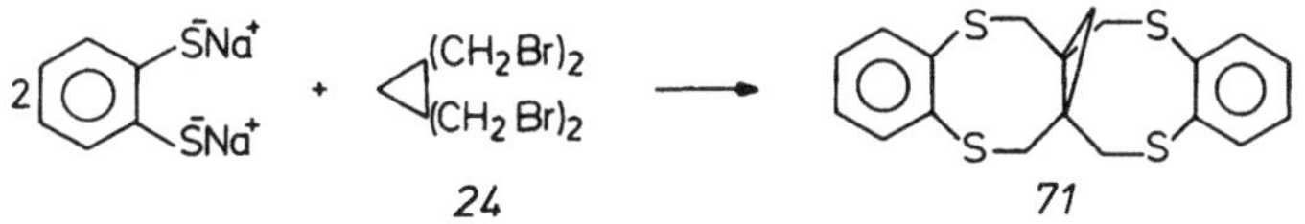

24 71

3,6,10,13-Tetrathia[6.6.1]propellane *72* has been prepared [21]. This is the parent of the dibenzo compound *71* just mentioned [20]. The same tetrasubstituted cyclopropane *24* was used, this time simply with 1,2-ethanedithiol, under conditions of high dilution in ethanol in order to avoid the formation of polymers. Carbene addition to the double bond in the bicyclic compound *73* afforded the same product [21].

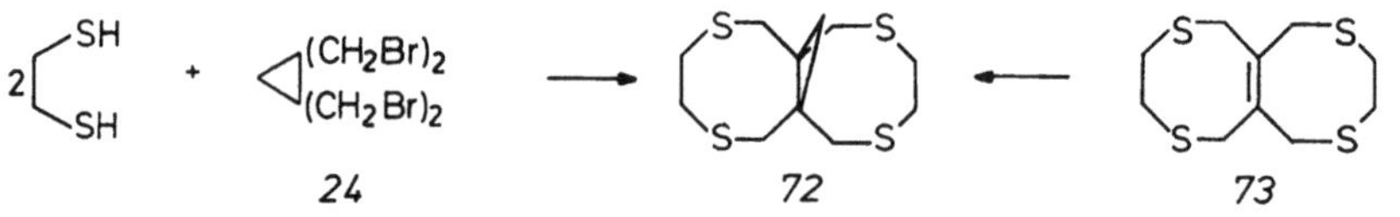

Jamrozik summarizes his work by pointing out that while 1,2-dithiols give [6.6.1]propellanes, the corresponding 1,2-diols may give [6.0.6.1]dispirans [21]. This is explained in terms of conformations of the rings in the products, in which there are different transannular effects for sulfur and oxygen atoms. NMR data is used to support this contention [21]. Such a difference in an eight-membered ring, of course, cannot be generalized so as to apply for the smaller ring propellanes and dispirans discussed earlier.

In the Baeyer-Villiger oxidation of [4.3.3]propellane-8,11-dione *74* the propellane-bis-lactones formed, *75* and *76*, are accompanied by a dispirolactone *77* [22]. Different product mixtures result when different (acidic or more basic) reaction conditions are employed but it has been shown experimentally for the head-to-tail propellane bis-lactone *75*, vis-à-vis the isomeric dispiran *77*, the latter appears to be the thermodynamically more stable product, resulting from the former under acidic conditions (p-TsOH/C_6H_6, 7 days, r.t.). The structures were established by means of X-ray diffraction and ^{1}H- and ^{13}C-NMR spectroscopy.

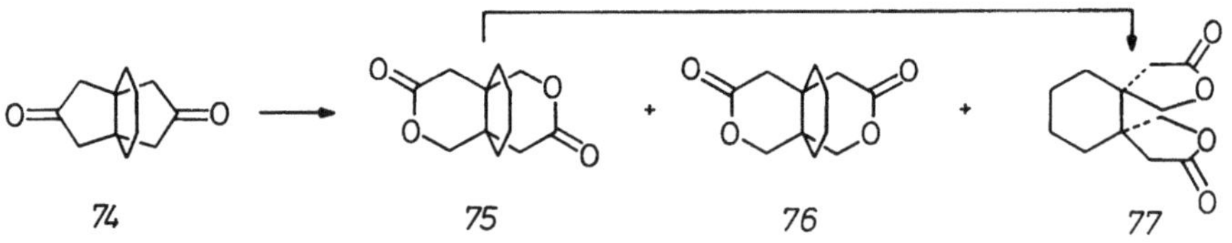

4 Polyspirans

The work which casts the most light on our subject, is, I think, that of the Conia and the Fitjer groups. It appears from several joint papers that the younger man had worked in the laboratory of the elder, and perhaps thus increased his interest in spirans of great beauty[1].

---

[1] Indeed, the most beautiful polyspirans are the rotanes. I have alredy taken issue with my friend, Don Cram, in a seminar I gave at UCLA. In a beautiful chapter [23], in his capacity as the real George Washington of cyclophanes he stated: "It seems that the selection of certain research problems is dominated by a *subliminal* (my underlining, D.G.) aesthetic judgment. Certainly, the author (D.J.C.) would not have selected cyclophanes as a field for investigation had they not possessed beautiful symmetry properties." The second sentence I fully understand but I take issue with the first. "*Subliminal*, not so!", said I to Don after reading the offending sentence from my first slide. "Not at all subliminal but quite conscious." I'm not saying that Monsieur Conia and Herr Fitjer are not interested in the chemistry of spirans. Of course they are and have contributed much to it (with some propellanes thrown in). But I am sure, without having asked them, that their inspection, in paper, of rotanes, inter alia, was quite consciously, a major factor in their beautiful work. The same for Buchta in his polyspirans [4b].

The Conia group synthesized [4]rotane *78* from a dispiro[2.0.2.2]oct-7-ene derivative. This and nearly all other syntheses of rotanes include one or more steps of carbene additon to an exocyclic double bond [24].

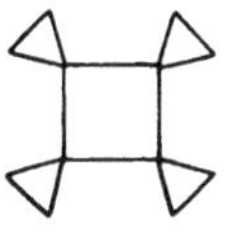

78

One exception as to the previous sentence is the synthesis of [4]rotane by thermal dimerization of bicyclopropylidene *79* [25].

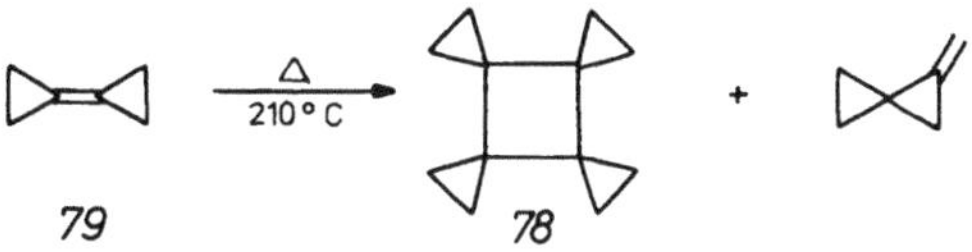

79 *78*

The preparation of [5]rotane *81* (pentaspiro[2.0.2.0.2.0.2.0.2.0]pentadecane), again includes one step of carbene addition, by the Simmons-Smith reaction, to 13-methylene-tetraspiro[2.0.2.0.2.0.2.1]tridecane *80* [26]. A full paper on the subject appeared later [27].

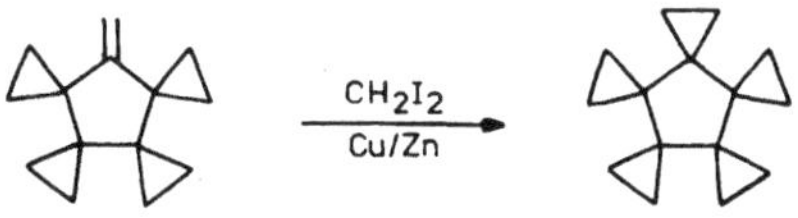

80 *81*

[3]Rotane *83* was similarly prepared from the olefin *82*, albeit by a newly modified Simmons-Smith reaction with silver replacing the copper in the above reaction scheme [28]. An alternative method from *79* was also reported [28–30]. Deductions have been made as to the strain energy of [3]rotane [29].

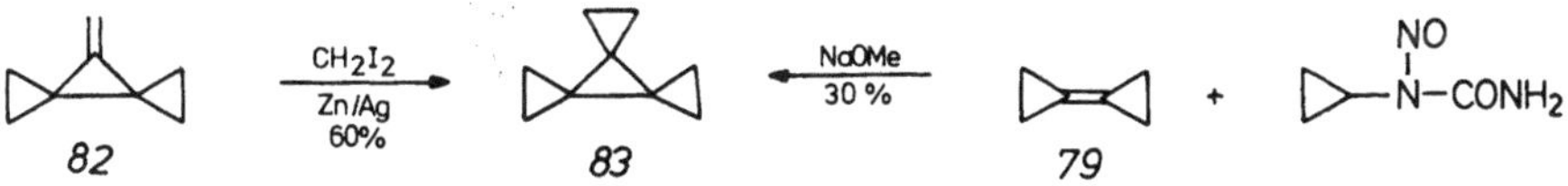

82 *83* *79*

A short paper, the beginning of whose title is designed to catch the eye, "A universal rotane synthesis" announced the synthesis of [6]rotane *85* for the first time via carbene addition to *84* [31]. An alternative method was to add carbene to a starting material *86*, containing two double bonds [32].

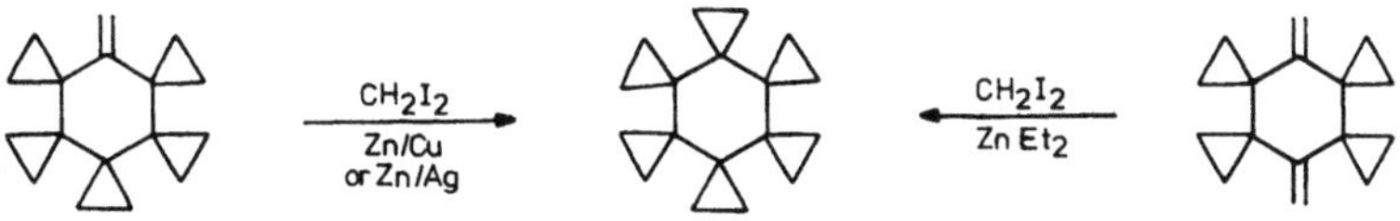

84 *85* *86*

Thermal stability of the rotanes has been mentioned in most if not all of the above papers [25-30]. But note that none of them mention acid-catalyzed rearrangements of rotanes. The preparation of intermediates for such a process is first mentioned for the eventual possibility of obtaining the compound called [6.4] coronane *87* [33].

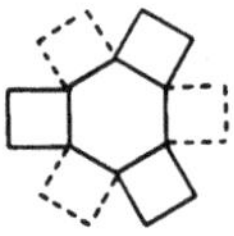

87

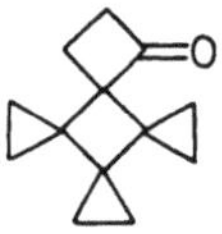

88

The syntheses of [4], [5], and [6]rotanes as well as precursors for those of [7] and [8]-rotanes have been recorded in a full paper [34]. In another synthetic study of rotane synthesis employing ring enlargement of certain ketonic intermediates, led perhaps by design, perhaps with the intervention of serendipity, to compounds such as *88* [35].

Thus appetite has (rightly) grown to obtain [m · n] coronanes, *87* (and their homologs) through rearrangement of the appropriate rotanes [36].

The point to note is that compounds such as *89*, give by a "cascade" of acid-catalyzed rearrangements products that are nearly the desired [m · n] coronanes, e.g. *89* → *90* [33-37].

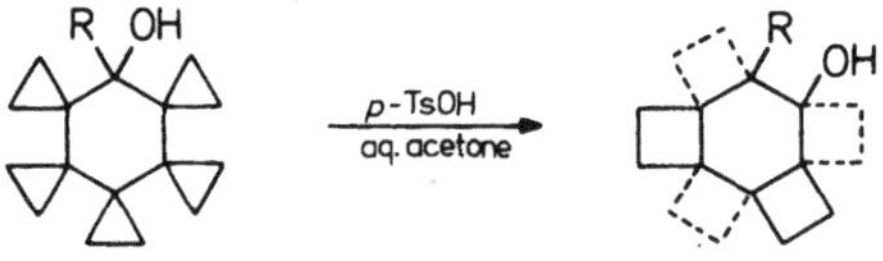

89 *90*

A very germane conformational discussion of sterically crowded cyclohexanes has appeared [38].

I have been following the work of the Fitjer group with bated breath because of a bee which has long been buzzing in my bonnet. I cannot fault Fitjer's interest in the beautiful [m.n]coronanes but it should surprise no one that I have a certain (vested?) interest in propellanes.

Since our work on the preparation of a dispiran rather than the isomeric propellane [7], I have lived with the feeling, alas, the feeling, by no means the certainty, that propellanes should be available from dispirans when the relative stability permits it. (Cf. Ref. 3b).

Indeed, propellanes have been obtained from some of Fitjer's new compounds, e.g. *91* → *92* [39].

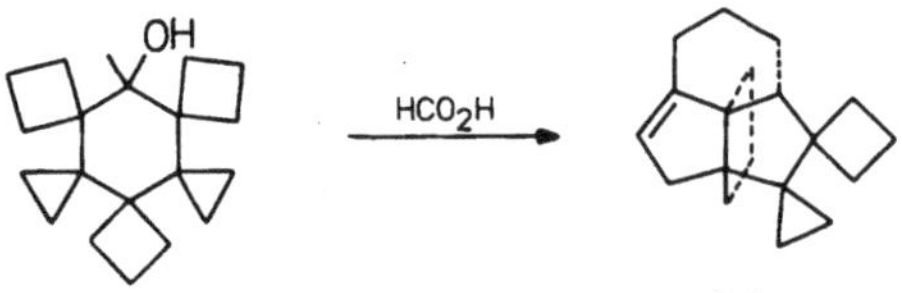

91 *92*

Further, a very recent paper has reported a cascade rearrangement, under acidic conditions of the simpler educt, a dispiro[3.0.3.3]undecane derivative *93* to the dehydrated isomeric propellane *94* [40]. It is somewhat reminiscent of the analogous case of *26* where silver ion is the catalyst [10]. Treatment of the dispiro-alcohol *78* when heated for 2 hrs at 70 °C with *p*-toluenesulfonic acid in benzene gives in quantitative yield the [3.3.3]propellene *94*. The following cascade is proposed to explain the rearrangement.

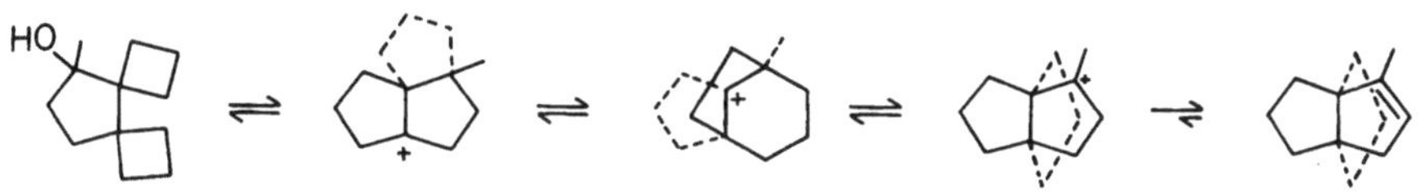

93 94

The corresponding dispiroketone *95* gives the isomeric propellanone *96* after 10 hrs at 70 °C by an analogous cascade which is not reproduced here.

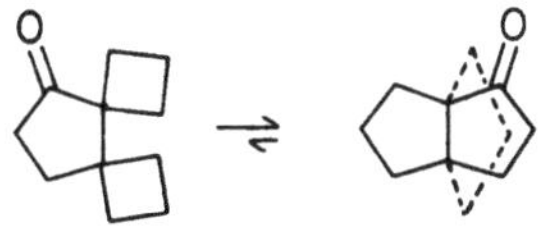

95 96

Molecular mechanics calculations show the [3.3.3]propellanes to be thermodynamically favored over the other, intermediate, structures proposed above [40], and presumably over the respective dispiran derivatives albeit we have not conducted the calculations for the abovementioned starting materials.

5 Conclusion

It appears that if one wishes to anticipate whether a 1,2-dispiran or an isomeric propellane would be formed, it would be wise, before beginning trial and error, to carry out simple molecular mechanics calculations as to which product experiment is likely to give.

On the basis of the extant evidence, the organic chemist finds that use of his intuition with respect to strain energies of ring systems would usually bring him a priori to postulate that, say, dispiran *22* is more likely to form from *97* than propellance *98*. This is so even for the cases cited in six-membered analogs, although this may be regarded as a border-line case for intuition. Not all chemists have the intuition of a K. B. Wiberg but nevertheless he conducts calculations. In cases such

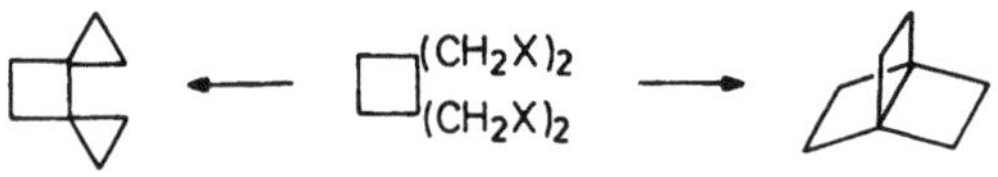

22 97 98

as *75*, if any equilibration exists under acidic conditions the equilibrium is all on the side of *77*.

Once we leave the realm of small-ring propellanes[5] and post-1975 reports[41,42], it appears that it should be possible to convert suitably disposed sets of dispiran moieties into their respective propellane counterparts[39,40] because of the relative stabilities of the two families. By the same token, there are plenty of examples of acid-catalyzed rearrangement of propellanes, through which systems of yet greater stability may be obtained[5,41,42].

Apparently Fitjer likes to work with *p*-TsOH in aqueous acetone. Since "as of the day of the destruction of the Temple, prophecy has been withdrawn from the prophets and has been given to fools and to babes"[43] I shall not prophecy; but I shall hazard a prediction. I belive that there are acidic conditions, yet to be found experimentally, which will permit the synthesis of beautiful molecules, by malice aforethought, such as, say, *99*, from the properly constituted derivatives of certain rotanes and their homologs. May Fitjer be successful in his quest for [m.n]coronanes but it behooves him, simultaneously, to seek optimal conditions for the preparation of, say, *99* and its propellane homologs. I would even settle for *100* or *101*!

99

100

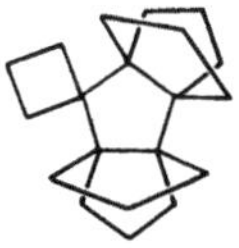
101

It should be possible to develop a general route from di-, tri, etc. — spirans of suitable ring size to afford the respective mono-, bis-, tris-, etc. — propellanes.

6 References

1. *Licentia poetica*: The second "Of" in the title may be considered by some to be superfluous. It was added only in honor of Lewis Carroll and to be compatible with his use of Iambic meter.
2. a) Yalpani, M., Köster, R., Wilke, G.: Chem. Ber. *116*, 1336 (1983); b) Yalpani, M., Schmöller, W., Henneberg, D.: Z. Naturforsch. *B39*, 1241 (1984)
3. a) Stanger, A., Apeloig, Y., Ginsburg, D.: Helv. Chim. Acta *68* 1179 (1985); b) I am indebted to A. Stanger for the contents of this table; c) Spanget-Larsen, J., Gleiter, R., Gubernator, K., Ternansky, R. J., Paquette, L. A.: J. Org. Chem. *47*, 3082 (1982)
4. a) Buchta, E., Merk, W.: Lebigs Ann. Chem. *695*, 34 (1966); b) Idem.: ibid. *716*, 106 (1968); c) Idem.: Chimia *22*, 193 (1968); d) Buchta, E., Billenstein, S.: Liebigs Ann. Chem. *702*, 38 (1967); e) Idem.: ibid. *702*, 51 (1967); f) Buchta, E., Kröniger, A.: Chimia *22*, 430 (1968) g) Idem.: ibid. *23*, 225 (1969); h) Dolbier, Jr., W. R., Lomas, D., Tarrant, P.: J. Am. Chem. Soc. *90*, 3594 (1968)
5. For a review of these compounds until 1975, *cf.* Ginsburg, D.: "Propellanes. Structure and Reactions", Verlag Chemie, Weinheim, 1975, pp. 30–53
6. a) Eaton, P. E., Kyi, N.: J. Am. Chem. Soc. *93*, 2786 (1971); b) Eaton, P. E., Temme III, G. H.: ibid. *95*, 7508 (1973)
7. a) Magrill, D. S., Altman, J., Ginsburg, D.: Isr. J. Chem. *7*, 479 (1969); b) Winkler, T., von Philipsborn, W., Altman, J., Ginsburg, D.: Helv. Chim. Acta *52*, 1603 (1969)
8. Altman, J., Babad, E., Pucknat, J., Reshef, N., Ginsburg, D.: Tetrahedron *24*, 975 (1968); b) Petersen, J. B., Ginsburg, D.: Isr. J. Chem. *6*, 843 (1968); c) Altman, J. Cohen, E., Maymon, T., Petersen, J. B., Reshef, N., Ginsburg, D.: Tetrahedron *25*, 5115 (1969)

9. De Meijere, A.: Chem. Ber. *107*, 1684 (1974)
10. a) Landheer, I. J., de Wolf, W. H., Bickelhaupt, F.: Tetrahedron Letters 2813 (1974); b) Idem: ibid. 349 (1975); c) van Straten, J. W., Landheer, I. J., de Wolf, W. H., Bickelhaupt, F.: ibid. 4499 (1975); d) *Cf.* Peelen, F. C., Rietveld, G. G. A., Landheer, I. J., de Wolf, W. H., Bickelhaupt, F.: ibid. 4187 (1975)
11. Wiberg, K. B., Burgmaier, G. J., Warner, P.: J. Am. Chem. Soc., *93*, 246 (1971)
12. Tenud, L., Farooq, S., Seibl, J., Eschenmoser, A.: Helv. Chim. Acta *53*, 2059 (1970)
13. Weinges, K., Wiesenhütter, A.: Liebigs Ann. Chem. *746*, 70 (1971)
14. Weinges, K., Klessing, K., Kolb, R.: Chem. Ber. *106*, 2298 (1973)
15. Jamrozik, J.: J. prakt. Chem., *321*, 437 (1979)
16. a) Nakazaki, M., Yamamoto, K., Maeda, M., Sato, O., Tsutsui, T.: J. Org. Chem., *47*, 1435 (1982); b) Tobe, Y., Fujita, H., Wakaki, I., Terashima, K., Kobira, K., Kakiuchi, K., Odaira, Y.: J. C. S. Perkin Trans. II 2681 (1984); c) Sarma, K., Witt, W., Schröder, G.: Chem. Ber. *116*, 3800 (1983)
17. Jamrozik, J.: Monatsh. Chem. *112*, 785 (1981)
18. Smolinski, S., Jamrozik, J., Jamrozik, M.: ibid. *108*, 1145 (1977)
19. Jamrozik, J.: Montash. Chem. *111*, 643 (1980)
20. Idem.: J. prakt. Chem. *322*, 909 (1980)
21. Idem.: Monatsh. Chem. *116*, 229 (1985)
22. Ashkenazi, P., Kapon, M., Piantini, U., von Philipsborn, W., Ginsburg D.: Helv. Chim. Acta *68*, 614 (1985)
23. Cram, D. J.: in: "Cyclophanes", Vol. 1, (Ed. Keehn, P. M., Rosenfeld, S. M.) p. 5. Academic Press, New York, 1983
24. Conia, J. M., Denis, J. M.: Tetrahedron Letters 3545 (1969)
25. Le Perchec, P., Conia, J. M.: ibid. 1587 (1970)
26. Ripoll, J. L., Conia, J. M.: ibid. 979 (1969)
27. Ripoll, J. L., Limasset, J. C., Conia, J. M.: Tetrahedron, *27*, 2431 (1971)
28. Fitjer, L., Conia, J. M.: Angew. Chem. Int. Ed. Eng. *12*, 761 (1973)
29. Idem.: ibid. *12*, 334 (1973)
30. Fitjer, L.: ibid. *15*, 762 (1976)
31. Idem.: ibid. *15*, 763 (1976)
32. Proksch, E., de Meijere, A.: Tetrahedron Letters, 4851 (1976)
33. Fitjer, L.: Chem. Ber. *115*, 1035 (1982)
34. Idem.: ibid. *115*, 1047 (1982)
35. Fitjer, L., Wehle, D.: ibid. *115*, 1061 (1982)
36. Fitjer, L., Giersig, M., Clegg, W., Schormann, N., Sheldrick, G. M.: Tetrahedron Letters *24*, 5351 (1983)
37. Fitjer, L., Wehle, D., Noltemeyer, M., Egert, E., Sheldrick, G. M.: Chem. Ber. *117*, 203 (1984)
38. Fitjer, L., Klages, U., Kühn, W., Stephenson, D. S., Binsch, G., Noltemeyer, M., Egert, E., Sheldrick, G. M.: Tetrahedron *40*, 4337 (1984)
39. Fitjer, L., Kuhn, W., Klages, U., Egert, E., Clegg, W., Schormann, N., Sheldrick, G. M.: Chem. Ber. *117*, 3075 (1984)
40. Fitjer, L., Kanschik, A., Majewski, M.: Tetrahedron Letters *26*, 5277 (1985)
41. Ginsburg, D.: "Propellanes. Sequel I", Technion, Haifa, 1981
42. Idem.: "Propellanes. Sequel II", Technion, Haifa, 1985
43. Rabbi Yochanan: Bava Batra, *12*, 2 (ca. 80)

A New Treatment of Chemical Reactivity: Development of EROS, an Expert System for Reaction Prediction and Synthesis Design

Johann Gasteiger, Michael G. Hutchings[1], Bernd Christoph, Leopold Gann, Christian Hiller, Peter Löw, Mario Marsili[2], Heinz Saller and Kazumi Yuki[3]

Organisch-Chemisches Institut, Technische Universität München,
D-8046 Garching, FRG

Present addresses:
1 Imperial Chemical Industries plc, Organics Division, Hexagon House, Blackley, Manchester M9 3DA, England.
2 Computer Chemistry Lab., CNR, Area della Ricerca, Monterotondo S., I-00016 Rome, Italy
3 Sumitomo Chemical Co. Ltd., Osaka, Japan

"*A theory has only the alternative of being right or wrong. A model has a third possibility — it may be right but irrelevant.*"

Manfred Eigen

Table of Contents

An expert system for the prediction of the course of organic chemical reactions and for the design of organic syntheses has been built. It does not depend on a database of chemical reactions. Instead, the system generates reactions from first principles by formal bond- and electron-shifting processes. The reaction sites are found by application of quantitative models for the prediction of chemical reactivity. This approach is founded on procedures that allow rapid calculation of the various physicochemical effects that influence the course of chemical reactions. The extent to which these factors influence chemical reactivity has been studied by statistical methods. Examples of the prediction of quantitative data on chemical reactivity, and of the course of complex organic reactions, are described.

1 Introduction

Much of the work of the practicing organic chemist is centered on the two questions implied by Fig. 1.

— Given a target molecule, how can it be synthesized to best effect?

— Conversely, given a substrate molecule, how will it react under certain conditions?

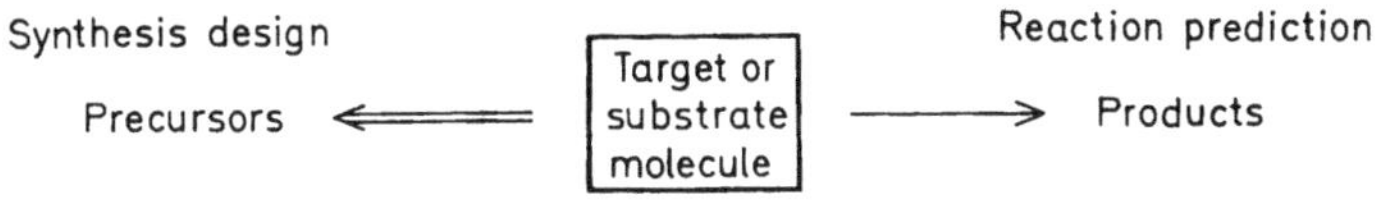

Fig. 1.

The problems are of both industrial and purely scientific importance, and they can also be approached qualitatively and quantitatively. The "reaction prediction" question, for example, might be directed not so much at what the products might be, but rather how likely the reaction might be in comparison with an analogous system. The scope of the problems is indeed vast, and Figs. 2 and 3 give some indication of the width of the spectrum of organic synthesis and reactivity.

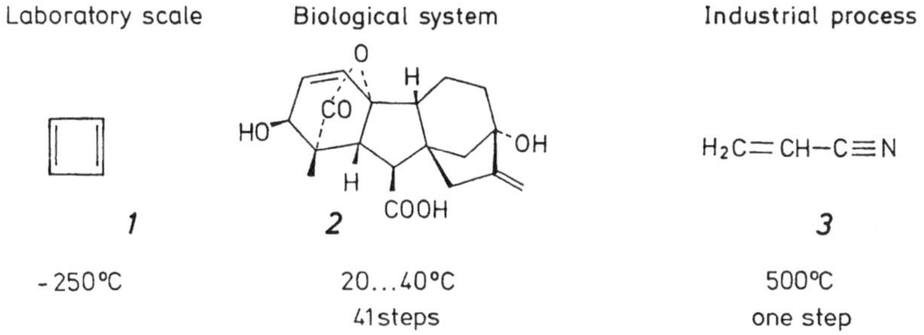

Fig. 2. Examples of targets for organic synthesis

The synthesis of a highly reactive compound like cyclobutadiene, *1*, requires rather esoteric reaction conditions. Economic considerations, for instance, played no role in the design of its synthesis; the aim was to synthesize this compound — at any price. Nature makes complicated structures through highly specific reaction sequences, and at ambient temperature and pressure. Chemists have taken up some of the challenging goals set by nature, but competing with nature can be quite laborious: a total of 41 steps was required to synthesize gibberellic acid, *2*, from readily available starting materials [1]. On the other hand, economic considerations are a major factor in the synthesis of bulk chemicals on an industrial scale. Frequently, these syntheses involve only a few steps, but the reaction conditions can be quite drastic as illustrated by the synthesis of acrylonitrile, *3*. Any approach intending to help with the synthesis design question will have to face the complexities given by the wide range of reaction conditions, and the varying degrees of importance given to economic and other considerations.

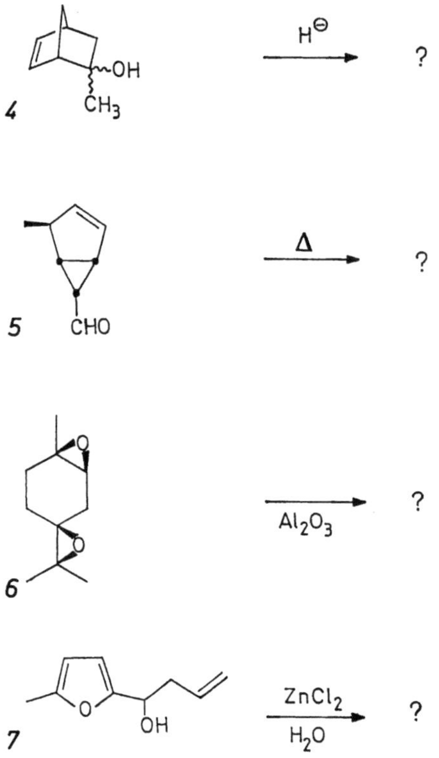

Fig. 3. Typical reaction prediction problems

The problem of reactivity is no less extensive. Readers may care to test their organic chemical expertise against the problems posed in Fig. 3. It is just such chemistry that will later become the central focus of this paper.

The extent of the interaction of synthesis design and reactivity prediction is even more extensive than implied above, in that both interrelate with a third very important element of organic chemistry — that of molecular design. The factors which control and are responsible for reactivity are precisely those that dictate how a particular

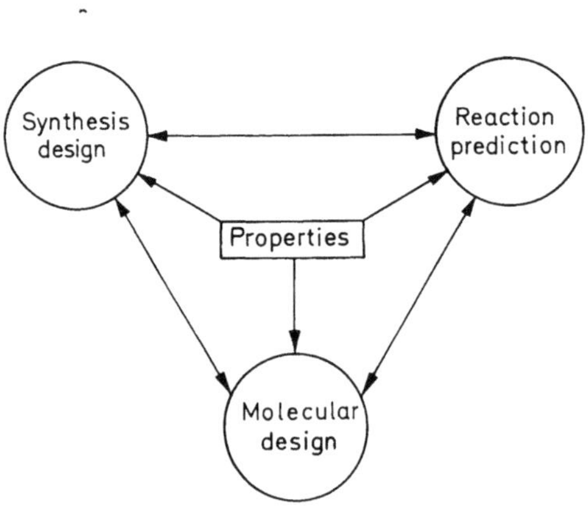

Fig. 4. The central importance of molecular property to synthesis and molecular design, and reactivity prediction

property should be built into a molecule. The manifestation of the property could be physical or chemical, or correspond to a biological activity. Fig. 4 summarizes the way in which we picture this symbiosis.

Organic chemistry has benefited most in its historical development from advances in the conceptual understanding of chemical principles[2]. Milestones were provided by the structural theory of Kekulé, the mechanistic perception of organic reactions, conformational analysis, and the disconnection and synthon-approach [3,4]. These developments have helped to put the more intuitive — and often failing — approach to organic chemistry on a more rational and effective basis. In the electronic age, with computers infiltrating many disciplines, it seems quite natural to keep an eye open for applications in chemistry. The huge arsenal of synthetic reactions, and the many different points to be considered in synthesis design and reaction prediction require the processing of a large amount of information, and decision-making between many alternatives. It is in just such circumstances that computers should be of use [5]. Computers have several features that render them particularly attractive for assisting in the problem areas defined by Fig. 1:

1. Rapid calculations: these may be useful in evaluating many different reactions and synthetic pathways in a short time.
2. Large storage space: this can be used to build a database.
3. Logical operations: ideal for development of strategies and decision-making processes concerning alternative reactions and pathways.
4. Graphical manipulations: these are important for the rapid input and output of molecular structures, and their correct 3D manipulation.
5. Interactive: ideally, the computer can respond to interrogation and outline the reasons for its decisions.

Synthesis design and reaction prediction can draw benefits from all these features of a computer. Our own work in this area began in 1974, and in 1978 the computer program system EROS (*E*laboration of *R*eactions for *O*rganic *S*ynthesis) was first presented [6]. Since then, several reports on certain aspects of the system development have appeared, but sometimes in less easily available journals or books[7]. Moreover, there has been no description of the overall system as it now stands. This article is intended to rectify this situation.

In developing EROS we have several objectives in mind:

1. To develop a practical tool that can help the chemist to design organic syntheses.
2. To develop a system that can predict the products of the reaction of given starting materials.
3. To provide a framework for testing quantitative models of chemical reactivity.
4. To provide a framework for studying strategies of organic syntheses.
5. To provide quantitative data which can be applied at the design stage of organic molecules, by modelling chemical, physical and biological properties.

As we discuss in the next section, we considered a system which is based largely on a library of reactions to have limitations. The route followed was therefore quite different, in that we intended the program to work out its own chemistry. This could be applied to prediction of reactivity and reactions, and also to suggesting retrosynthetic pathways. In fact, the two problems are related in that a knowledge of likely reaction path is necessary to know whether a potential synthetic precursor will indeed lead to its intended target, or alternatively follow a different path to give an unwanted product.

Recent developments have been aimed at facilitating EROS's chemistry, and use of the latter in the various evaluation steps necessary to the problem. This has led to a version of EROS which is capable of solving some quite sophisticated reactivity prediction questions, such as those posed in Fig. 3. Further developments of the synthesis design capabilities have intentionally been kept in abeyance, and are following on in the light of our experience with reaction prediction.

The main aim of this article is to describe the advances of the last 7 years, and show how they have been brought together to give a working reactivity prediction version of EROS. Secondly, we shall indicate where appropriate how the reactivity work impinges on the problem of synthesis design. To conclude, we touch on the question of EROS's relationship to artificial intelligence and expert systems in general.

We are concerned only with EROS, a unique system for such studies, and this article is in no way an overall review of computer-aided synthesis in general.

2 The EROS Philosophy

In order to understand the conceptual framework of EROS, it is desirable to consider first the approach to computer-aided synthesis planning which is based on a library of known reactions.

2.1 EROS — What It Isn't

One of the principal strategies of a chemist for designing an organic synthesis is a stepwise retrosynthetic approach. The target molecule is analyzed for structural features for which synthetic reactions are known. Taking the hydroxycyclopentenone *8* of Fig. 5 as an example, it must first be recognized that a hydroxy group is in a position β to a carbonyl group. Secondly, the chemist must recall that such a substructural unit can be obtained through an aldol condensation between an aldehyde and another carbonyl group. The reverse of the changes that occur in an aldol condensation are applied to the target structure, leading to the synthetic precursor *9*. Usually, there will be several reactions that conform to various substructures in the target molecule.

It is not surprising that computer programs have been developed that imitate such an approach. To this end, the program must be able to recognize certain structural features for which synthetic reactions are known. It must also contain a database of retroreactions which when called perform these structural changes. (The retroreactions that give the structural changes, like those contained in the frame of Fig. 5, are frequently called transforms.) Such an approach is attractive as it models a basic strategy used by the organic chemist in designing syntheses.

However, a database of reactions brings with it problems. The most crucial question concerns the size of the database. To cover the entire range of organic chemistry certainly requires very many reactions. Some classes of compounds have their own, rather specific synthetic reactions; some technical processes apply to one compound only; each week new reactions are being discovered. Are all these reactions to be

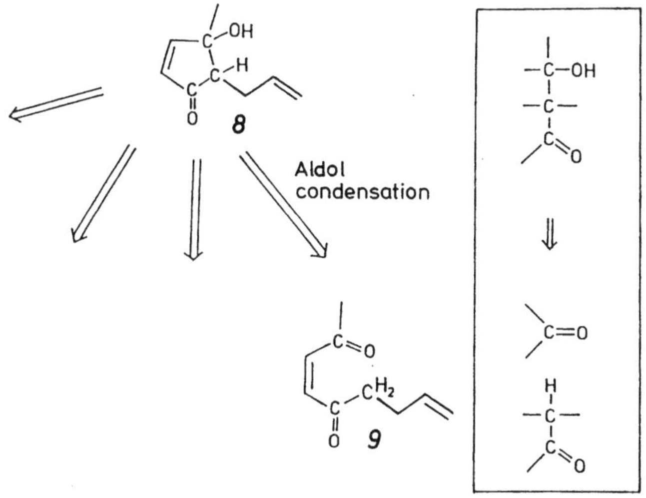

Fig. 5. Retrosynthetic analysis of an aldol condensation

stored in the database? Systems have been developed with 5,000 transforms in the database, but there is as yet no end to be seen. Furthermore, a program designed around a database will only give access to known reactions, since obviously only these will be stored.

Conventional thinking can be similarly inhibiting and limited when applied to reactivity and reaction prediction. Thus, when a chemist is faced with the task of predicting the reaction products for given starting materials, he usually looks for the functional groups present, which in turn suggest reactions that he had learnt previously. For example, when an aldehyde group is found in one molecule, the known reactions of aldehydes will be scanned mentally and those compatible with functional groups elsewhere in the molecule or in reaction partners will be selected and further investigated. A more detailed analysis of the possible course of a reaction and the expected products will elaborate on the mechanism of the reactions of these functional groups. This approach is very powerful in rapidly directing the chemist to the actual reactions between given starting materials. However, the concept of functional groups also has its limitations and pitfalls. Thus, the reactivity of chloral hydrate, *10*, in the presence of hydroxide ion cannot be completely understood from a knowledge of the two functional groups present in this molecule: an aldehyde hydrate, and an alkyl chloride. Whereas formation of the free aldehyde could take place in a rapid equilibrium, no substitution of chlorine by hydroxide ion occurs. Rather, by a haloform reaction, formation of chloroform and formate is observed. In this case, a CC single bond is broken (Fig. 6). As another case, 3-chloro-3-methylbutanol, *11*, reacts with hydroxide ion to give neither a 1,3-diol, nor an allyl alcohol or an oxetane. Instead, in a fragmentation reaction again a CC single bond is broken, giving formaldehyde, isobutene and a chloride ion (Fig. 6). These examples indicate that bonds beyond those contained in conventional functional groups must be considered when determining chemical reactivity.

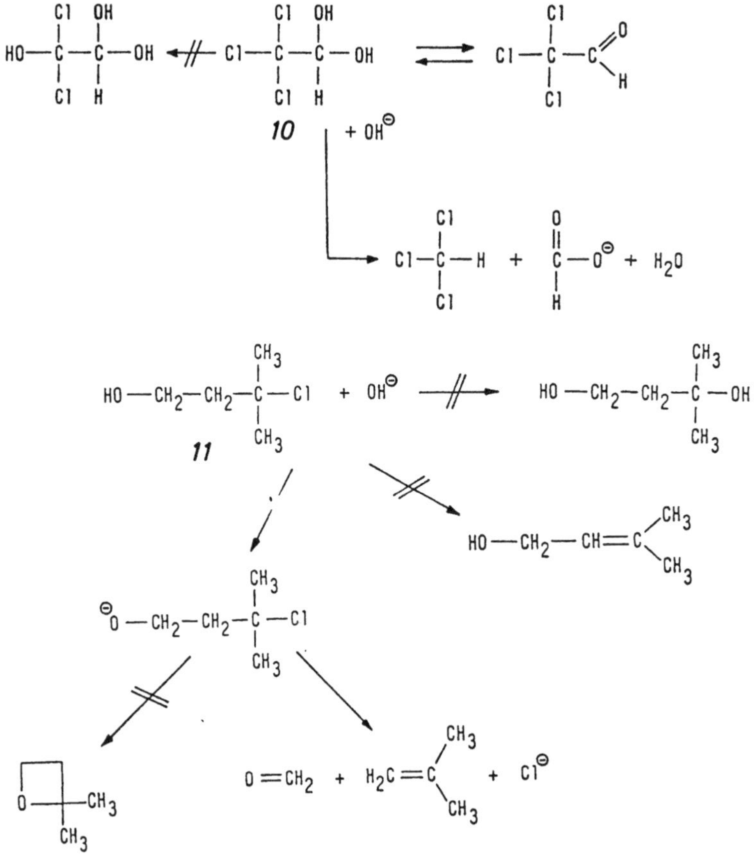

Fig. 6. Breakdown of the concept of functional groups

The concept of EROS was devised in order to avoid the problems inherent in a database of reactions, as well as to avoid the treatment of chemistry from the standpoint of functional groups. This work was stimulated by a mathematical model of constitutional chemistry [8,9]. Central to the approach is a formal handling of organic reactions, where they are treated as bond-breaking and -making and electron-shifting processes. Furthermore, the approach is applied to all atoms and bonds in a molecule, regardless of any preconceptions of functional groups.

2.2 Reaction Generators, an Introduction

For explanation, the example *8* of Fig. 5 is again used (Fig. 7). The essential features of an aldol condensation in its retro-form are the breaking of a CC- and of an OH-bond, and the making of a CO- and of a CH-bond. Alternatively, if this process is considered in an even more general manner, two bonds between atoms I, J and K, L are broken and two new ones between the four atoms involved are made.

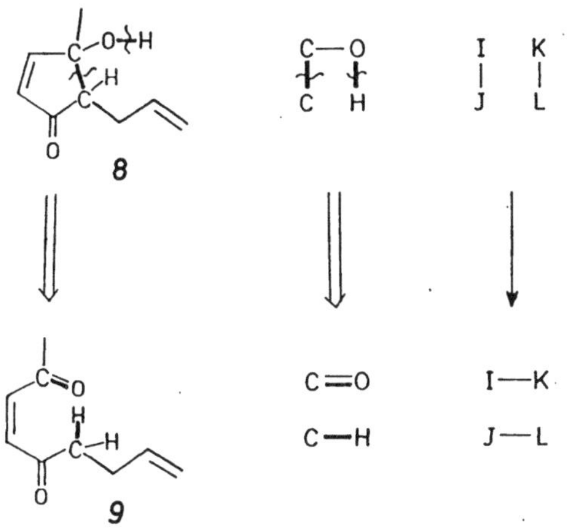

Fig. 7. Bond changes in a retro-aldol condensation

As we shall see in the next section, other formal reaction schemes play their part in describing other chemical reactions. However, this reaction scheme, breaking two bonds and making two new ones, is of paramount importance in organic chemistry. In fact a majority of organic reactions, of different mechanistic types, follows the scheme. Figure 8 gives some representative examples. Applying the scheme to various combinations of bonds of the molecule *8* gives the reactions of Fig. 9

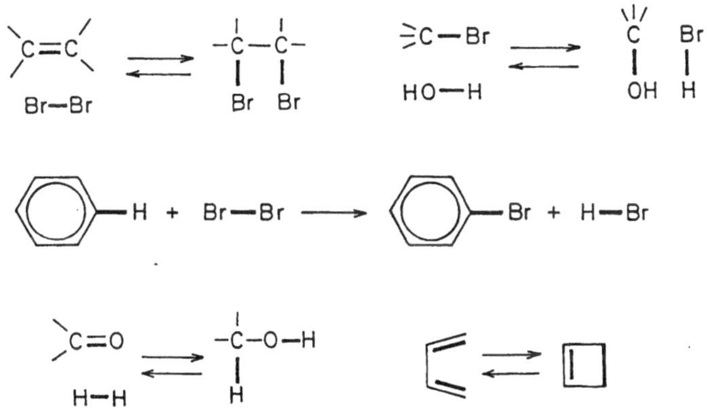

Fig. 8. Reactions breaking two bonds and making two new ones, as indicated

For each set of two bonds broken there are two alternatives for making two new ones as indicated with reactions 9.1 and 9.2 as well as 9.4 and 9.5. The bonds made can be contained in two different molecules (reactions 9.3 and 9.6), or the bonds broken

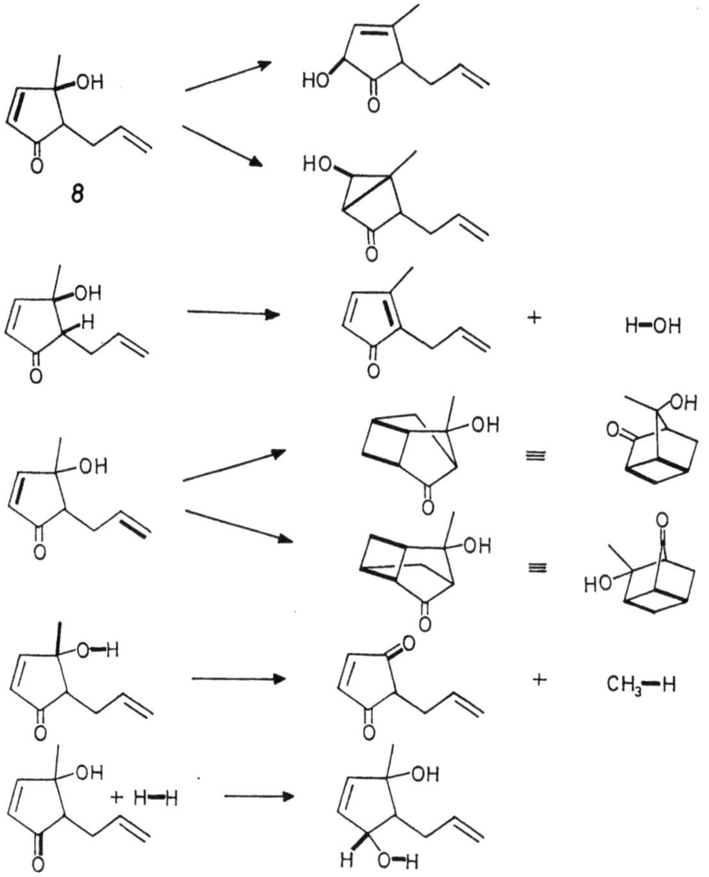

Fig. 9. Some reactions obtained by breaking two bonds of molecule *8* and making two new ones (bonds broken and made are emphasized)

can come from two different molecules (reaction 9.7). Naturally, all these reactions can also be generated by pencil and paper. A recent example of such an approach, in effect, is the retrosynthetic analysis of α-methylene-γ-butyrolactones [10]. However, the human generator may get tired or overlook certain reactions, so in this situation the potential of the computer to do the job exhaustively becomes quite helpful.

Observe that to generate the reactions of Fig. 9 no information was necessary on whether such a reaction is known: *no database of reactions is necessary*. The problems in building, updating and maintaining a reaction library are thus avoided. The formal treatment of reactions as bond and electron-shifting processes allows the generation, in principle, of all conceivable reactions, and can be seen as a method to deal freely with molecular architecture. The program's result could be a known reaction, but equally a new, as yet undiscovered reaction which could be realised in the laboratory.

But a price has to be paid for this potential advantage, because the number of reactions that could be obtained by applying a formal reaction generator scheme could be very high. Furthermore, most of the suggestions could be chemically meaningless. One cannot just pick any bonds of a molecule, break them and make new ones, and always get a good reaction. For example, reaction 9.6 of Fig. 9 seems to be rather unlikely. Thus, either selections have to be made among the formally conceivable reactions to find the chemically interesting ones, or steps must be taken to ensure that only sensible ones are generated. *This is the major task in program development.* The selection is based on various evaluation procedures that are conceived to model chemical reality. Approaches and solutions to this problem are the main theme of this article.

The high number of reactions that can be obtained by the formal reaction generators have forced us from the very beginning to work on an automatic evaluation package that allows the selection of chemically feasible reactions. With continuing program development the evaluation process will become increasingly better and more selective, thus discarding poor reactions and offering only a few, realistic, good reactions. There is convergence in this development.

A system for synthesis design working with a database of reactions will initially contain only a few reactions and thus produce only a few alternatives. Thus, evaluation and selection are not very important. However, as the size of the database is expanded, more and more alternatives will be obtained for a given target structure. Here again, the development of a general evaluation and selection package will be required to handle the various transforms of the database.

There is a further advantage in working with formal reaction generators. The reaction scheme of Figs. 7 and 8 has two bonds on both sides of the reaction equation, and moreover implies nothing about the directionality of the reaction represented. In other words, the scheme is formally reversible, and the reactions generated by such a scheme can be interpreted to proceed in either direction. If it is assumed that the reactions occur physically in the direction as generated by the computer, then a forward search, reaction prediction, can be performed (Fig. 10a). From given starting materials, the system proceeds through various intermediate structures until it ends up with the products of a reaction. Alternatively, the reactions as obtained by the computer can be considered as retro-reactions (Fig. 10b). The system proceeds from the target molecule through synthetic precursors until arriving at available starting materials. Again, various alternatives are explored.

Naturally, there must be differences in the two types of search procedures. However, they do not originate in the mechanisms for generating reactions; the latter is achieved for both types of searches via the formal reaction schemes. Instead, the differences come from the way in which the reactions are *evaluated* and *selected*. The various

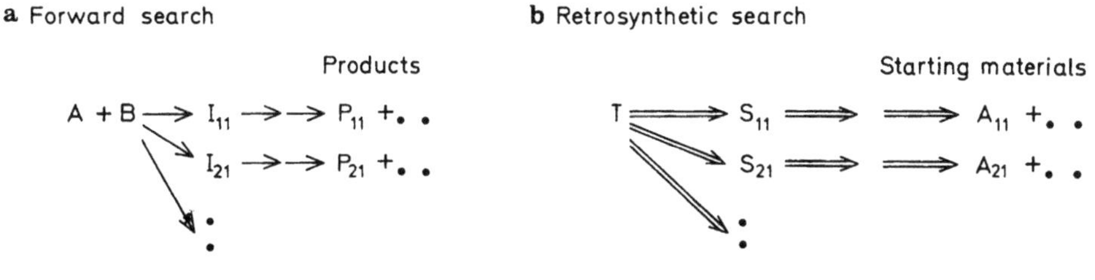

Fig. 10. Forward and retrosynthetic search by formal reaction generators

evaluations are used with different weight and emphasis depending on whether a problem is one of reaction prediction or of synthesis design.

To summarize the features which we regard as characteristic and crucial in EROS:

- EROS contains no database of known reactions;
- EROS does not recognize or work on functional groups;
- EROS is based on formal reaction generators which regard reactions as bond- and electron-shifting processes;
- the chemistry generated can be forward or backward;
- evaluations and selections will restrict the formally possible reactions to the chemically realistic.

3 Representation of Molecules and Reactions

Before continuing with the discussion of how reactions are represented in EROS, it is more sensible to describe in outline some of the formal features associated with the representation of molecules.

3.1 Molecules

In the first phase of the development of EROS it was decided to consider only the constitutional aspects of molecules. By and large the most important factors influencing chemical reactivity are dominated by the nature of the atoms of a molecule and the way in which they are bonded. Steric effects were regarded to be of lesser influence. As the prediction of chemical reactivity has progressed, work has been initiated on the stereochemical features of molecules and steric effects in reactions. Some of the logic for the treatment of the stereochemistry of molecules and reactions was presented some time ago [11]. This, and the Cahn-Ingold-Prelog rules [12], form the basis of the algorithms for the treatment of stereochemistry.

The constitution of molecules is given by lists of atoms and bonds (connectivity lists; "topological representation") [6]. In addition, the number of free electrons for each atom is also carried in a separate vector. This is necessary as some reaction generators may transform free electrons into bonds, or vice versa. Thus, by working with free electrons and the electrons involved in bonds, all valence electrons of a molecule are explicitly specified.

In representing a molecule by a connectivity list, the atoms have to be numbered. For a molecule with n atoms this can be performed in n! ways. Thus a molecule can be represented by n! bond lists which are all different provided that there is no symmetry in the molecule. Several tasks in synthesis design require the automatic comparison of molecules for identity. For example, this is necessary when the molecules generated in a retrosynthesis study have to be compared with those contained in a database of available starting materials, or with molecules already previouly generated in the synthesis tree. As n! increases rapidly with n ($10! \simeq 3.6 \cdot 10^6$), comparison of all the different n! bond lists has to be rejected out of hand. Rather, we have developed an algorithm for numbering the atoms of a molecule in a unique, canonical manner, to give an unambiguous code [13].

Breaking equivalent bonds of a molecule would result in the same products, and redundant reactions or retrosynthetic pathways would be generated. For example, the three CH-bonds of the methyl groups of acetaldehyde are equivalent and therefore it suffices to break only one of them in an aldol condensation. During the process of canonically numbering the atoms of a molecule, constitutionally equivalent *atoms* are detected[13]. In order to define two constitutionally equivalent *bonds*, the atoms at the ends of the two bonds must be pairwise equivalent. However, this is a necessary but not a sufficient condition as the example of triphenylene, *12*, shows: the six atoms of the central six-membered ring are all constitutionally equivalent, but they split into two groups of constitutionally equivalent bonds: 1–2, 3–4, 5–6, and 2–3, 4–5, 6–1.

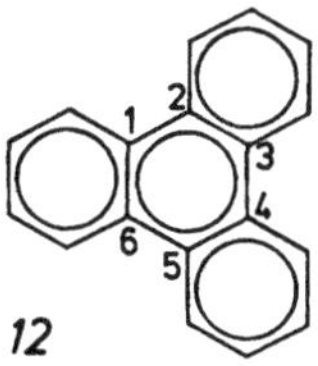

12

New algorithms have been developed in order to detect such constitutionally equivalent bonds in a molecule [14]. Various spanning trees having different atoms as a root are grown for each molecule. These spanning trees are compared with each other to give both constitutionally equivalent atoms and bonds. Extension of these algorithms provides the full automorphism group of a molecule, which in turn gives even deeper insights into the constitutional symmetry of a molecule.

The para-disubstituted benzene derivative, *13*, of Fig. 11 has two groups of equivalent bonds to hydrogen: the ones to hydrogens 1 and 4, and those to hydrogens 2 and 3. If only one C—H bond from the first group in this molecule is to be broken, it

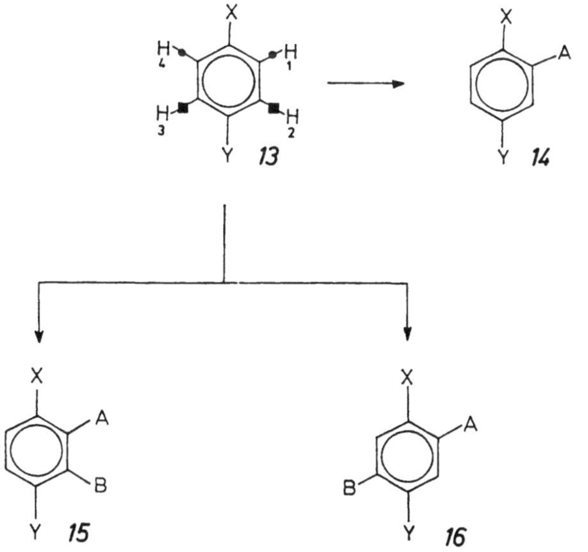

Fig. 11. Relationship between constitutionally equivalent bonds

suffices to select only that to atom 1, leading to structure *14*. The bond to atom 4 would obviously give the same result, *14*. If, however, two bonds are to be broken, one from each of the two equivalence classes, both choices have to be made as the two alternatives *15* and *16* are now different. These symmetry relationships can be found in the automorphism group of a molecule. (Incidentally, these relationships also express themselves in the fact that the hydrogen atoms 1 and 4, as well as 2 and 3, respectively, are chemically equivalent, but the corresponding protons are magnetically non-equivalent. Thus, the ^1H nmr spectral type, AA'BB' or AA'XX' for *13*, can be obtained automatically from the automorphism group.)

A further aspect of the constitution of a molecule, the presence of rings, is of great importance. Rings have a profound influence on the physical and chemical properties of molecules. To mention just a few: small-membered rings are highly reactive; aromaticity is intimately connected to the presence of rings; special care has to be given to the construction of rings in a synthesis. Algorithms for the perception of rings in molecules have been developed [14, 15], and on the basis of this program a smallest set of smallest rings (SSSR) of a molecule is determined. An SSSR completely describes a ring system. Additional rings can be obtained from the SSSR by linear combination, if needed in specific evaluations.

3.2 Reactions and Reaction Generators — Further Discussion

In the previous section, we described how the generation of reactions within EROS is performed by formal reaction schemes for breaking and making bonds, and intimated that some also involve free electron pairs. At that stage, discussion was restricted to just the one type. In fact, there are many formal possibilities, encompassing different combinations of bonds and free electron pairs. In order to keep the problem tractable, a selection was made from amongst the set of all these possible reaction schemes. Firstly, only those schemes were taken which comprise changes in the electron distributions that the chemist considers to be an entire reaction. Those representing single mechanistic steps, but incomplete reactions (e.g. bond homolysis) were excluded, as the main objective in program development was the representation of complete reaction steps. Furthermore, those reaction schemes most commonly met in organic chemistry were preferred. Our own conclusions as to those schemes which are most common found support in studies of reactions reported in the literature [16]. The presently available reaction schemes are given in Fig. 12. (The notation RG*xyz* is merely a simple identifier for a reaction generator, *x* giving the number of bonds broken, *y* the number of bonds made. *z* is a consecutive number where *xy* alone is not unique.)

We emphasize that the exclusion of other feasible reaction generators should not be construed as a limitation of the system — others could be included easily. In fact, our current thinking is that the chemistry of the system should drive the reaction generators, and not vice versa. In this approach, reactions are obtained from mechanistic steps, which allows the automatic generation of *any* reaction scheme. Work on implementing this approach is in progress (see Section 10).

When bonds are counted as electron pairs, the same number of electrons is found on both sides of a reaction equation. In order to ensure that both forward and retro-

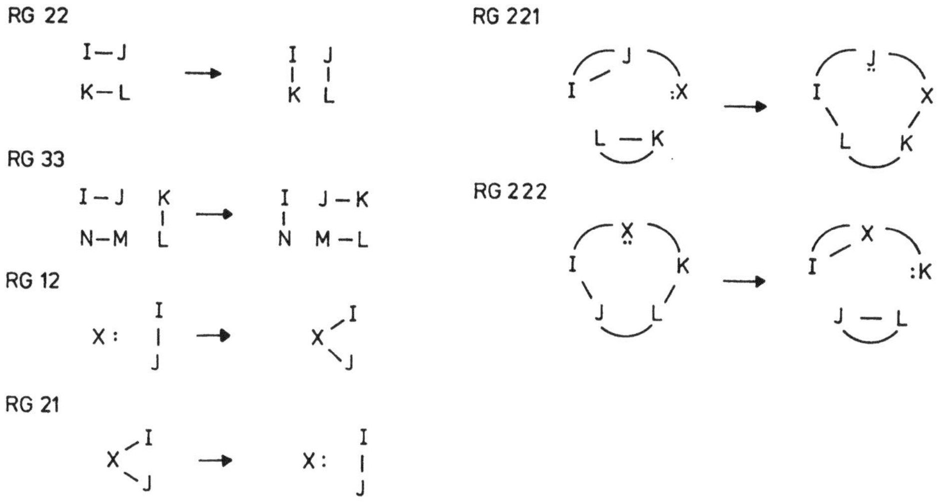

Fig. 12. Reaction schemes contained in EROS; bent lines indicate bonds that are necessary, but are not involved in the bond and electron rearrangement

synthetic processes have the same formal characteristics, care was taken in the selection of reaction generators to include the reverse process for each scheme.

In the following, the individual reaction schemes selected for inclusion in EROS are briefly discussed. The schemes only give the overall changes in the bond and electron distribution of a reaction. As such, they make no assumption regarding the timing of bond-breaking and -making, which could be in a concerted or a stepwise manner. Nor do they imply any other physical details on what is happening during a particular transformation.

RG22: As already pointed out and illustrated with Fig. 7–9 this scheme is of great importance in organic chemistry.

RG33: This is also a very important scheme, comprising many organic reactions.

RG12: Valence electrons are transformed from a free electron pair into bonding electrons and a change in the valence state of atom X occurs. Before this scheme is applied in EROS, a table of valence states for each atom is scanned to determine whether this change in the valence for atom X is allowed. The scheme has importance in representing oxidations at atom X as exemplified with the change $S^{II} \rightarrow S^{IV}$ (Fig. 13).

RG21: This is the formal reverse process of RG12, but here the valence of atom X is reduced. Obviously, this scheme should not be applied indiscriminately, (for instance to each tetravalent carbon, changing it to a carbene) since many reactions with no chemical significance would result. Care has therefore been taken in the evaluation phase to find the appropriate sites for its application.

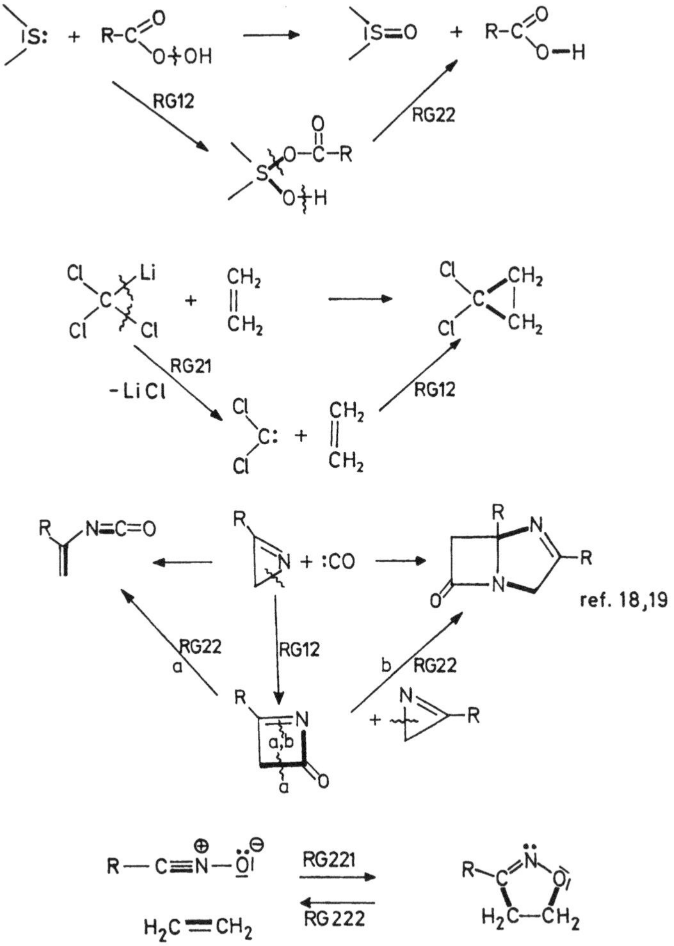

Fig. 13. Reactions described by application of the formal reaction schemes.

RG221 and RG222: Both reaction schemes involve the breaking and making of two bonds and the shifting of a free electron pair:

$$X{:}\quad I-J\quad K-L \rightarrow L-I\quad J{:}\quad X-K$$

However, before such a scheme is applied, additional bonds are required between the five atoms involved. These, however are not changed on bond rearrangement. These constraints fall into two types, thus breaking the general scheme up into the two reaction generators RG221 and RG222. RG221 represents 1,3-dipolar cycloadditions[17]. RG222 represents retro-1,3-dipolar cycloadditions when applied in a forward search, or 1,3-dipolar cycloadditions in a retrosynthetic search (Fig. 13).

Figure 13 gives examples of reactions obtained by single and consecutive application of these schemes.

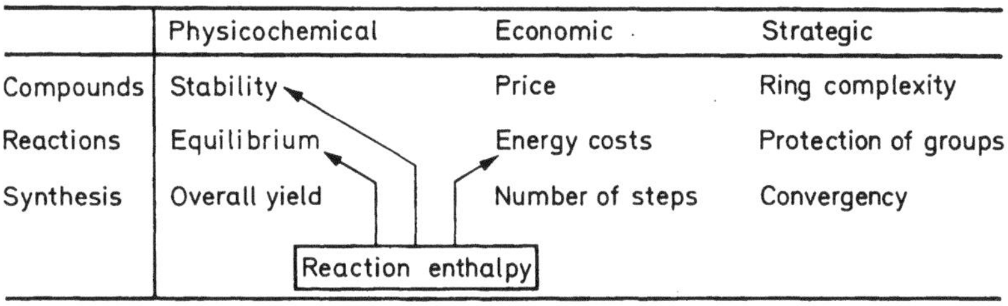

Fig. 14. Classification of evaluations

4 Evaluations

We have already emphasized our view that the *evaluation* of chemical reactions and synthetic pathways is of preeminent importance in any system for computer-assisted synthesis design or reaction prediction. The quality of the evaluation process will determine to a large extent the overall quality of such a system.

The various evaluations which form the basis for selecting the interesting reactions and synthetic pathways fall into several categories: physicochemical, economic, strategic, hazard or toxic evaluations spring to mind. Furthermore, the evaluations can be classified according to whether they are to be performed on molecules, reactions, or overall synthetic or reaction pathways. Figure 14 gives one example each for the first three types of evaluations.

The intention is to keep the different types of evaluations strictly separate from each other within EROS. Thus, the system is designed to be flexible, and can respond to the different types of problems mentioned in Figs. 2 and 3. As an example, by raising the weight assigned to the economic evaluations, more emphasis can be put onto such considerations when studying industrial processes. However, although the various categories of evaluations are distinct, some might draw their information from a common source. For instance, values of reaction enthalpies determined in EROS (see next Section) can be used for estimating the relative stabilities of products and for determining the equilibrium of a reaction. Additionally, these enthalpy values can also be employed for deriving an estimate of the energy costs — an economic parameter. By giving different weightings to these various flows of information — i.e. the extents to which reaction enthalpies are used for physicochemical and for the economic evaluations — the system can be made responsive to the needs of the user.

We now address the all-important question of how the evaluations themselves are to be represented and enabled within EROS. Various possibilities exist, but our approach is unequivocal: we have devised and applied wherever possible *models of chemistry* to the evaluation processes. The use of models is common throughout chemistry, as has been very effectively discussed by Suckling, Suckling and Suckling [20]. They can take many general forms, but the intention is always the same — for the model to represent as closely as possible an aspect of chemical reality, without it necessarily being applicable outside its intended context. Eigen's warning [21] quoted at the beginning of this article is wholly appropriate. It is well known, for instance, that Dreiding molecular models can represent static spatial relationships between atoms in a molecule relatively accurately. On the other hand, the same models are notoriously unsuccessful if used to try to predict energy barriers to conformational interconversions in the molecule, even qualitatively. The same type of restriction applies to the models we shall discuss in the following sections. Models, of whatever type, must be used with care; we believe they are in the present context.

The models which we have developed can be classified as follows. Some are intended to represent *physicochemical* processes and properties by mimicking quantitatively concepts which have become accepted by chemists in general. A simple example would be the transfer of electronic charge between two atoms of differing electronegativities. Other models are *statistical* in nature. We have applied parameters quantified by the physicochemical models to series of chemical data. The relationships thus derived by various statistical techniques, and their form, is such that they are readily applicable to the task of quantifying the evaluation process in EROS. Further discussion of these points is a major feature of this article.

In developing the discussion of evaluations, physicochemical parameters are the only ones necessary for the prediction of reaction products (forward search). Thus once the effects influencing the reactivity of a system are correctly modelled, the course of a reaction and its products can be predicted. Economic and strategic considerations only come into play in retrosynthetic searches, along with the physicochemical evaluations. In synthesis design, selections have to be made among various pathways that could all be chemically amenable, but which nevertheless could be ranked according to some economic or strategic reasoning. The physicochemical evaluations should also reflect whether a forward or retrosynthetic search is being performed. To an extent, the same types of parameters can be used but with weights that portray the direction of a reaction. The discussion here will center initially around the development of physicochemical evaluations for a forward search. The modifications necessary for retrosynthetic searches will be briefly considered later.

We use the term "chemical reactivity" quite loosely throughout this article, and it is intended to imply reactivity in both the kinetic and thermodynamic senses. When several reaction sites are present in a molecule, it must be decided which site is the most reactive in order to determine the predominant reaction path and whether reaction at some of the other sites might lead to side reactions. Basically, this requires an evaluation of chemical reactivity at each reaction site. This requirement led to the conclusion that a global approach to chemical reactivity should assign a *reactivity value to each bond of a molecule*. However, there is no general method available for performing this task — there is no general theory of organic chemical reactivity. In this situation we embarked on a program to develop just such a general scheme which is able to assign reactivity values throughout a molecule. The working hypothesis in this endeavor was that the concepts used by an organic chemist in discussing reaction mechanisms have matured over the last several decades to a point that allows a rationalization of the individual steps of many organic reactions. These concepts include bond dissociation energies, atomic charges, inductive and resonance effects, polarizability, hard and soft character, etc. However, in many cases the concepts can only be used in a qualitative or, at best, a semi-quantitative manner. We believed that quantification of these concepts could provide a firm basis for a quantitative access to chemical reactivity values.

The computer plays a vital role in this undertaking at several stages. Foremost, it enables the efficient handling of all the numerical calculations for estimating the magnitude of the various energetic and electronic effects. As the objective is to perform them on each bond of a molecule, these calculations can be quite extensive. Beyond this, the computer can also serve as an invaluable tool in the *development* of the methods for calculating the magnitude of the various effects on chemical reactivity. In this process, a numerical model of the effect under consideration is designed which should reflect chemical observations that are under the influence of this effect. As a next step the model is put into an algorithm. Clearly, this requires an exact definition of the model since only then can it be rigorously programmed. In this way, the requirement of an algorithm enforces a clear logical conception of any model on the chemist. Once such an algorithm has been implemented as a program, the model can be tested for its validity, scope, and limitation to an extent that can hardly be achieved by any other means. The EROS system provides the ideal environment. The formal reaction schemes can generate from input molecules a multitude of reactions and product molecules, and all these can serve as test cases for the newly developed model in one run of the program (Fig. 15). Comparing the results of such a test run against chemical reality can lead to modifications and extensions of the initial model thus improving its performance. In effect, a tunable feedback loop can be established that leads to new insights, and to an optimization of the model and the associated program.

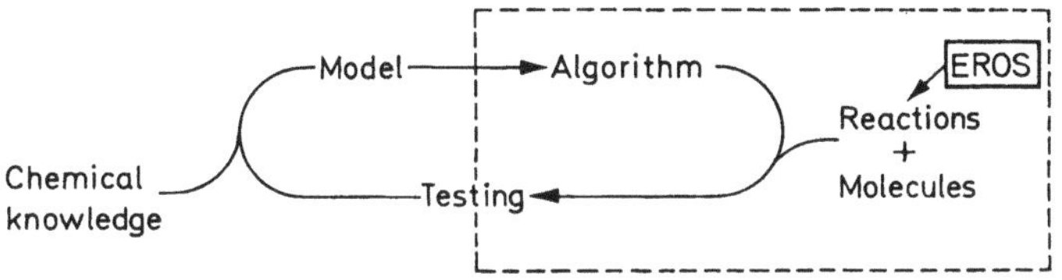

Fig. 15. Iterative process for development of a model (the large frame contains the tasks that are performed in a program run)

This setup for the development of a model is of particular merit for those situations where the performance of several models is being simultaneously tested with a single run of the EROS program. One of the complicating features associated with deciphering and modelling chemical reactivity is that it is simultaneously influenced by many of the various energetic, electronic and geometric effects, which can counteract or reinforce one another.

Particular care is taken in the design of a model to define it in such a way that it can be converted to a procedure characterized by short computation times. This is deemed essential for the evaluation of the large number of molecules which can be generated during a synthesis study or in reaction prediction. Since studies on molecules with up to 70 atoms are quite often performed with EROS, rapid evaluations of many large molecules has to be realistic.

The various evaluations can also be classified into two different types based on the way in which they are invoked. Some are performed only on the starting molecules. For example, this is the case when determining the dissociation energies of the bonds in a molecule. Other evaluations need a knowledge both of the starting materials and the products of a reaction, as in the calculation of heats of reaction (see next

Section). Clearly the ideal situation is to employ as many evaluations of the first type as possible, as these detect those bonds that actually take part in a reaction. A preselection is thus possible, guiding the reaction generation phase in the right direction. In the second type of evaluation, the reaction products must first be generated, with the possibility that they are ultimately evaluated as not being feasible for some reason. The formal reaction generation step would then be wasted time.

The next two sections deal with physicochemical models for the thermochemical and electronic effects which are used in the evaluations. This work is fairly well progressed. However, there are other effects which we are fully aware influence reactivity appreciably, but which we have so far investigated to only a limited degree. These include solvent and steric effects. We refer to these topics only briefly later in this article.

5 Thermochemistry

In this section we deal with the first of the physical effects which impinge on reactivity — the influences which heats of reaction and bond dissociation energies have on the course of chemical reactions. Both heats of reaction and bond dissociation energies are enthalpy values that are experimentally determined by thermochemical methods, in the first case usually by direct calorimetric methods, in the second by more indirect techniques [22].

5.1 Heats of Reaction

Heats of reaction give the enthalpy part of the overall thermodynamic expression of chemical reactions. The entropy contribution will not be discussed here, apart from noting that estimates of reaction entropies can also be obtained by an additivity scheme along similar lines to those used for estimating reaction enthalpies. Work in this direction, which also has to consider the changes in the number of molecules in a reaction and the symmetry numbers, has given promising results and w.ll soon he included in the EROS system [23].

The heat of a reaction, ΔH_r, is given by the difference in the heat of formation of starting materials and products (Eq. 1).

$$\Delta H_r = \Sigma\, \Delta H_f{}^\circ(\text{products}) - \Sigma\, \Delta H_f{}^\circ(\text{reactants}) \qquad (1)$$

In developing an empirical model for calculating reaction enthalpies [24, 25] resort was made to experience gained with additivity schemes developed for estimating heats of formation of organic compounds. These can be obtained by summing parameter values assigned to the various substructures in the molecule, the accuracy being improved by increasing the number of parameters considered. The Allen scheme [26] was chosen as a reasonable compromize between the number of parameters and estimation accuracy. Parameters for each molecule are assigned to 1,2-interactions (bonds) and those 1,3-interactions not involving hydrogen atoms. This scheme is numerically equivalent to the widely used Benson group additivity scheme [27] but offers advantages over the latter for our purposes. An increase in the accuracy of the estimate has been achieved [28, 29] by reparameterization of the scheme through multilinear regression analyses of experimental heats of formation. Only parameters for those substructures involving the bonds broken and made in a reaction need be considered in calculating heats of reaction. Thus, only the reaction site need be scanned, rendering computation times independent of the size of the molecules involved in the reaction. Due to the parameterization the values refer to the gaseous state at 293 K.

Ring strain energies [30] and aromatic delocalization energies must also be parameterized and calculated. This is handled automatically in EROS and makes use of the ring perception routine mentioned earlier [15].

The standard deviation between experimental and calculated heats of reaction are between 0.5 and 1 kcal/mol for those classes of compounds where enough experimental heats of formation are available to allow a full parameterization. For those classes of compounds where insufficient heats of formation are known to allow the determination of all parameters for 1,2- and 1,3-interactions, an estimate can be given for the bond energy terms which are the dominating parameters. Even here, therefore, a reasonable value for the reaction enthalpy is available.

Table 1. Products and heats of reaction of C_3-species from propane, oxygen and water

$CH_3{-}CH_2{-}CH_3 + O_2 + H_2O \rightarrow$	ΔH_r(exp)	ΔH_r(calc)
	(in kcal/mol)	
O_0-species		
$CH_3CH = CH_2 + H_2O_2 + H_2O$	− 2.82	− 2.72
$CH_2{-}CH_2$ (ring via CH_2) $+ H_2O_2 + H_2O$ \ / CH_2	+ 5.03	+ 5.66
$CH_3{-}C{\equiv}CH + 3H_2O$	−46.38	−45.87
$CH_2{=}C{=}CH_2 + 3H_2O$	−45.14	−43.93
$CH{=}CH$ $+ 3H_2O$ \ / CH_2	−24.57	−24.15
O_1-species		
$CH_3CH_2CH_2OH + H_2O_2$	−11.07	−10.53
$CH_3CHCH_3 + H_2O_2$ OH	−15.02	−14.44
$CH_2{=}CH{-}CH_2OH + 2H_2O$	−62.52	−63.72
$CH_3CH_2CH{=}O + 2H_2O$	−78.42	−77.13
$CH_3C{-}CH_3 + 2H_2O$ ‖ O	−84.87	−85.03
$CH_3CH{=}CHOH + 2H_2O$	—	−64.11
$CH_3C{=}CH_2 + 2H_2O$ OH	—	−65.55
$CH_3CH{-}CH_2 + 2H_2O$ \ / O	−55.60	−53.01

Table 1 (continued)

$CH_3-CH_2-CH_3 + O_2 + H_2O \rightarrow$		ΔH_r(exp) (in kcal/mol)	ΔH_r(calc)
CH_2-CH_2 CH_2-O	$+ 2H_2O$	−52.22	−50.42
CH_2-CH_2 CH–OH	$+ 2H_2O$	—	−59.25
CH_2-CH_2 C O	$+ 2H_2O + H_2$	—	−46.77
$CH_2=CH-CH=O$	$+ 2H_2O + H_2$	—	−48.97
$CH_3CH=C=O$	$+ 2H_2O + H_2$	—	−51.37
O_2-species			
CH_3CH-CH_2 OH OH	$+ H_2O$	−79.80	−75.40
$CH_2CH_2CH_2$ OH OH	$+ H_2O$	—	−71.53
$CH_3CH_2CH_2OOH$	$+ H_2O$	—	−18.30
CH_3CHCH_3 OOH	$+ H_2O$	—	−22.21
CH_3CH-CH_2 O—O	$+ H_2O + H_2$	—	+18.21
$CH_2-CH_2-CH_2$ O–O	$+ H_2O + H_2$	—	+ 2.22
$CH_3CH-CHOH$ O	$+ H_2O + H_2$	—	−42.04
CH_3 C–CH_2 HO O	$+ H_2O + H_2$	—	−42.95
CH_2CH-CH_2 OH O	$+ H_2O + H_2$	—	−30.94
CH_2-CHOH CH_2-O	$+ H_2O + H_2$	—	−39.53
CH_3C-CH_2 O OH	$+ H_2O + H_2$	—	−62.96
$CH_3CH-CH=O$ OH	$+ H_2O + H_2$	—	−58.97
$CH_2-CH_2-CH=O$ OH	$+ H_2O + H_2$	—	−55.06

Table 1 (continued)

$CH_3{-}CH_2{-}CH_3 + O_2 + H_2O \rightarrow$		ΔH_r(exp)	ΔH_r(calc)
		(in kcal/mol)	
$CH_3{-}C({=}O){-}CH{=}O$	$+ H_2O + H_2$	−39.97	−43.03
$CH_3CH_2C({=}O)OH$	$+ H_2O + H_2$	−83.57	−82.86
$CH_2{-}C{=}O$ / $CH_2{-}O$ (ring)	$+ H_2O + 2H_2$	−42.77	−39.76
$CH_2{=}CH{-}C({=}O)OH$	$+ H_2O + 2H_2$	—	−54.78
O_3-species			
$CH_2(OH)CH(OH){-}CH_2(OH)$	$+ H_2$	−56.67	−53.45
$CH_2(OH)C({=}O){-}CH_2(OH)$	$+ 2H_2$	—	−40.97

As an example, reaction products generated from propane, oxygen, and water are contained in Table 1 together with experimental and calculated heats of reaction. All products were obtained with the reaction generators RG22 and RG33, either directly from the starting materials, or by consecutive application of these two reaction schemes. Only those reactions were generated that did not involve the breaking of C—C bonds. This restriction was imposed, and the example chosen, to allow comparison with a sizeable number of experimental data. Within the C_3 series, experimental heats of formation are available for quite a few compounds of varying oxidation state. The values for the experimental heats of reaction were obtained with Eq. 1 by using the experimental heats of formation of the species involved.

As can be seen, experimental and calculated values for the reaction enthalpy are in rather good agreement. The compounds contained in Table 1 comprise quite a variety of functionalities, illustrating that the method for calculating heats of reaction is of general applicability. It can also predict values for compounds for which heats of formation have not yet been determined or which are unstable, like the vinyl alcohols. The formulae of the starting materials and products of the reactions contained in Table 1 always add up to the empirical formula $C_3H_{10}O_3$. This directs attention to an important application of the EROS system: the formal reaction schemes allow the generation of various sets of species for a given empirical formula. The method for calculating heats of reaction can then give an estimate of the enthalpy of these various ensembles of molecules. Thus, those points on an energy hypersurface can be determined which correspond to a given empirical formula and which represent stable molecular species or reaction intermediates, and their relative position with respect to each other can be fixed. Clearly this approach alone is not sufficient for predicting the course of chemical reactions, since no predictions of the energies of transition states can be made. However, it provides a powerful means for determining the thermochemical framework within which the actual course of a reaction has to be searched.

The values of reaction enthalpies in a forward search can be of use in predicting the products of a reaction: the more exothermic a reaction, the more it should be preferred. The situation will be different in a retrosynthetic search where retroreactions should be calculated to be endothermic to some degree. This underlines the point previously made (Sect. 2, Fig. 9) that the differences between a forward and a retrosynthetic search do not reside in the way reactions are generated — in both cases in EROS by the formal reaction schemes — but in the way they are evaluated.

So far we have not touched on the fact that the important topic of solvation energy is not yet taken into account. The extent to which solvation influences gas-phase energy values can be considerable. As an example, gas-phase data for fundamental enolisation reactions are included in Table 1. Related aqueous solution phase data can be derived from equilibrium constants [31]. The gas-phase heats of enolisation for acetone and propionaldehyde are 19.5 and 13 kcal/mol, respectively. The corresponding free energies of enolisation in solution are 9.9 and 5.4 kcal/mol. (Whether the difference between gas and solution derives from enthalpy or entropy effects is irrelevant at this stage.) Despite this, our experience with gas-phase enthalpies calculated by the methods described in this chapter leads us to believe that even the current approach is most valuable for evaluation of reactivity.

5.2 Bond Dissociation Energies

Homolysis of a bond is an elementary reaction that is of profound influence on reactivity in many processes. The enthalpy of such a step, the bond dissociation energy (BDE), can be calculated from Eq. 1 with the products now being atoms or radicals.

$$A - B \rightarrow A. + B.$$
$$BDE(AB) = \Delta H_r$$

Since homolytic or radical processes are largely governed by the effects of bond dissociation energies, a knowledge of BDE is required for the evaluation of chemical reactivity in such reactions. However, we have found, as we mention later, that BDE's are also an important factor influencing other types of reactions involving bond heterolyses.

The same program as used for calculating heats of reaction can also be used for estimating BDE when parameters for radicals are included. Such parameters have been determined for the important types of bonds in molecules and incorporated into tables in the program, so that the BDE is calculated automatically for each bond of a molecule [29]. Table 2 gives the results of a calculation on 2,2,5-trimethylheptane. The small changes in C—H and C—C BDE observed in going from CH_3 to a primary, secondary, or tertiary carbon atom are well reproduced. Equally good results are obtained for those classes of molecules where enough data on experimental heats of formation of radicals are available to allow full parameterization. For other cases higher uncertainties in the BDE's are inevitable.

Table. 2. Bond dissociation energies calculated for 2,2,5-trimethylheptane compared with corresponding experimental value; (a) Ref. 32 (b) Ref. 33

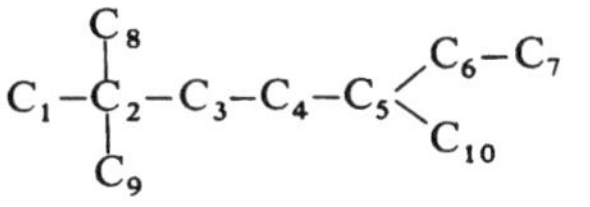

BDE(kcal/mol)	calc		exp (a)	exp (b)
C_1—H	98.67	C_{prim}—H	98	98.2 ± 1
C_3—H	94.93	C_{sek}—H	95	95.1 ± 1
C_5—H	92.53	C_{tert}—H	92	93.2 ± 1
C_3—C_4	82.56	C_{prim}—C_{prim}	81.8	82.2 ± 1
C_4—C_5	79.86	C_{prim}—C_{sec}	80.4	81 ± 1
C_3—C_2	78.44	C_{prim}—C_{tert}	77.7	79.1 ± 1
C_6—C_7	86.54	CH_3—C_{prim}	84.8	85.8 ± 1
C_5—C_{10}	83.84	CH_3—C_{sec}	84.2	85.7 ± 1
C_1—C_2	82.42	CH_3—C_{tert}	81.8	84.1 ± 1

In calculating the BDE, the program automatically accounts for effects of ring strain, aromatic delocalization energies, and stabilization through allylic or benzylic type resonance. Figure 16 gives the values of the BDE's in structure 7, a molecule that has a variety of structural features, and which incorporates allylic and benzylic resonance effects, as well as the resonance energy of the furan system when a ring bond is broken.

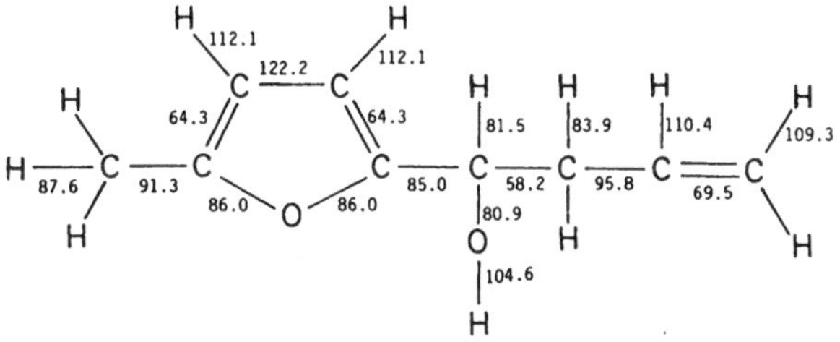

Fig. 16. Bond dissociation energies (in kcal/mol) calculated for 2-(but-3-en-1-olyl)-5-methylfuran, 7 (for double bonds, the value for the π-bond only is given)

6 Electronic Effects

Heats of reaction and bond dissociation energies allow the estimation of the feasibility of homolytic processes, as these are largely — but not solely — governed by thermochemical effects. The quantitative treatment of heterolytic processes, however, presents a far more difficult problem. Basic electrostatic considerations indicate that the dissociation of a covalent bond into positive and negative ions is inherently a highly endothermic process. It will be facilitated by any mechanism that allows dissipation or stabilization of the incipient charges. Chemists have come to differentiate these stabilization mechanisms according to various effects: charge, electronegativity, inductive, resonance, polarizability, hyperconjugation, hydrogen-bonding, etc. and, last but not least, solvent interactions. In most cases, several of these effects are simultaneously operative making a clear conceptual separation difficult. Indeed, two or more of the different effects may sometimes be representing the same physical phenomenon, and differ only in semantics. Additionally, these effects are frequently based to a large extent on qualitative reasoning only, or are applied qualitatively. Progress in understanding chemical reactivity can be best expected when the various factors of influence are put on a quantitative basis. Development and application of quantitative models for electronic effects is the theme of this section.

One of the concepts most widely used in discussing chemical reactivity and reaction mechanisms is the notion of *partially charged atoms* in reflecting the *polarity* of bonds. Thus, nucleophilic substitution at a saturated carbon atom is ascribed to an attack of a nucleophile at the carbon atom which bears a partial positive charge (Fig. 17). The nature of X in Fig. 17 will affect the positive charge on the carbon atom, but the overall reactivity does not correlate well with this charge value alone. Other factors beyond charge distribution are responsible for reactivity in nucleophilic aliphatic substitution. These factors have been termed nucleofugicity, polarizability, or soft character, and again are usually only applied in a qualitative manner.

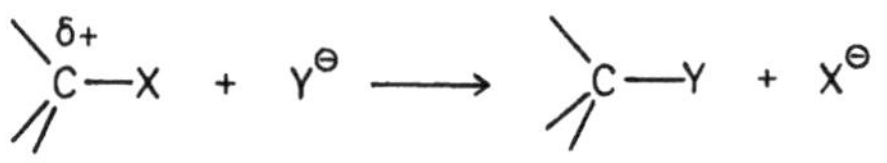

Fig. 17. Charge distribution in nucleophilic substitution

Looking at nucleophilic aliphatic substitution in more mechanistic detail reveals a further complicating feature. On heterolysis of the carbon-halogen bond in tertiary alkyl halides by an S_N1 process, electronic reorganization through the three CC-bonds occurs so as to decrease the partial positive charge on the central carbon atom. On the other hand, formation of a unit positive charge at carbon is avoided altogether in an S_N2 process by simultaneous formation of a new bond to the nucleophile during heterolysis of the carbon-halogen bond.

This example serves to make the point that in a quantitative treatment of chemical reactivity, electronic effects additional to those observed in the ground states of molecules have to be considered. In the following these extra effects will be comprehensively called transition state effects although they might also refer to electronic effects in other intermediate structures, including true reaction intermediates. Having stressed the importance of transition state effects we nevertheless aim to predict some of them from calculations performed on the ground states of molecules. In other words, can calculations on the starting materials of chemical reactions give information on the extent of electronic rearrangement occurring when these molecules react with each other? This approach is in line with our intention, stated in section 4, to assign a reactivity value to each bond of a molecule from calculations performed on that molecule. The reaction sites of molecules should thus be found without having to perform calculations on the products or transition states of a reaction, as well as starting materials. It will be shown later, for example, how data on the gas phase acidity of alcohols can be quantitatively modelled by parameters obtained from cal-

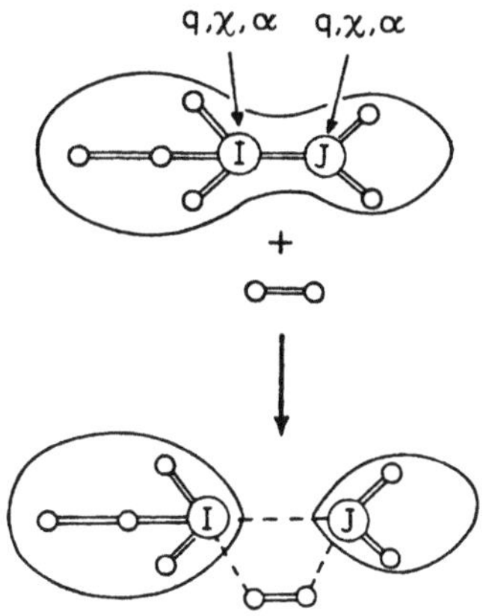

Fig. 18. Basic model for treating chemical reactivity (q = atomic charge, χ = electronegativity, α = polarizability)

culations on the alcohols themselves, without the necessity for calculations on the alkoxide ion products.

Our objective has been to develop methods that allow the calculation of various electronic parameters such as partial atomic charge, q, electronegativity, χ, polarizability, α, for each atom of a molecule. In this way, the values assigned to an atom not only reflect the type of the atom, but also the particular molecular environment into which this atom is embedded (Fig. 18). The electronic parameters assigned to the atoms of a bond will then be used to arrive at a quantitative value for this bond which reflects its reactivity. A detailed description of a reaction will also have to include parameters characteristic of the reagent in order to account for its influences on bond breakage and formation.

In attempting to evaluate electronic effects in organic molecules, we perform calculations on each molecule as an integral entity, that is, it is not separated into individual groups (e.g. substituent, skeleton, reaction site) that are assigned characteristic parameters transferable from one molecule to another. (This, incidentally, contrasts with the transferability of thermochemical parameters; see Sect. 5.) The interaction of a certain group with the rest of the molecule is dependent on the particular molecule considered, as for instance, with a nitro group, which not only interacts differently with aryl and alkyl groups, but also differently with various alkyl groups.

A common feature of the various methods that we have developed for the calculation of electronic effects in organic molecules is that they start from fundamental atomic data such as atomic ionization potentials and electron affinities, or atomic polarizability parameters. These atomic data are combined according to specific physical models, to calculate molecular descriptors which take account of the network of bonds. In other words, the constitution of a molecule (the topology) determines the way the procedures (algorithms) walk through the molecule. Again, as previously mentioned, the calculations are performed on the entire molecule.

Concepts like electronegativity, inductive, resonance effects, etc. have been developed by the chemist to bring order into a wealth of experimental observations. They are of an empirical nature defying an unequivocal theoretical derivation. Any attempt to put these concepts onto a quantitative basis has to face up to this situation. Values calculated for these electronic effects must be compared with experimental data in order to demonstrate their own validity, as well as the overall validity of the methods derived for their calculation. A further complication is the fact that in nearly every chemical reaction several of these electronic effects are operating simultaneously, and they do so to varying extents. A chemical reaction is a multiparameter event. It is therefore not simple to establish the significance of the individual values calculated by a particular method, and to demonstrate that they do indeed reflect the electronic effect for which they were designed.

In the absence of clear-cut guidelines on the mathematical form of the influence of a given electronic effect on chemical reactivity, we took the simplest form, a linear equation. If several effects were thought to be of importance, each of them was considered separately in a linear fashion. The significance of the calculated values was then tested, if possible, against experimental data through direct correlation, or via multilinear regression analysis (MLRA). Statistical techniques of this type, generally termed correlation analysis, are becoming increasingly important for analyzing quantitative data throughout organic chemistry[34]. However, such techniques bring certain dangers deriving from their misuse. For this reason, certain standard precautions were taken. For instance, the number of parameters applied at a time was always kept to a minimum, and a particular parameter was only included in a MLRA study if a definite indication of its relevance existed.

Electronic effects pertaining to the ground state of molecules were tested as far as possible against *physical* data of molecules. If feasible, several independent physical measurements were investigated, and primarily those data were selected where a clear understanding was available of the dependence of the physical observation on the electronic effects being modelled. Investigation of transition state effects necessarily depended on chemical reactivity data. Here, data on gas-phase reactions were investigated in the first instance. Recent progress in experimental techniques has given access to many data of known and high accuracy on reactions in the gas phase. These data are of particular interest as they reflect the *inherent* reactivities of individual molecules. In particular, the complicating effects of solvent which still largely elude a quantitative treatment could be excluded. Only later have we directed our attention to more generally interesting organic reactions, where the experimental conditions are less well defined. It is significant that the same models can be applied to such problems, as we describe below.

6.1 Partial Atomic Charges

The notion of a molecule as consisting of partially charged atoms is widely used in organic chemistry. It is clear, however, that it can only be a rough approximation to reality as in such an approach the continuous electron distribution of a molecule is split up and assigned to individual atoms. There is no unequivocal theoretical criterion for performing this separation. The most widely used method is Mulliken population analysis [35] of wave functions obtained from quantum mechanical calculations. However, the results of such an analysis are heavily dependent on the quantum mechanical method used and the basis set chosen. Furthermore, it is known that a Mulliken population analysis has basic deficiencies, primarily in its treatment of the overlap population. Therefore, results from quantum mechanical calculations cannot provide an unambiguous definition of partial atomic charges.

In our context one of the main prerequisites was a procedure for calculating atomic charges which was very rapid, as it is sometimes necessary to process many molecules of a fairly large size. The method we developed starts from the electronegativity concept, and uses electronegativity data, derived from atomic ionization potentials, IP, and electron affinities, EA (Eq. 2) [36].

$$\chi = 0.5(\mathrm{IP} + \mathrm{EA}) \tag{2}$$

Electronegativity was considered to be dependent both on the hybridization state of an orbital, and also on its electron occupation: an empty orbital must be more electronegative than a singly, and even more so than a doubly, occupied orbital. Equation 3 quantifies this dependence on occupation number, or more generally, on charge, q.

$$\chi = a + bq + cq^2 \tag{3}$$

The coefficients a, b, and c (with $a > 0$, $b > 0$) for this charge dependence can be derived from the electronegativity values of a given atomic orbital in the neutral state, and in the positive and negative ions, which in turn are derived from the relevant IP's and EA's [37]. Thus, the latter are the fundamental data on which the whole method is based.

On bond formation the more electronegative atom attracts electron density from the less electronegative atom. The former atom thus becomes partially negatively charged and consequently (Eq. 3) its electronegativity decreases. For the less electronegative atom the reverse applies. Thus, the orbital electronegativities of the atoms in a molecule tend to adjust, that is, to equalize. However, contrary to other approaches [38, 39], the orbital electronegativities of the free atoms are not completely equalized by our method, since the electrostatic potential that is created on charge transfer acts against further charge transfer. An iterative procedure was developed to quantify this Partial Equalization of Orbital Electronegativities (PEOE) [40, 41]. The form of the algorithm ensures rapid convergence. Consideration of changes in hybridization enables the peculiar bonding situations encountered in systems containing three- and four-membered rings to be calculated [42]. The conceptual basis of the PEOE procedure has been analyzed and compared with other definitions of electronegativity [43].

As there is no unambiguous definition of partial atomic charges, a host of physical data was investigated to show the utility of the charge values. The data studied included C-1 s ESCA chemical shifts [41], ^{1}H-NMR chemical shifts [44], dipole moments [42, 45], $^1J_{CH}$ coupling constants [42], and $^2J_{HH}$ coupling constants [46]. Calculated values for these data are in good agreement with experiment, thus demonstrating that the values of partial atomic charges obtained by our method are of physical significance and reproduce the phenomenon for which they were designed. Thus they can be confidently used to study the effects of charges on chemical reactivity.

An extension of the above method was developed for conjugated π-systems: Partial Equalization of Pi-Electronegativity (PEPE) [47, 48]. After calculation of the charge distribution in the σ-skeleton, the various resonance structures of a π-system are generated. The π-charge distribution is obtained by assigning weights to these resonance structures. In this case too, physical data such as C-1s ESCA shifts [48], ^{13}C-NMR chemical shifts [47,48], and dipole moments [48] served to establish the importance of the charge values thus obtained. A successful correlation with σ_R substituent constant values [47] was an early indication of the potential of these charge values for predicting chemical reactivity data.

In the course of these studies it became clear that the merit of the PEOE and PEPE methods lies not only in providing for each atom of a molecule a uniquely defined value of its σ- and π-charges. Beyond that, an electronegativity value can be obtained for each atom from the charge value through Eq. 3. This electronegativity value is not only characteristic of the atom type, but also of the molecular environment into which the atom is embedded. We call this value the *residual electronegativity*, since it reflects the remaining potential of the atom to attract further electron density [49]. In this sense, the residual electronegativity constitutes the property for which electronegativity was conceived: "the power of an atom *in a molecule* to attract electrons to itself" [50]. However, a dependence on molecular environment is now explicitly incorporated into the electronegativity values.

6.2 Inductive Effect

The nature of the inductive effect has been the subject of much controversy. Through-bond (inductive effect in its restrictive sense) and through-space (field effect) mechanisms have been suggested and experiments have been designed in attempts to differentiate between them. In our conceptualization of the inductive effect, we stay clear from these two limiting situations, noting that both types of mechanisms are operative and that, in general, the two effects parallel each other. However, it should be kept in mind that deviations could occur in those situations where two atoms are near each other in space, although separated by several bonds.

In attempting to quantify the inductive effect, two aspects have to be modelled: firstly, the magnitude of the inductive effect when changing the atom or the group that is exerting such an effect; and secondly, the attenuation of the inductive effect with distance, or, equivalently, with the number of bonds. Both aspects of the inductive effect as it influences ground state properties are quantitatively modelled by the PEOE method as reflected by the charge values. This can be monitored by the various correlations of the charge values with physical properties. For instance, the C-1s ESCA shifts of a substituted alkyl chain are attenuated the further they are from the substituent. It is found that the correlation holds up well in these cases [41].

Inductive effects in chemical reactions demonstrate their influence when charges are formed during bond-breaking or -making. A measure of the inductive effect is sought here which goes beyond that occurring in ground state charge distribution. In these situations residual electronegativity values provide an excellent quantitative measure of inductive effects on reactive states. Unfortunately, no satisfactory chemical reaction was found which is solely under the influence of inductive effects, and for which a sizeable amount of experimental data is available. The nearest test set we could devise contains only four points, and concerns the gas phase proton affinities (PA) of ethylamine and its three β-fluorinated derivatives [51]. In the light of other investigations on such protonation reactions (Sect. 6.4), we were satisfied that the magnitude of the PA should be influenced by inductive effects alone, since the polari-

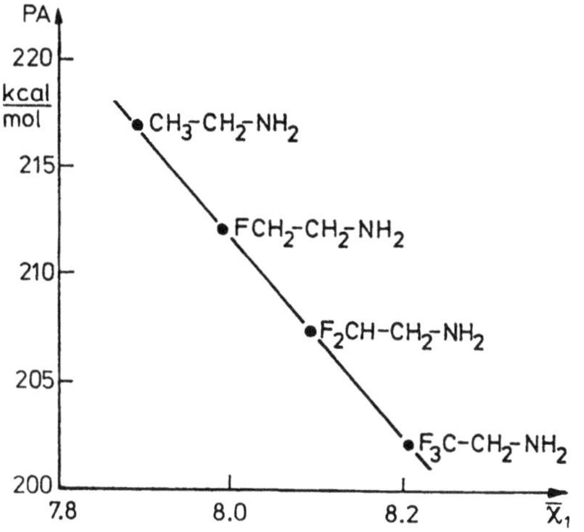

Fig. 19. Plot of experimental PA of ethylamine and fluoro derivatives against residual electronegativity, a measure of the inductive effect

zabilities of fluoroethyl groups are the same as that of ethyl itself. Consistent with this, a good correlation between PA and χ was found (Fig. 19). The data set can be extensively enlarged, but the effect of polarizability then becomes significant (Sect. 7).

Further discussion of the inductive effect is deferred until other effects have been described.

6.3 Resonance Effect

As with the inductive effect, resonance effects on ground state properties have already been included in the procedure, PEPE, for calculating partial atomic charges. This has been achieved by generating and weighting the various resonance structures of a molecule. The significance and quality of the results has been shown by correlations and calculations of physical data [47,48,52].

However, as for the inductive effect, models must again reflect the influence of resonance effects on reacting states (transition states or reaction intermediates), as well as just ground states. The breaking and making of bonds are strongly facilitated when the charges thereby generated are stabilized by delocalization. In fact, this effect is probably the single most important influence on chemical reactivity, as reflected by the range of reactions which are predominantly governed by this influence: electrophilic and nucleophilic aromatic substitutions, aldol, Claisen and related condensations, reactions of allylic and benzylic systems, etc. (Fig. 20).

Both positive and negative charges are generated in a heterolytic bond cleavage, and in both cases resonance stabilization mechanisms can be envisaged. The $-$M effect pertains to stabilization of negative charges, and the $+$M effect to stabilization of positive charges (Fig. 21). For the former, the negative charge resulting from a free electron pair is delocalized into an empty orbital located on an atom E, or into an antibonding orbital of a double bond, C=D, with the result that D has the largest orbital coefficient. Thus, the electronegativity of the π-orbital on E or on D is of predominant importance, and we take this π-electronegativity, obtained from the PEPE procedure (see 6.1), as a quantitative measure, R^-, of the $-$M effect.

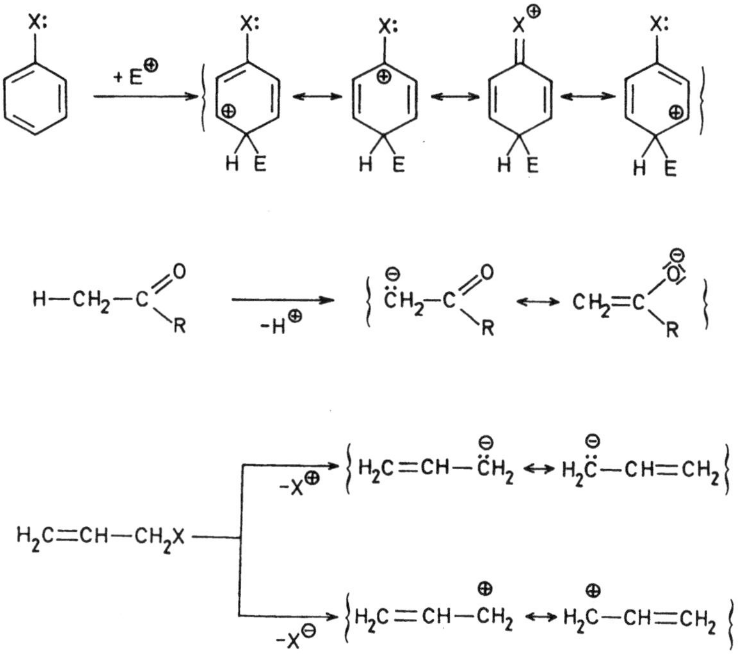

Fig. 20. Reactions that are influenced by the stabilization of charges through resonance effects

A positive charge in an empty π-orbital can be stabilized by the +M effect exerted by the free electron pair on an adjacent atom X, or by a filled π-orbital of a double bond C=D. In delocalizing a positive charge into a double bond C=D, the larger orbital coefficient is again on atom D. The higher the electronegativities of the orbitals on X or on D, the less they are available for donation of electron density into the

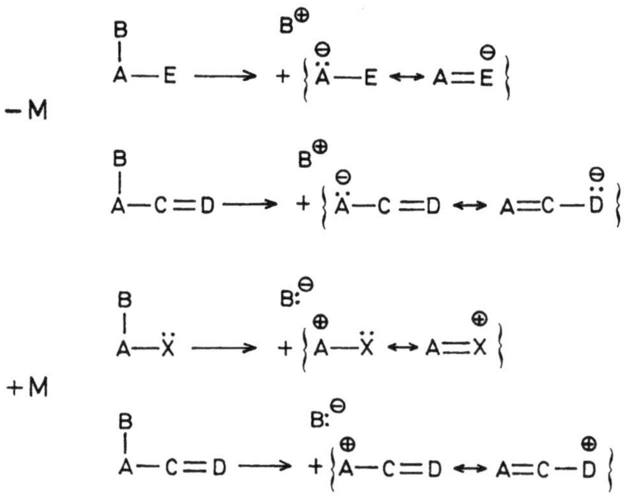

Fig. 21. Situations where the −M and the +M effect are operating

empty orbital on A. The reciprocal of the π-electronegativity of X or of D is taken as a measure of the +M-effect, R^+. Values of R^+ and R^- could only be studied in conjunction with other effects, in multiparameter analyses of various systems. These are discussed in Sect. 7.

6.4 Polarizability Effect

An external electric charge induces a dipole in a molecule such as to stabilize the presence of that charge. This induced dipole is the higher, the more polarizable the molecule, as measured by its mean molecular polarizability, $\bar{\alpha}$. The stabilization energy, E_{ci}, due to interaction between the external charge and the induced dipole is given by Eq. 4, with r being the distance between charge, q, and induced dipole, and ε the dielectric constant of the medium

$$E_{ci} = -\bar{\alpha}q^2/2\varepsilon r^4 \quad (4)$$

However, this formula loses its significance when the charge is introduced into the very same molecule that is being polarized. This is the situation most commonly met in chemical reactivity, where the charge results from the attack of an electrophile or nucleophile. In particular, the distance between the charge and the induced dipole is no longer defined, and *mean* molecular polarizability is no longer the appropriate property to use. In any case, different atom types have different ease of electronic distortion and their contribution will be strongly distance-dependent. To account for these two phenomena we have defined an *effective polarizability*, α_d, and introduced empirical methods for its calculation [53, 54]. α_d is to be taken as a measure of the po-

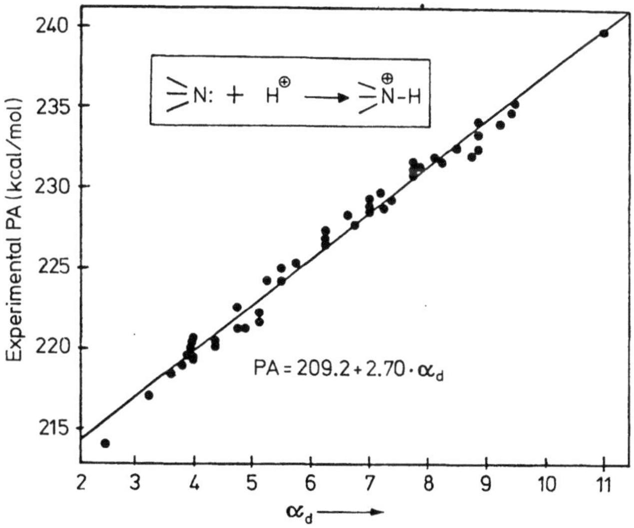

Fig. 22. Proton affinity of 49 unsubstituted alkylamines (n 49; r 0.984; s 1.0 kcal/mol) [55]

larizability effect and in this sense is proportional to, and represents, the *stabilization energy resulting* from the charge-induced dipole interaction.

The significance of the values calculated for the effective polarizability was first established with physical data, among them relaxation energies derived from a combination of X-ray photoelectron and Auger spectroscopy, as well as N-1s ESCA data [53, 54]. From our point of view, however, the most important applications of effective polarizability are to be found in correlating chemical reactivity data. Thus, the proton affinity (PA) of 49 unsubstituted alkylamines comprising primary, secondary and tertiary amines of a variety of skeletal types correlate directly with effective polarizability values (Fig. 22).

This is a case where α_d suffices to reproduce the experimental data. For most of the chemical reactions investigated, which were primarily gas-phase reactions for reasons given previously, effective polarizability had to be used in combination with other electronic effects, especially measures of the inductive effect. This becomes necessary, for instance, when studying the PA of amines that contained species with heteroatoms in the substituent. This and other reactions are discussed further in Sect. 7. In all gas-phase reactions investigated, effective polarizability proved to be of importance. It also provided a quantitative explanation [56] of the puzzling result that gas-phase protonation of alcohols (positive charge development) and gas-phase acidity of alcohols (negative charge development) are both favored in the order Me < Et < i-Pr < t-Bu. The polarizability effect results in stabilization of both positive and negative charges.

In solution, stabilization due to polarizability is of less importance, as the solvent will provide other charge stabilization mechanisms. However, recent work shows that polarizability is still of influence in solution [57, 58, 59].

6.5 Hyperconjugation

It now seems well established that hyperconjugation provides an efficient means of stabilization for positively charged species where the charge results from an empty p-orbital, but is of less importance in neutral molecules. Alternatively, in MO terminology, orbitals of appropriate symmetry of C—H and C—C bonds of alkyl groups can overlap with adjacent empty orbitals and thus donate electron density. We have modified [60] a simple procedure introduced by Kreevoy and Taft [61] for estimating hyperconjugation empirically. Here again, there is no physical or chemical phenomenon which allows a direct confirmation of the method, via correlation analysis, However, the values have been found acceptable in multiparameter treatments (see Sect. 7), where we found that hyperconjugation contributions of C—H and C—C bonds are the same. Such contributions are calculated in EROS itself simultaneously with resonance effects. The two are of course conceptually related.

6.6 Frontier Molecular Orbital Approaches

Frontier molecular orbital (FMO) theory [62] has provided new insights into chemical reactivity. This, and the simplicity of its application, has led to its widespread use, particularly in the treatment of pericyclic reactions [63]. An FMO treatment depends on the energy of the highest occupied (HOMO) and lowest unoccupied molecular

ε_{HOMO} = IP	χ_j = 0.5 (IP+EA)
ε_{LUMO} = EA	
c_{ij}^2 = q_{ij}	q_j

Fig. 23. Comparison between FMO-parameters and charge/electronegativity values. c_{ij} refers to the coefficient of orbital i at atom j

orbitals (LUMO) of the reaction partners, as well as their orbital coefficients. According to Koopmans' theorem, the energy of the HOMO can be set equal to the first ionization potential, IP, and the LUMO energy corresponds to the electron affinity, EA. The definition of electronegativity (Eq. 2) indicates some correspondence between an FMO treatment and our charge/electronegativity treatment (Fig. 23).

We have only just started to explore empirical access to FMO parameters based on these similarities [52)]. Recently, others have reported empirical equations for calculating IP's and EA's for a variety of π-bonded systems [64)]. This approach used a large number of parameters for the underlying π-system, heteroatom substitution, and the substituents on the π-system. However, we aim at calculating FMO parameters from fundamental atomic data while taking due account of the bond structure of a molecule.

In line with expectations suggested by the comparison of Fig. 23, electronegativity data do indeed prove valuable for correlating ionization potentials. Thus, the IP's of a variety of dienes, enynes, diynes and heterodienes comprizing molecules like butadiene, styrene, acrylonitrile, hexatriene, dihalodiynes, furan, naphthalene, phenol, and anthracene, could be reproduced by three parameters: the mean of the σ- and of the π-residual electronegativities, as well as the mean of the R^+ values [52)]. The R^+ parameter (Sect. 6.3) is necessary because on ionization a positive ion is generated that can be stabilized by resonance effects. The statistics of the correlation with these three parameters is reasonable. A similar result could be obtained with these same three parameters for the IP's of dienophiles. In fact, the coefficients for the three parameters in these two correlation equations are rather similar; a sizeable difference exists only in the constant term of the two equations. However, the difference of 1.46 eV corresponds nearly exactly to the difference between the IP's for the two basic systems of the two correlations, butadiene and ethylene. Thus, the two data sets can be combined into a single correlation equation. This result is remarkable in that it covers both electron-rich and electron-deficient systems, including acetylenes, ethylenes, dienes, and aromatic compounds with a variety of substituents, as well as systems with heteroatom substitution. The ionization potentials for these systems vary from 7.29—17.70 eV. They can now be estimated through the electronegativity and resonance effect parameters, directly from the constitution of a molecule, without recourse to M.O. methods.

Frontier orbital approaches are not yet implemented in EROS. Nor does EROS take account of the features of reactivity which are controlled by orbital symmetries. This will follow the current work on stereochemistry and conformation.

7 Chemical Reactivity — A Multiparameter Event

In the preceding two sections we have introduced and described models which quantify various physicochemical effects. Wherever possible we have demonstrated their validity by direct correlation with appropriate data. However, as was implied in those sections, in many cases the parameters are applicable *in combination* to descriptions of less well resolved situations. It is widely accepted that chemical reactions are frequently under the influence of more than one effect. Thus, when developing quantitative descriptions of such reactions it becomes necessary to use several parameters which describe the different effects: the treatment becomes *multiparameter*. In this section we develop the discussions of the previous two. However, we continue to concentrate on well-defined, accurate data wherever possible, in a gradual approach toward building up a general treatment of reactivity. In the following, MLRA is applied to the various problems.

The proton affinities (PA) of two restricted subsets of amines were correlated directly with inductive and polarizability effect parameters, respectively (Figs. 19 and 22). These can be combined with data on other hetero-substituted amines to give a set of 80 amines of different skeletal and substitution types (e.g. Fig. 24). In this and all other systems (below), a residual electronegativity value, $\bar{\chi}_{12}$, (Eq. 5) derived from those of the atoms of the first, $\bar{\chi}_1$, and second, $\bar{\chi}_2$, sphere neighbors of the nitrogen atom is preferred as a measure of the inductive effect [49].

$$\bar{\chi}_{12} = 0.5(\bar{\chi}_1 + 0.25\bar{\chi}_2) \tag{5}$$

Thus, the PA data (in kcal/mol) on the 80 amines can be reproduced quantitatively with Eq. 6 (r 0.998; s 1.33 kcal/mol) [55], $\bar{\chi}_{12}$ being a measure of the inductive effect and α_d that of the polarizability effect.

$$\text{PA(amines)} = 343.0 - 27.79\bar{\chi}_{12} + 2.99\alpha_d \tag{6}$$

The signs of the coefficients in this equation are consistent with the interpretation of the two factors: the negative sign of the coefficient for the $\bar{\chi}_{12}$ parameter indicates that an increase in the inductive effect destabilizes the corresponding ammonium ion and thereby leads to a decrease in the proton affinity. On the other hand, an increase in the effective polarizability, α_d, stabilizes the ammonium ion and therefore

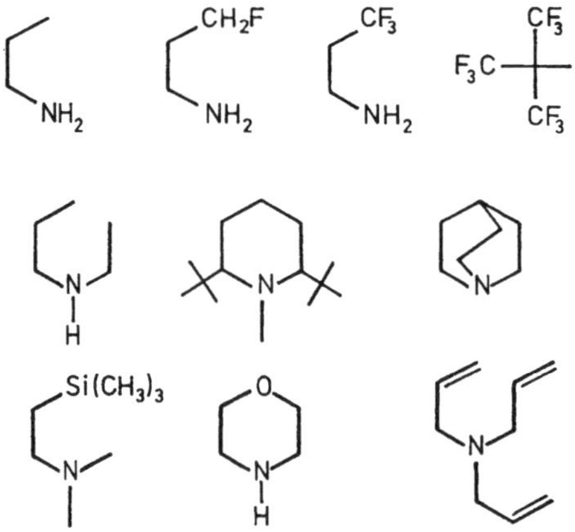

Fig. 24. Example amines included in the proton affinity study (Eq. 6)

leads to increased PA. Note that the influence of the inductive effect, as well as that of polarizability, on stabilization or destabilization of the ammonium ion has been derived from calculations on the neutral amine. This is in line with our intention, mentioned previously, to predict data on chemical reactivity from calculations on the ground state of reactants.

Extensive studies on other gas-phase reactions support the interpretation of a residual electronegativity value, e.g. calculated as in Eq. 5, as being a quantitative measure of the inductive effect for reacting species. Thus, two-parameter equations similar to Eq. 6 using $\bar{\chi}_{12}$ and α_d values were developed to reproduce PA data of alcohols and ethers, as well as those of thiols and thioethers [56] (Fig. 25). In addition, the same two parameters were used for correlating data on gas phase acidity of alcohols (Fig. 25) [56]. In this case the coefficients of the $\bar{\chi}_{12}$ and the α_d parameters have the same sign, consistent with the physical picture that an increase of both the inductive effect and of polarizability stabilizes the negative alkoxide ion, leading to increased acidity.

Our study of these elementary reactions was extended to protonation reactions of unsaturated species (Fig. 26) [60]. For protonation of ketones and aldehydes the

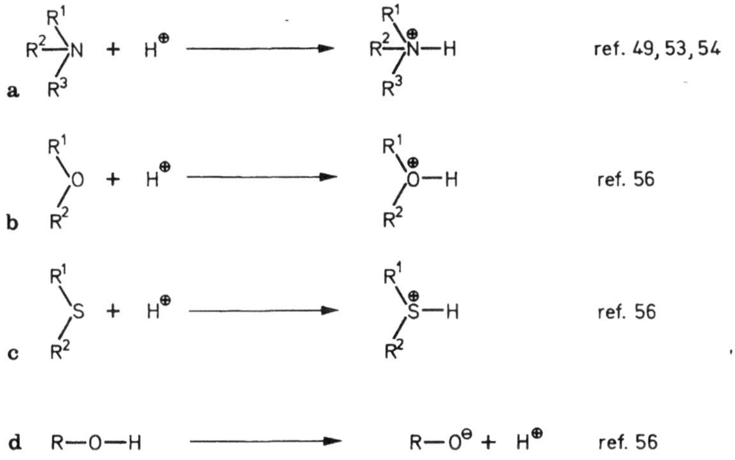

Fig. 25. Gas-phase reactions for which quantitative 2-parameter models have been developed

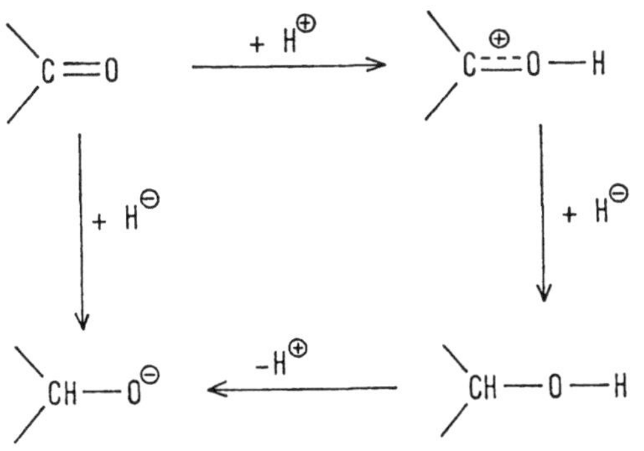

Fig. 26. Reaction cycle between carbonyl compounds and alcohols

same inductive effect and polarizability terms can be used, but it is found that a hyperconjugation parameter is also needed. This latter effect is a feasible source of stabilizing interaction between alkyl substituents and the positively charged hydroxycarbenium center. Values for hydride ion affinities of carbonyl compounds can be established by considering the differences in the heats of formation of carbonyl compounds and of alcohols (Fig. 26), and combining with PA and acidity values. Correlation equations with the above parameters have also been obtained for these hydride ion affinities.

These models refer to reactions with the simplest nucleophile, H^-, both under neutral conditions and in the protonated form. Chemical reactivity can be strongly altered by catalytic effects; acid/base catalysis is of particular importance. We regard the studies on gas phase acidities and on proton affinities discussed in the above sections to bear special significance for quantitative modelling of acid/base catalysis in the future.

Yet further extension of these multiparameter models brings in not just hyperconjugation effects of simple alkyl groups, but also resonance stabilization effects of hetero and unsaturated substituents. One such example is the rate of gas phase elimination of HCl (or HX) from substituted alkyl chlorides (or from bromides and alcohols). These reactions proceed via rate-determining cleavage of the C-X bond, to give transition states which resemble substituted carbenium ions. Correlation analysis between log k data (or activation energies) and three parameters which describe inductive, resonance, and polarizability effects, gave good quantitative models [65].

One of the underlying reasons for the intensive study of gas-phase reactions by workers in this area has been the belief that only by eliminating the influence of solvent can inherent effects be studied. Our approach is consistent with this, in that once chemical reactivity in the gas phase can be modelled, a foundation has been laid for investigating the additional influence of solvent. In fact, a study along these lines has already related gas phase alcohol acidity to the equivalent process in aqueous solution [59]. The underlying trend in pK_a variation is dominated quantitatively by inductive effect considerations. In contrast to the gas phase, polarizability influences are not generally apparent, and only enter into the model when particularly polarizable groups such as CCl_3 and CBr_3 are present. On this basis, a good two-parameter model can be derived for polar-substituted derivatives. The major difference between acidity in the gas phase and aqueous medium is the opposite influences of nonpolar alkyl and aryl groups in the two phases. Via a polarizability effect, such groups stabilize negative charge in the former, while their dominant acid-weakening effect in solution is ascribed to their different interaction with solvent structure, compared with polar-substituted derivatives. This results in differences in entropy (rather than enthalpy) contributions to free energy. Such an effect can be modelled by quantifying the number of such groups in the alcohol, and the good overall correlation derived includes 45 alcohols [59].

This chapter has outlined specifically how quantitative data on somewhat idealized reaction systems can be used as a basis for demonstrating the validity of our empirical electronic models in the field of reactivity. The multiparameter statistical models derived for the systems studied (PA, acidity, etc.) have limited direct application in EROS themselves. The next section develops the theme of applying the models in a much more general way, leading up to general reactivity prediction in EROS itself.

8 A General Model of Chemical Reactivity

More than just a few parameters have to be considered when modelling chemical reactivity in a broader perspective than for the well-defined but restricted reaction sets of the preceding section. Here, however, not enough statistically well-balanced, quantitative, experimental data are available to allow multilinear regression analysis (MLRA). An additional complicating factor derives from comparison of various reactions, where data of quite different types are encountered. For example, how can product distributions for electrophilic aromatic substitutions be compared with acidity constants of aliphatic carboxylic acids? And on the side of the parameters: how can the influence on chemical reactivity of both bond dissociation energies and bond polarities be simultaneously handled when only limited data are available?

Nevertheless, we are aiming at a global treatment of reactivity that incorporates all the various effects mentioned in Sects. 5 and 6 and is applicable to any type of chemical reaction. This is most certainly an ambitious goal. To progress in this direction we are balancing the various effects against each other by a combination of efforts. A brief outline of one approach that determines a function for calculating a value for chemical reactivity is given here.

8.1 The Reactivity Space

The breaking or making of a bond is an event influenced by many parameters. In order to represent these influences on a system, a multi-dimensional space can be constructed. Each coordinate represents a certain effect and has one of the parameters calculated by the procedures mentioned in the previous sections as its quantitative measure. Each bond of a reacting molecule is represented in this space by a single point having a specific value for each parameter. For example, Fig. 27 shows such a reactivity space spanned by using the differences in charge and in electronegativity of a bond, as well as bond polarizability, as three coordinates. Two bonds of differing reactivity are shown in this space. The I-Br bond is characterized by high polarizability but low polarity and electronegativity differences. The opposite is true for the H–F bond.

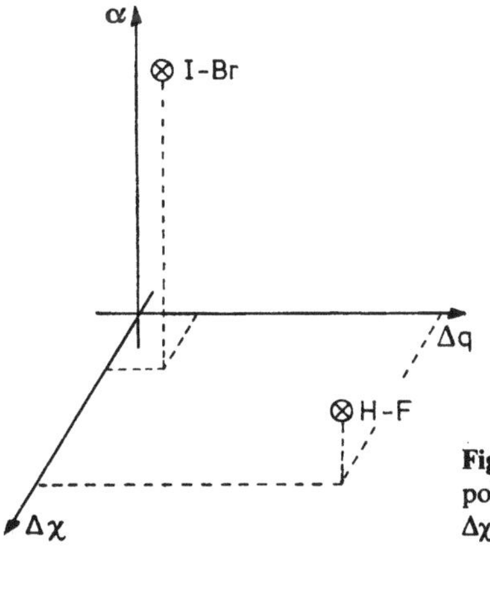

Fig. 27. A three-dimensional reactivity space defined by polarizability, α, and charge, Δq, and electronegativity, Δχ, differences

In general, reactivity spaces of higher dimensionality have to be considered as more physicochemical effects become influential. Although such spaces can no longer be given pictorially, they represent no problem to investigation by pattern recognition methods. We have applied both supervised and unsupervised techniques such as linear discriminant analysis, the K-nearest neighbor method, and principal component and cluster analyses. These methods have shown that chemical reactivity is a property well represented in such spaces, bonds being further separated from each other the more they differ in reactivity. When the bonds are classified as either breakable or non-breakable a good clustering of the two types is observed.

This classification of bonds allowed the application of *logistic regression analysis* (LoRA), which proved of particular benefit for arriving at a function *quantifying* chemical reactivity. In this method, the binary classification (breakable or non-breakable, represented by 1/0, respectively) is taken as an initial probability P_0, which is modelled by the following functional dependence (Eqs. 7 and 8) where f is a linear function, and x_i are the parameters considered to be relevant to the problem. The coefficients c_i are determined to maximize the fit of the calculated probability of breaking (P) as closely as possible to the initial classification (P_0).

$$P = 1/(1 + e^{-f}) \quad (7)$$

$$f = c_0 + c_1x_1 + c_2x_2 + \ldots \quad (8)$$

An important feature of the logistic regression method is that although the input modelling data (P_0) are binary, the calculated probability (P) is a continuous function.

In application, the method involves the following: from a series of molecules, assign a selection of bonds as definitely breakable or definitely non-breakable. These binary data are input, along with their calculated physical properties (e.g. BDE, Δq, $\Delta\chi$, α_d, etc.) into the LoRA. This in turn calculates the optimum functions, f and P, to describe the input data set. This calculated f is defined as the *reactivity function*.

In one analysis aimed at determining the reactivity of single bonds, 29 molecules were chosen as a training set for representing a wide variety of aliphatic chemistry [66]. From the 770 bonds contained in these 29 molecules, 111 were selected and specified as either breakable (36) or non-breakable (75). Four of these molecules are shown in Fig. 28, and the bonds selected as breakable or non-breakable are indicated. Note that not all bonds of a molecule have to be specified as either breakable or non-breakable. Thus the LoRA study can be concentrated on a particular field of chemistry. In fact, this data set contained a sizeable number of multiple bonds that are of course potentially reactive. However, in the present study no specification of breakable or non-breakable was made for multiple bonds in order to focus the investigation on the reactivity of single bonds. It was found by LoRA that with three variables (the polarity Δq, BDE, and resonance effect, R), a probability, P, is calculated that reproduced well the classification of bonds into breakable and non-breakable. Of even further significance is the fact that with the corresponding reactivity function, f, a numerical estimate of the ease of breaking of each bond is obtained, that is, a quantitative measure of its reactivity. By using logistic regression analysis we have managed to use the *qualitative* information of whether a bond is breakable or not, to arrive at a function that predicts chemical reactivity *quantitatively*.

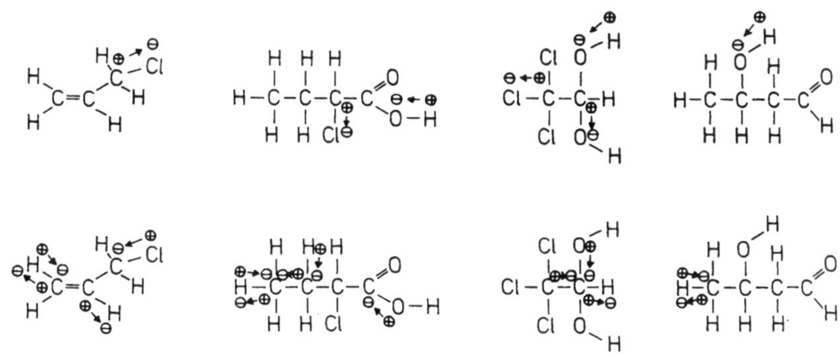

Fig. 28. Example molecules used in a training set for LoRA. Assignments of breakable and non-breakable bonds are shown

Although the reactivity function obtained above is of quite general validity, we do not believe that just one such simple, singular function will describe chemical reactivity in a universal manner. Rather, we are presently working with several such functions. Thus, in a similar manner to that described above, additional functions were obtained by LoRA for the reactivity of multiple bonds, and for bonds in charged species.

8.2 Heuristics in EROS

However, there are still important reactivity features which have so far been neglected by the reactivity functions, but yet which must be accounted for even at this stage of development if a sensible overall approach is to result. An important case concerns the special position of the hydrogen atom, and its ion, the proton. Its peculiar role in chemistry is reflected particularly in the way that even weakly basic solvents are able to interact with, and stabilize, it to a degree sufficient to render it a common and feasible independent entity in chemical reactions. This is in marked contrast to simple alkyl group ions, such as the methyl cation, whose electronic properties in many respects are very similar to those of the proton. Our current level of model development does not reflect this difference, and so specific allowance must be made artificially for the proton.

This is achieved by the inclusion of "heuristic rules". These are general rules-of-thumb which owe nothing to theory or an understanding of basic underlying principals, but which emanate from empirical observations of relevant systems (i.e. chemical reactions). (For further discussion of heuristics, see Sect. 11.)

The most important heuristics relate to reaction conditions, in particular, to acid-base catalysis. Depending on whether acidic or basic conditions are specified, the reactivity of certain bonds is changed. As an example, under basic conditions the breakability of H–X bonds is increased in comparison with other bonds. In fact, the relative acidities of all H–X bonds (X = any other element) can be rapidly calculated in EROS, and this allows further distinction within this class of bonds.

EROS also contains other less important, more formal heuristics which together with the reactivity functions comprise the chemical knowledge from which the program works out reactivity in general.

The approach which is now available is sufficiently general to be applied to organic reactivity in a much broader sense. The examples in the next section show how the reactivity functions perform in EROS when confronted with reactivity problems of several types.

9 Examples for Reaction Predictions

The important question is how good the reactivity functions are when applied in EROS to real problems, centered on molecules other than those whose parameters and properties led to its derivation. For this purpose, we now return to the problems

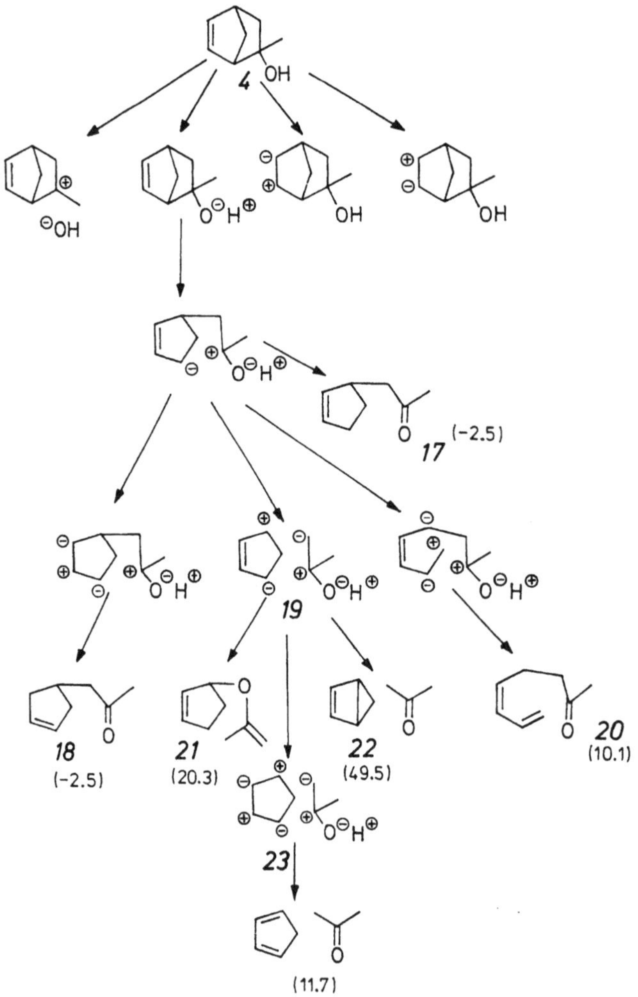

Fig. 29. Bond breakages calculated by EROS for the reaction of hydroxynorbornene *4* with base. Enthalpies are in kcal/mol in parentheses, relative to *4*

posed in Fig. 3. The following studies were made by combined application of the three reactivity functions mentioned above, the two pertaining to the reactivity of single and double bonds in neutral species, as well as the reactivity function valid for charged species. These functions determine a series of preferential heterolytic bond cleavages giving intermediate structures. It is emphasized that the charges shown on the intermediate structures in Figs. 29 to 32 are not to be taken literally. They represent bond polarization and direction of bond-breaking, and do not imply that species containing multiple charge centers are actually present as intermediates or transition states! At each level in these figures, the intermediate structures are arranged from left to right by decreasing value of breakability as calculated by the reactivity functions. Some are not developed further into the next level, when rejected by a heuristic judgement, e.g. that based on acidic or basic reaction conditions. Values for the reaction enthalpies are given in parentheses and are in kcal/mol with respect to the heat of formation of the starting molecule.

A summary of EROS's results for the reaction of the hydroxynorbornene, *4*, under basic conditions is given in Fig. 29. In practice, the experiment was performed with the potassium salt in HMPT solution [67].

The reactivity function calculates four potential bond breakages. (Each bond is represented twice in EROS, once for each direction of polarization on heterolysis, thus the C=C bond can break in two ways.) However, in the presence of base, only one of these is considered likely by EROS and developed further. A reactivity function derived for charged species is then applied to this "intermediate", with the result that at the next stage, equivalent to the reaction generator RG22 (Sect. 2.2 and 3.2), only one further bond is considered breakable. Bond reformation according to RG22 leads to the ketocyclopentene, *17*, as a possible product. The process of applying the reactivity function is then repeated, to give the RG33 level containing three possible 3-bond combinations. One of these corresponds to *18*, an isomer of *17*.

In practice, fragmentation of exactly this kind is observed under basic conditions, to give *18* in 45% isolated yield along with less than 10% *17* [67]. (With alkyl groups other than the methyl substituent the ratio of 2- to 3-cyclopentenyl isomers can rise to 40:60.) [67]. The reactivity functions in EROS evaluate these products quite highly. The enthalpy of reaction is also realistic at —2.5 kcal/mol for both products. This value is more favorable than those for the other possible products suggested.

What of the other suggested products in Fig. 29? The RG33-type fragmentation leading to *20* is still formally permitted by EROS. However, orbital symmetry constraints suggest that such a reaction would not occur in practice. As yet, EROS pays no heed to this type of orbital overlap consideration, which depends on knowledge of stereochemistry and conformation.

The third alternative product precursor in Fig. 29, *19*, can lead to two possible products by RG33-type bond formation, *21*, and *22* plus acetone. Each is rather unlikely, a fact which is also reflected by the unfavorable reaction enthalpies. However, further development of *19* gives *23* which results in cyclopentadiene and acetone. (When starting from *4* this sequence corresponds to an RG44 type, a reaction generator not used directly in EROS. Rather, it is currently obtained by consecutive application of RG33 and RG22 generators.) In fact, evidence for a retro-Diels-Alder reaction was found when the reaction was performed on the anion of *4* containing a benzyl or phenyl group in place of methyl. For the phenyl derivative, the product corresponding to *18* was isolated in only 8% yield, accompanied by 62% acetophenone, which was assumed to be formed in a retro-Diels-Alder reaction [67]. Furthermore, the bond-breaking scheme of *19* does suggest an alternative pathway to the observed reaction products, *17* and *18*, as follows. Could the alkoxynorbornene in the reaction mixture undergo retro-Diels-Alder reaction to these intermediates (enolate anion rather than acetone), which then recombine in a Michael-type reaction to give *17* and *18*? This pathway in practice is quite different from that implied by Fig. 29, but is sensible enough to warrant consideration. After all, Fig. 29 suggests carbanionic intermediates which would not normally be considered particularly favorable products to emanate from less basic alkoxides.

The starting hydroxynorbornene, *4*, contains 22 bonds, of which 18 are constitutionally distinct. There are therefore 36 distinct single bond heterolyses. Application of RG22 to all possible bond combinations leads to at least 864 possible reactions [68]. RG33 can give at least 10,000 possible reactions [68]. Thus, to summarize, EROS's evaluations not only reduce a totally intractable number of formal possibilities to five feasible products, it also suggests, correctly, which of these are the most likely.

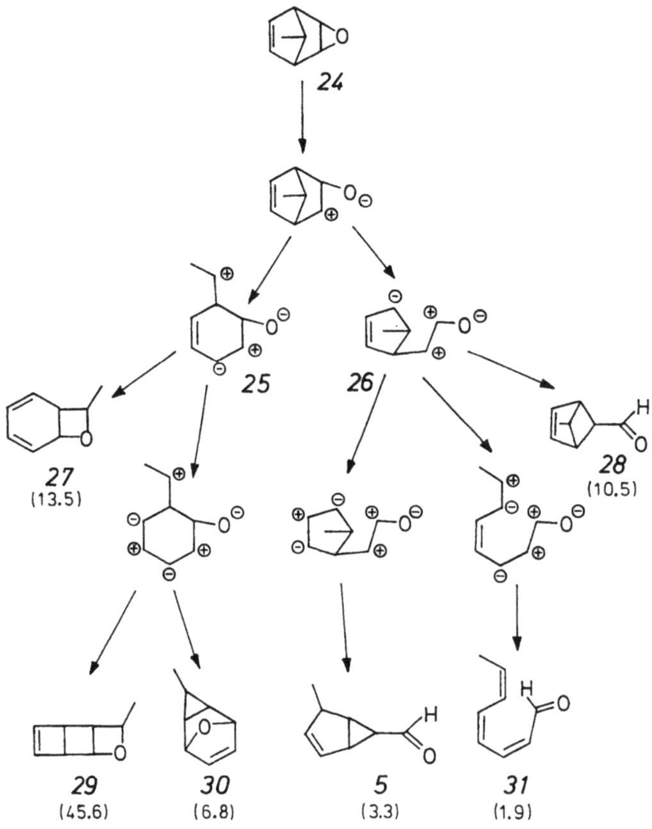

Fig. 30. Acid-catalyzed rearrangement of 7-methyl-norbornadieneoxide, *24*. Values in parentheses are enthalpies relative to *24*, in kcal/mol

Furthermore, it has also suggested a less obvious pathway for their formation. And finally, it points to a side reaction actually observed in a slightly modified system.

A second interesting reaction sequence is presented in Fig. 30, which displays the intermediate structures and possible products as predicted by the reactivity functions of EROS.

In practice, during epoxidation of norbornadiene with peracid, *24* could not be isolated but rearranged to give *5* as product [69-72]. Apparently, passing through *26* is favored over a pathway involving *25*, an effect not yet modelled by our reactivity functions. However, it is significant that the two intermediates *25* and *26* differ only slightly in their evaluation. In discussing the structure of the final product, *27* was considered as a possible product, but was rejected on the basis of spectral data [70]. This illustrates that EROS can be of help in structure elucidation work by suggesting structures that should be considered when analyzing the spectral data. As can be seen from the values of the reaction enthalpies, *5* is thermochemically preferred over *27*.

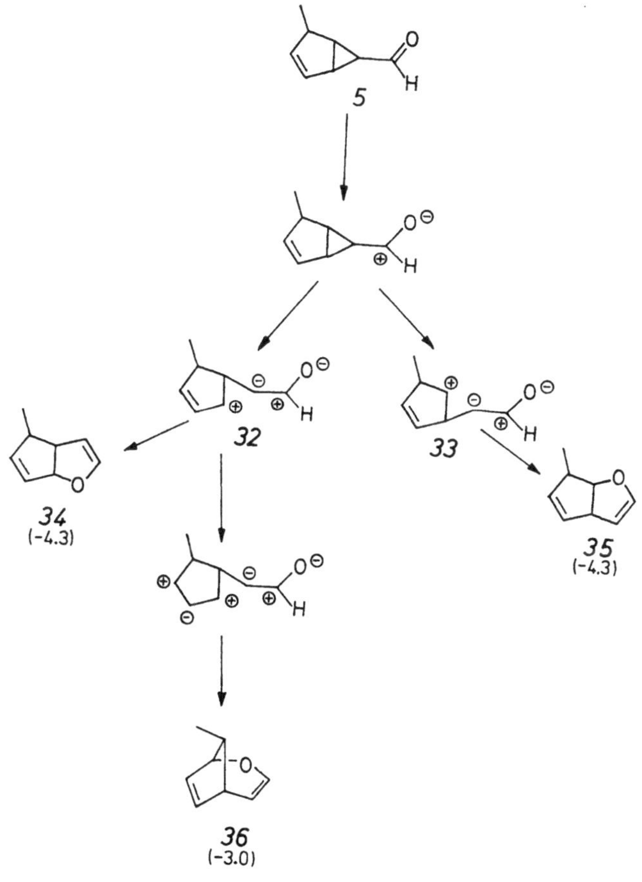

Fig. 31. Rearrangement of the formylbicyclo[3.1.0]hexene, *5*. Values in parentheses are enthalpies relative to *5* in kcal/mol

Formation of *31* should be disfavored by the same orbital symmetry constraints mentioned for the sequence *4* to *20* (see above and Fig. 29).

The chemistry of the system can be extended further as presented in Fig. 31. Application of the reactivity functions to the product *5* of the previous reaction shows that an additional rearrangement is possible as indicated by the high values obtained with the reactivity functions for the relevant reaction steps. An RG33 process breaking and making the bonds indicated in Fig. 31 was evaluated as the highest rating for further reaction of *5*. This leads to product *36*, again as observed experimentally [71]. The RG22 process to *34* is slightly less favorable. This structure *34* has been considered as a possible reaction candidate for the reaction product but was excluded on the

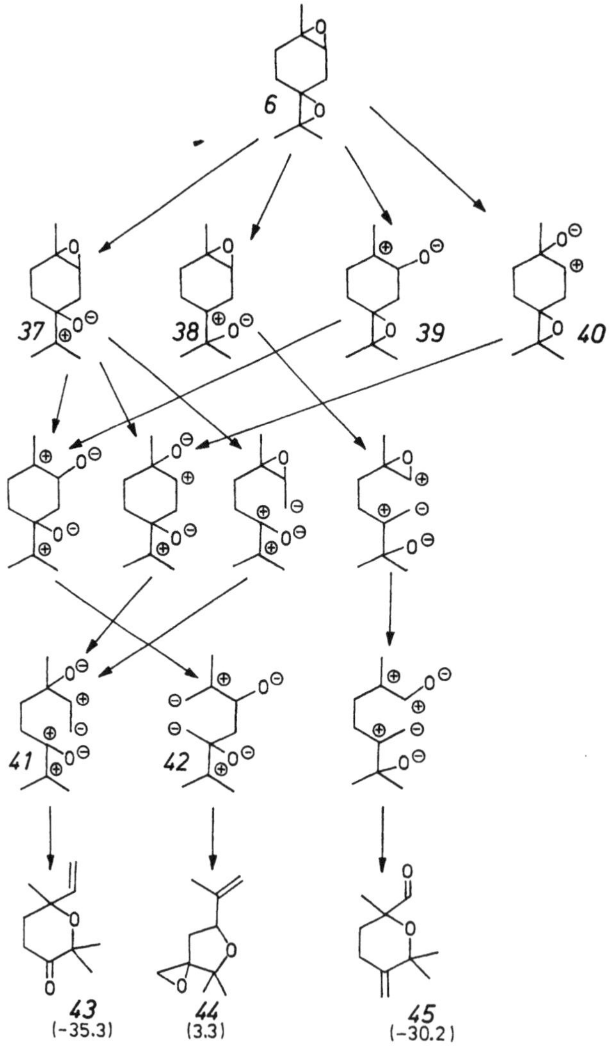

Fig. 32. Acid-catalyzed rearrangement of 1,2:4,8-diepoxy-p-menthane, *6*. Values in parentheses are enthalpies relative to *6* in kcal/mol

basis of the NMR data [71]. The RG22 process to the isomeric structure *35* is rated even less, as *33* lacks the allylic cation stabilization encountered in *32*. Interestingly, the reaction of *5* to *36* is predicted to be only slightly exothermic ($\Delta H = -3.0$ kcal/mol). In fact, *5* and *36* exist in a mobile solvent-dependent equilibrium [71]. Thus the reactivity function correctly predicted both steps of this reaction sequence, and the values of the reaction enthalpies rationalize the observed equilibrium.

1,2:4,8-Diepoxy-p-menthane, *6*, rearranges when heated with alumina in toluene [73]. What is the product of this reaction? An organic chemist would predict that acid treatment of the diepoxide *6* (Figs. 3 and 32) would induce one or other of the oxirane rings to open. But which of the two will be the more reactive, and would overall reaction necessarily involve such an initial step? Furthermore, for each oxirane there are two possible C—O cleavages.

The reactivity functions in EROS give rise to a reaction network which is reproduced in part in Fig. 32. The first three oxirane cleavages, *37* to *39*, receive the same rating, since they all lead to tertiary carbenium ions which the current evaluations do not distinguish. The initial rating of intermediate *40* is lower, as the carbenium ion is at a secondary center. Nevertheless, this oxirane cleavage is evaluated to be of importance in the best *overall* bond rearrangement process leading to *43* via intermediate *41*. At the third level of application of the reactivity functions, it is found that *41* is more likely than *42*, and any other intermediate structure. Significantly and surprisingly, a C—C bond of the saturated six-membered carbocycle is also found to be breakable. This illustrates the power of the EROS approach which *simultaneously* considers the breakability of all bonds involved in a reaction. Thus, in this example, an intuitively less likely oxirane ring opening (to the secondary carbenium ion) and the breaking of an unstrained C—C single bond are predicted to occur on the basis of evaluating the overall electron shifting pattern. The evaluations led to the prediction that structure *43* should be the product of acid treatment of *6*. Indeed, this is also the experimental finding [73]. Interestingly, this is the most exothermic reaction path.

Other predicted reactions, not shown in Fig. 32, include rearrangements of the oxiranes to give a cyclohexanone, and various allyl alcohols. These predicted products are entirely consistent with the type of by-product to be expected under such reaction conditions.

Figure 33 gives the results of a more extended study. Intermediate structures are no longer shown. Only the molecules involved in the scheme which result from the application of the RG22 and RG33 generators are drawn. The circled numbers above the arrows give the ranking that was assigned by EROS to the corresponding reaction. In the vertical direction the compounds are arranged to reflect their enthalpy content.

An entry to this network was obtained by starting from *7*. The reaction most favored by the reactivity functions leads to *46*, but this reaction is endothermic by 17.4 kcal/mol. The second best reaction leads from *7* to *47*. In order to explore the chemistry of these systems further, structures *46* and *47* were also submitted to EROS. Starting from *46*, the reaction predicted to be the best leads back to *7*, the second best reaction to *47*. Thus overall, rearrangement of *7* to *47* is expected, with *46* as a potential intermediate of this reaction. The most favored reaction of *47* leads to the cyclopentenone *8*, a reaction which is also favored by its exothermicity (—14.8 kcal/mol). The second best reaction leads from *47* to *48*. In a further study, *48* was also submitted to EROS. The two reactions evaluated to be best by the present reactivity functions lead to *49* and

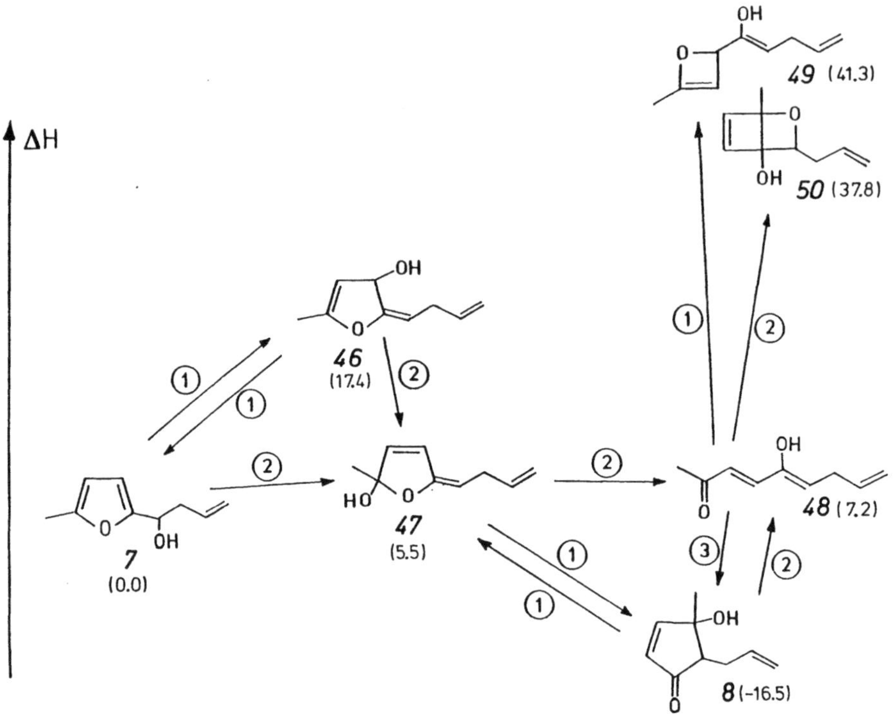

Fig. 33. Reaction network for furan to cyclopentenone interconversion, predicted by EROS

50. However, these are highly endothermic processes and may therefore be disregarded. (These reactions are also restricted by orbital symmetry considerations.) Thus, combining electronic and thermochemical evaluations, the preferred reaction of *48* should lead to *8*. The picture that emerges from this analysis is that *47* will rearrange to *8* with *48* as a possible reaction intermediate. Overall, the reaction pathway *7* → *47* → *8* is predicted.

These predictions correspond closely to the experimental observations and the mechanism suggested in the literature [74]. The conversion of furans to cyclopentenones is used industrially to obtain intermediates for the synthesis of insecticides, prostaglandins, perfumes, and compounds for energy storage [75].

We have intentionally selected example reactions (Figs. 29–33) that would not usually be immediately obvious to a chemist. The examples chosen have all been concerned with rearrangements of various types, since their courses are frequently difficult to predict. It remains to emphasize that the reactivity functions contained in EROS perform perfectly well with other types of reaction. This is true, for example, with reactions that a chemist could derive directly from an analysis of the functional groups in a molecule. Thus, EROS predicts addition reactions to carbonyl compounds, nucleophilic substitutions, and condensation reactions, to name just a few examples. In all these reaction types, the possibility of assigning a quantitative estimate to the reactivity at the various sites via the reactivity functions is of particular merit. It permits a decision on the extent of competing reaction pathways when several functional groups are present.

Throughout this section, and indeed the majority of this article, we have been concerned with developing and using evaluations in order to model reactions in a forward direction, that is, to predict the products from given starting materials. We now briefly consider how the reactivity functions could have application in *retrosynthesis* studies.

When the reactivity functions are applied to *8* in Fig. 33, the two reactions most favored by the reactivity function lead back to *47* and *48*. Both reactions are endothermic and thus unfavorable in the direction leading to *8*. In synthesis design, endothermic retroreactions should be preferred. Therefore, in searching for a synthesis of *8*, the compounds *47* and *48* are attractive precursors: based on considerations of electronic effects, favorable reaction mechanisms for the conversion of *47* or *48* to *8* can be established. Furthermore, these conversions are thermochemically favorable.

Thus, applying the reactivity functions onto a target molecule has led to interesting synthetic precursors. This suggests that reactivity functions could have more general use in retrosynthesis studies than we had previously thought. Of course, other strategic considerations will also become important in any general approach to this problem, but further discussion of these in the context of EROS's current development would be premature.

10 Current Developments and Future Prospects

Throughout this article we have indicated areas where further development is continuing. This section elaborates on work which is incomplete but in progress.

The full scope and potential of reactivity functions has yet to be explored. Particular attention is presently being given to the differences between aliphatic and aromatic chemistry. The role of acid-base catalysis in modifying the reactivity functions is being developed. Further in the future, the effects of different solvents, or solvent-classes, on reactivity will be built into the models. Perhaps the most promising development to come from the area of reactivity functions will be the possibility of shedding further light on the basic understanding of organic chemistry itself, as for instance, in the balance of the various effects on reactivity under different conditions.

Up until now, the results of the reactivity functions have been taken as one type of evaluation, while the calculated heats of reaction (Sect. 4.1) have been considered as a second. This parallels ideas based on kinetic and thermodynamic approaches. While it is possible to interpret the EROS results in terms of these two evaluations separately, it would be preferable to consider them together in some sort of combined function. Initial studies are indicating that this idea is feasible.

Experience with reactivity functions has led us to reconsider the manner of the generation of reactions altogether. As of now, various types of reaction generators (see Sect. 3.2) have been presupposed and then the chemistry has been built around them. We have now come to the conclusion that, instead, the evaluation phase should also decide which type of reaction generator should be applied. In other words, the chemistry, corresponding to the individual steps in Figs. 29–32, should drive the reaction generators and not vice versa. This has the added advantage that no selection of a subset of reaction generators, as previously necessary, now has to be made. Rather, the evaluation phase can select any pattern of bond-breaking and -making and electron shifting. This should lead to a closer modelling of the reaction mechanism. In effect, the new concept depends on just two "reaction generators": break a bond, and make a bond. Extension of the original scheme to include many more formal reaction generators is therefore superfluous. A preliminary version of EROS based on this new concept is already functioning, and giving highly promising results.

Recent work has concentrated on the reactivity prediction version of EROS described in this article, and less effort has gone into synthesis design. Further efforts are now being made in this direction. The extent to which the reactivity functions already derived could be applied retrosynthetically has already been touched on in the preceding section. Of course, this alone will not suffice to give sensible retrosynthetic routes, and a major goal is to define and build into EROS the necessary strategies. Many of these depend on formal information already calculated by EROS procedures based on constitutional information. For instance, factors relevant to the symmetry of a molecule, ring-combinations and bridgehead bonds, and molecular complexity as defined by Bertz [76], are all available from routines already functioning.

Both synthesis design and reactivity prediction require a knowledge of the 3-dimensional structure of molecules which is so far lacking in EROS. Procedures for the perception of molecules as 3D objects are being developed. Programs for handling the stereochemical features of both molecules and reactions are now available [77], some of the logic being based on concepts derived some time ago [11]. The formal aspects of stereochemistry also take account of the Cahn-Ingold-Prelog rules [12]. The program package that generates 3D coordinates works from constitutional information on molecules, plus stereochemical descriptors [77]. A direct application is to steric and conformational influences on chemical reactivity. However, this work also has significance for the whole area of molecular modelling in general. Thus, the interrelationships between synthesis design, reaction prediction and molecular design summarized in Fig. 4 are again stressed.

To date, EROS development has concentrated very much on the system itself and its chemistry. Far less attention has been paid to the interface between EROS and user. To rectify this situation, work has begun on implementing procedures for graphical input and output of molecular structures.

11 EROS, Artificial Intelligence and Expert Systems

Several chemists unrelated to the EROS project have recently described it publicly as an example of artificial intelligence (AI), or as an expert system (ES). In fact, amongst computer chemists in general there seems to be a slide towards referring to many programs, usually their own, as having "artificial intelligence". This is regardless of the fact that the program might not conform to any of the accepted criteria applied to such systems, and despite the fact that a few years ago the chemist would have been perfectly satisfied to accept classification of his system as a conventional program. It is almost as if some sort of late twentieth century symbol of scientific or technological virility attaches to the magic words "artificial intelligence".

In view of the current interest in AI and ES across many disciplines [78], as well as in chemistry, it seems appropriate to give our views on how EROS fits into the scheme of things.

We have described in an earlier section how EROS is able to predict reactions given starting materials and conditions, at a level sufficiently advanced to tax professional organic chemists. Surely, then, EROS must be deserving of some accolade in the AI/ES assignment stakes? It would be best to consider first how professionals in this field are currently defining the terms AI and ES. Unfortunately, even the professionals seem to be having difficulty. For instance, in their superb book [78] Michie and Johnston state that there is no rigorous definition of AI, but make it clear that it has to do with the "construction of a mechanizable logic of commonsense reasoning". Our view is that an acceptable transposition of this statement to the area of organic chemical reactivity is the following. A system would be definable as chemically "intelligent", if it made use of *chemical* "commonsense reasoning" in the sense that it could accept individual observations on known chemical reactions, draw whatever inferences and conclusions it could from these data, and thereby make a stab at answering other reactivity problems. In the light of correction by more chemically astute assistants, and the provision of more information on known reactions, it would then reformulate its opinions such that it became more successful in tackling further problems posed to it. The same would apply to retrosynthetic planning, or any other chemical problem.

EROS is not able to do this. In fact, it falls down in its inability to extract its own rules from input organic reactivity data sets. It is this information to which organic chemists have applied *their* intelligence and experience, and then, in the case of our research group, *built into* EROS. Thus, in the sense which we believe was intended when the term "artificial intelligence" was coined, EROS is not yet an example of AI.

Expert systems (ES) are offshoots from earlier development work towards AI systems. They are characterized by containing a database, or "knowledge base", which stores "rules", interfaced to a "rule interpreter" (or "inference engine"). The "rules" are frequently, but not necessarily, wholly heuristic in nature — that is, they are rules-of-thumb which derive from pure empiricism, or belief, or folklore, applying to the area of expertise in question. The "rule interpreter" then works out the logical consequences of these rules taken together, given a starting proposition (i.e. some sort of question). A further desirable characteristic is that at any time during a consultation, the ES can explain its line of reasoning to an interrogator, just as a human expert should be able to do when requested. The ES, then, can be regarded as that part of an AI system after the latter has been divested of the requirement to work out its own rules.

In the context of ES, EROS is close to the definition we have outlined above. In the jargon of AI/ES, EROS's "rules" are based largely on a "top-down" approach, using "causal models". That is, the rules have been derived by some sort of treatment which is based more on a fundamental understanding of the system of interest (organic chemistry, and reactivity, in particular), than to pure empiricism. The latter approach is a "bottom-up" one, and depends on the heuristic models mentioned above. Although EROS is based largely on "causal models", some purely heuristic models, or rules, are included in EROS's evaluations, as noted in Sect. 8.2. In the computational sense, most of the models on which the evaluations are built are to be found in specific procedures of the program, immodestly titled BRAIN's. The application of the models is, however, purely procedural, according to algorithms we have developed. An inference engine, in the sense applied to true ES, is not used (nor, we believe, needed).

We are therefore inclined towards the opinion that EROS can be classified as an ES, based as it is on *distilled knowledge of organic chemistry* in the form of our quantitative models (Sect. 8). EROS calculates its results based on these models, in contrast to the organic chemist who depends on a sort of mental qualitative pattern recognition when considering functional groups and their interrelationships. Despite this difference, EROS can provide answers to questions that are not always straightforward even for an expert in the field.

We have dwelt on the semantics of AI and ES merely because of the current interest in such matters. However, we conclude this article by emphasizing the fact that EROS functions successfully as a tool to aid chemists, wherever it fits into the taxonomy of computer software. We believe that this is far more important than whether the system conforms, or not, to any particular definition of AI or ES.

12 Acknowledgements

The support of the Deutsche Forschungsgemeinschaft for this research over nearly a decade is greatly appreciated. This allowed us to embark on such an ambitious project, which was recognized from the outset to be a long term effort.

The support of Imperial Chemical Industries, plc, United Kingdom, came at a decisive time both through secondment of Dr. M. G. Hutchings to the T. U. Munich for two years, and through funding via an I. C. I. Joint Research Scheme. We thank Dr. C. W. Greenhalgh, Dr. P. Bamfield and Dr. B. Langley for initiating and furthering this collaboration. Comments and suggestions by Dr. D. B. Baird are much appreciated, and M. G. H. in particular wishes to acknowledge the invaluable part played by Dr. Baird in mounting, maintaining, and further extending the capabilities of the I.C.I. versions of EROS.

Sumitomo Chemical Co. Ltd., Japan helped through secondment of K. Yuki and financial contributions. In this context we wish to thank Prof. S. Sasaki, K. Ito, Dr. I. Dohgane, Dr. H. Yamachika, Dr. M. Takahashi, and Dr. S. Asao.

We also thank Tecnofarmaci SpA., Pomezia, Italy and Prof. L. Caglioti, CNR for supporting our research.

Dr. M. D. Guillen developed some of the procedures for dealing with charge calculations in small rings, and programming assistance was provided by R. Buschsieweke, G. Feissel, K. Hartl and W. D. Ihlenfeldt; to all of these we extend our thanks.

Computation time was provided by the Leibniz Rechenzentrum. München.

13 Appendix — Current Versions of EROS

The EROS program system is written in PL/I and implemented on various IBM, Amdahl, CDC, and Siemens machines. Presently, two versions are supported, the 3.2 and 4.1 versions, consisting of about 9,000 and 13,000 statements, respectively. The essential difference between these two versions is that in EROS 3.2 the various methods for calculating electronic effects discussed throughout this article have not been implemented, whereas they are included in EROS 4.1.

In this article we have said little specific about the 3.2 version. This represents a lower stage in the conceptual development of the system, but is still supported for various reasons. Its main strength is that it is capable of stimulating lateral thinking in the design of synthetic routes, and can suggest new reactions. It is being used successfully and routinely in industry.

EROS 4.1 is what has been described in this article. It too is running routinely, but, as outlined, is mainly of use only for reaction prediction. EROS 4.1 is still undergoing development and extension as outlined in Sect. 10.

14 References

1. Corey, E. J., Dankeiser, R. L. Chandrasekaran, S., Keck, G. E., Gopalan, B., Larsen, S. D., Siret, P., Gras, J.-L.: J. Am. Chem. Soc. *100*, 8034—8036 (1978)
2. For a brief, enlightening discussion see: Turner, S.: The Design of Organic Syntheses, Elsevier, Amsterdam 1976
3. Corey, E. J.: Pure Appl. Chem. *14*, 19–37 (1967)
4. Warren, S.: Organic Synthesis, The Disconnection Approach, Wiley, Chichester 1982
5. Gasteiger, J.: Chapter: Syntheseplanung, in: Computer in der Chemie, (Ziegler, E., Ed.), Springer, Heidelberg 1984, p. 207—257
6. Gasteiger, J., Jochum, C.: Topics Curr. Chem. *74*, 93—126 (1978)
7. a) Gasteiger, J., Jochum, C., Marsili, M., Thoma, J.: Proceed. IVth Internat. Conf. on Computers in Chemical Research and Education, Novosibirsk, USSR 1978, p. 442—457
b) Gasteiger, J., Jochum, C., Marsili, M., Thoma, J.: Informal Commun. Math. Chem. *6*, 177–199 (1979)
c) Gasteiger, J., Jochum, C., Marsili, M., Thoma, J.: Proceed. IVA-Symposium "The Use of Computers in Organic Chemistry", Stockholm Sweden 1981, p. 51—68
d) Gasteiger, J., Marsili, M., Paulus, B.: in Data Processing in Chemistry, (Z. Hippe, Ed.) Elsevier, Amsterdam 1981, p. 229—246
e) Gasteiger, J.: Chim. Ind. (Milan) *64*, 714—721 (1982)
f) Gasteiger, J., Christoph, B., Gann, L., Hutchings, M. G., Saller, H., Yuki, K.: Royal Soc. of Chem., Residential School "Computer-Aided Organic Synthesis Design and Molecular Graphics", University of Leeds 1983
g) Marsili, M., Gasteiger, J., Carter, R. E.: Chimica Oggi, *1984*, (9) 11–18
8. Dugundji, I., Ugi, I.: Topics Curr. Chem. *39*, 19–64 (1973)
9. Ugi, I., Gillespie, P. D.: Angew. Chem. *83*, 980–981; 982–985 (1971); Angew. Chem. Int. Ed. Engl. *10*, 914, 915 (1971)
10. Hofmann, H. M. R. and Rabe, J.: Angew. Chem. *97*, 96–112 (1985); Angew. Chem. Int. Edn. Engl. *24*, 94–123 (1985)
11. Blair, J., Gasteiger, J., Gillespie, C., Gillespie, P. D., Ugi, I.: Tetrahedron *30*, 1845–1859 (1974)
12. Cahn, R. S., Ingold, C., Prelog, V.: Angew. Chem. *78*, 413–447 (1966); Angew. Chem. Int. Ed. Engl. *5*, 385–419 (1966);
Prelog, V., Helmchen, G.: Angew. Chem. *94*, 614–631 (1982); Angew. Chem. Int. Ed. Engl. *21*, 567–584 (1982)
13. Jochum, C., Gasteiger, J.: J. Chem. Inf. Comput. Sci., *17*, 113–117 (1977); ibid. *19*, 49–50 (1979)
14. Christoph, B.: unpublished results
15. Gasteiger, J., Jochum, C.: J. Chem. Inf. Comput. Sci. *19*, 43–48 (1979)
16. Bart, J. C. J., Garagnani, E.: Z. Naturforsch. *31b*, 1646–1653 (1976); ibid. *32b*, 455–464 (1977); ibid. *32b*, 465–468 (1977); ibid *32b*, 678–683 (1977)
17. Huisgen, R.: Angew. Chem. *75*, 604–637 (1963); ibid. *75*, 742–754 (1963)
18. Alper, H., Perera, C. P., Ahmed, F. R.: J. Am. Chem. Soc. *103*, 1289–1291 (1981)
19. Skakibara, T., Alper, H.: J. Chem. Soc. Chem. Commun. *1979*, 458

20. Suckling, C. J., Suckling, K. E., Suckling, C. W.: Chemistry Through Models, Cambridge University Press, Cambridge 1978
21. Eigen, M.: The Physicist's Conception of Nature, quoted in "A Dictionary of Contemporary Quotations", compiled by Green, J., Pan Books 1982
22. Cox, J. D., Pilcher, G.: Thermochemistry of Organic and Organometallic Compounds, Academic Press, London 1970
23. Caglioti, E., Marsili, M.: results presented at 7th Internat. Conf. on Computers in Chemical Research and Education, Garmisch-Partenkirchen, W. Germany 1985
24. Gasteiger, J.: Comput. Chem. *2*, 85–88 (1978)
25. Gasteiger, J.: Tetrahedron *35*, 1419–1426 (1979)
26. Allen, T. L.: J. Chem. Phys. *31*, 1039–1049 (1959);
Kalb, A. J., Chung, A. L. H., Allen, T. L.: J. Am. Chem. Soc. *88*, 2938–2942 (1966)
27. Benson, S. W., Buss, J. H.: J. Chem. Phys. *29*, 546–572 (1958);
Benson, S. W.: Thermochemical Kinetics, 2nd Ed. Wiley, New York 1976
28. Gasteiger, J., Jacob, P., Strauss, U.: Tetrahedron *35*, 139–146 (1979)
29. Gann, L., Löw, P., Sello, G., Yuki, K.: unpublished results
30. Gasteiger, J., Dammer, O.: Tetrahedron *34*, 2939–2945 (1978)
31. Guthrie, J. P., Cullimore, P. A.: Can. J. Chem. *57*, 240–248 (1979);
Guthrie, J. P.: ibid. *57*, 797–802 (1979)
32. Egger, K. W., Cocks, A. T.: Helv. Chim. Acta *56*, 1516–1536 (1973)
33. McMillen, D. F., Golden, D. M.: Ann. Rev. Phys. Chem. *33*, 493–532 (1982)
34. Chapman, N. B., Shorter, J.: Advances in Linear Free Energy Relationships, Plenum Press, London 1972;
Chapman, N. B., Shorter, J.: Correlation Analysis in Chemistry — Recent Advances, Plenum Press New York 1978
35. Mulliken, R. S.: J. Chem. Phys. *23*, 1833–1840 (1955)
36. Mulliken, R. S.: ibid, *2*, 782–793 (1934)
37. Hinze, J., Jaffe, H. H.: J. Am. Chem. Soc. *84*, 540–546 (1962); J. Phys. Chem. *67*, 1501–1506 (1963);
Hinze, J., Whitehead, M. A., Jaffe, H. H.: J. Am. Chem. Soc. *85*, 148–154 (1963)
38. Sanderson, R. T.: Science *144*, 670 (1964)
39. Huheey, J. E.: J. Phys. Chem. *69*, 3284–3291 (1965); ibid. *70*, 2086–2092 (1966)
40. Gasteiger, J., Marsili, M.: Tetrahedron Letters *1978*, 3181–3184
41. Gasteiger, J., Marsili, M.: Tetrahedron *36*, 3219–3228 (1980)
42. Guillen, M. D., Gasteiger, J.: ibid. *39*, 1331–1335 (1983)
43. Mortier, W. J., Van Genechten, K., Gasteiger, J.: J. Am. Chem. Soc. *107*, 829–835 (1985)
44. Gasteiger, J., Marsili, M.: Org. Magn. Resonance *15*, 353–360 (1981)
45. Gasteiger, J., Guillen, M. D.: J. Chem. Res. (S) *1983*, 304–305; (M) *1983*, 2611–2624
46. Knorr, R., Saller, H., Gasteiger, J.: unpublished results
47. Marsili, M., Gasteiger, J.: Croat. Chem. Acta *53*, 601–614 (1980)
48. Gasteiger, J., Saller, H.: Angew. Chem. *97*, 699–701 (1985); Angew. Chem. Int. Ed. Engl. *24*, 687–689 (1985)
49. Hutchings, M. G., Gasteiger, J.: Tetrahedron Lett. *24*, 2541–2544 (1983)
50. Pauling, L.: The Nature of the Chemical Bond, 3rd Edit., Cornell University Press, Ithaca, N.Y. 1960, p. 88
51. Lias, S. G., Liebman, J. F., Levin, R. D.: J. Phys. Chem. Ref. Data *13*, 695–808 (1984)
52. Saller, H.: Ph. D. thesis, TU München 1985
53. Gasteiger, J., Hutchings, M. G.: Tetrahedron Lett. *24*, 2537–2540 (1983);
Gasteiger, J., Hutchings, M. G.: J. Chem. Soc. Perkin II, *1984*, 559–564
54. Gasteiger, J., Löw, P.: unpublished results
55. n = number of data points; r = (multiple) correlation coefficient; s = standard deviation
56. Gasteiger, J., Hutchings, M. G.: J. Am. Chem. Soc. *106*, 6489–6495 (1984)
57. Wolf, J. F., Abboud, J. L. M., Taft, R. W.: J. Org. Chem. *42*, 3316–3317 (1977)
58. Bordwell, F. G., Drucker, G. E., McCollum, G. J.: ibid. *47*, 2504–2510 (1982)
59. Hutchings, M. G., Gasteiger, J.: J. Chem. Soc. Perkin 2, *1986*, in press
60. Hutchings, M. G., Gasteiger, J.: J. Chem. Soc. Perkin 2, *1986*, in press
61. Kreevoy, M. M., Taft, R. W., J. Am. Chem. Soc. *77*, 5590–5595 (1955)
62. Fukui, K.: Fortschr. Chem. Forsch. *15*, 1–85 (1970); Acc. Chem. Res. *4*, 57–64 (1971)

63. Houk, K. N.: Topics Curr. Chem. *79*, 1–40 (1979)
64. Burnier, J. S., Jorgensen, W. C.: J. Org. Chem. *49*, 3001–3020 (1984)
65. Gasteiger, J., Hutchings, M. G., Saller, H.: unpublished results
66. Saller, H., Löw, P., Hutchings, M. G., Gasteiger, J.: results presented at the 7th Internat. Conf. on Computers in Chemical Research and Education, Garmisch-Partenkirchen, W. Germany 1985
Gasteiger, J., Hutchings, M. G., Löw, P., Saller, H.: in: Artificial Intelligence Applications in Chemistry, ACS-Symposium Series, in press
67. Snowden, R. L.: Helv. Chim. Acta *66*, 1031–1038 (1985)
68. Not all 2-bond, or 3-bond, combinations are allowed in RG22, or RG33, reactions, respectively. e.g. two bonds emanating from a single atom cannot take part in an RG22 reaction. The numbers given are rough, but minimum, estimates, and include degenerate atom switches (e.g. $C^1-H^1 + C^2-H^2 \rightarrow C^1-H^2 + C^2-H^1$) which EROS would remove anyway.
69. Experimentally, this reaction sequence was observed with hydrogen [70] or a substituted octenyl group [72] in position 7 of the norbornadiene. In the calculations a methyl group was used as marker, to show the course of the rearrangement.
70. Meinwald, J., Labana, S. S., Chadha, M. S.: J. Am. Chem. Soc. *85*, 582–585 (1963)
71. Rey, M., Dreiding, A. S.: Helv. Chim. Acta *48*, 1985–1987 (1965)
72. Baxter, A. D., Roberts, S. M., Scheinmann, F., Wakefield, B. J., Newton, R. F.: J. Chem. Soc., Chem. Commun. *1983*, 932–933
73. Ho, T. L., Stark, C. J.: Liebigs Ann. Chem. *1983*, 1446–1447
74. Piancatelli, G., Scettri, A., David, G., D'Auria, M.: Tetrahedron *34*, 2775–2778 (1978)
75. Dohgane, I., Yamachika, H., Minai, M.: J. Synth. Org. Chem. (Japan) *1984*, 896–903
76. Bertz, S. H.: J. Am. Chem. Soc. *103*, 3599–3601 (1981); ibid *104*, 5801–5803 (1982)
77. Hiller, C., Christoph, B., Gann, L., Gasteiger, J.: results presented at 7th Internat. Conf. on Computers in Chemical Research and Education, Garmisch-Partenkirchen, W. Germany 1985
78. Michie, D., Johnston, R.: The Creative Computer, Pelican Books, England 1985

Transition-metal Catalyzed Decomposition of Aliphatic Diazo Compounds — New Results and Applications in Organic Synthesis

Gerhard Maas

Fachbereich Chemie der Universität Kaiserslautern,
D-6750 Kaiserslautern, FRG

Table of Contents

Abbreviations

EDA = ethyl diazoacetate; MDA = methyl diazoacetate; OTf = O_3SCF_3 (trifluoromethanesulfonate); acac = acetylacetonate; hfacac = hexafluoroacetylacetonate.

1 Introduction

Since the early observations by Loose [1] and Wolff [2], who noted a significant lowering of the decomposition temperature of diazoketones and diazoacetic ester in the presence of copper powder, cupric sulfate or silver salts, the catalytic decomposition of diazo compounds has become a standard procedure in organic synthesis. Generally speaking, nitrogen evolution from aliphatic diazo compounds occurs in the presence of several metals, metal salts as well as transition metal complexes. Among the reactions so initiated, one finds cyclopropanation and cyclopropenation, insertion into C—H, O—H and N—H bonds, ylide and dimer formation, in all of which the carbene moiety remaining after N_2 loss is formally involved. A number of reviews have covered the catalyzed decomposition of diazo compounds both from a preparative [3-15,15a] and from a mechanistic point of view [11,14,16]. The preparative value of the method is well supported by the fact that a plethora of several hundred aliphatic diazo compounds is easily accessible to the organic chemist [17,18].

Up to the late seventies, catalytic decomposition reactions were usually carried out in the presence of copper in different oxidation states, e.g. copper powder, copper bronze, Cu_2Cl_2, CuI · $P(OMe)_3$, CuI · Bu_2S, $CuCl_2$, $CuSO_4$ or copper bis(1,3-diketonates). If Wolff rearrangement products from α-diazocarbonyl compounds were desired, Ag_2O was the catalyst of choice. As a result of a systematic screening of transition-metal compounds in recent years, copper(I) or copper(II) trifluoromethanesulfonate as well as palladium and rhodium compounds have emerged as highly efficient catalysts which for many applications are superior to the traditional copper catalysts although the latter may still be considered to be more all-round catalysts. Detailed studies concerning the different aspects of selectivity of the new diazo compound/catalyst systems, as well as optimization of reaction conditions, have shed light on possible reaction intermediates and will help the experimentalist to draw optimum benefit from the correct choice of the catalyst. The majority of these systematic studies have been published, mainly by the research groups of Doyle in Michigan and of Noels and Hubert in Belgium after the last, very extensive review [14] was written. It seems, therefore, appropriate to collect these results in another account, together with some further applications of the new catalysts and comparison with traditional catalysts. Furthermore, decomposition of diazo compounds even with more traditional catalysts is involved in a number of novel contributions to organic synthesis which deserve recognition in a review article[1].

Not included in the present review is the fascinating new chemistry which results from reaction between diazo compounds and low-valent transition-metal complexes bearing easily displaceable two-electron ligands as well as with metal-metal multiple bonds and metal hydrides whereby a variety of novel organometallic molecules could be obtained. This field has been covered, in accord with its rapid development, by successive reviews of Hermann [19-22] and Albini [23].

[1] This survey covers the literature between 1978 and August 1985. Reference to earlier research papers has been made where it seemed appropriate.

2 Reaction with Olefins

2.1 Cyclopropanation with Diazoalkanes, Aryldiazomethanes and Diazocyclopentadienes

Methylene transfer from diazomethane to olefinic and aromatic double bonds has traditionally been carried out with Cu(I) halides [24]. However, other copper salts have occasionally been used.

Smooth and efficient cyclopropanation also occurs with copper(II) triflate and diazomethane. Intra- and intermolecular competition experiments show that, in this case, the less substituted double bond reacts preferentially [25]. The same is true for CuOTf and $Cu(BF_4)_2$, whereas with $CuX \cdot P(OMe)_3$ (X = Cl, I), $CuSO_4$ and copper(II) acetylacetonate, cyclopropanation of the more substituted double bond predominates. An example is given for cyclopropanation of *1*.

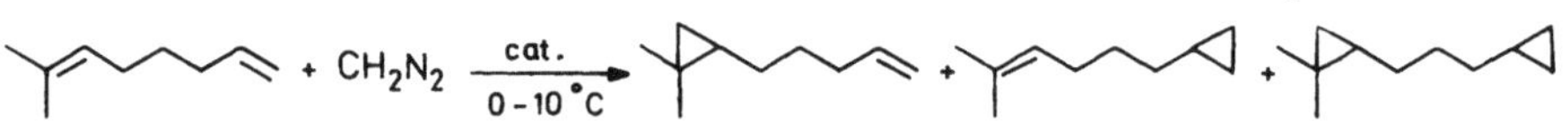

1

cat.	relative yields
$Cu(acac)_2$	1:0.1:0.6 (at 90% conversion)
$Cu(OTf)_2$	1:2.5:0.4 (at 50% conversion)

The same difference in regioselectivity holds for cyclopropanation with ethyl diazoacetate [25]. It is assumed that $Cu(OTf)_2$ or $Cu(BF_4)_2$ are reduced to the Cu(I) salts by the diazo compound; the ability of CuOTf to form stable complexes with olefins may then explain why, with these catalysts, cyclopropanation is governed by the steric environment around a double bond rather than by its electron-richness.

Copper(II) triflate has also been used for the carbenoid cyclopropanation reaction of simple olefins like cyclohexene, 2-methylpropene, *cis*- or *trans*-2-butene and norbornene with vinyldiazomethane *2* [26,27]. Although the yields were low (20–38 %), this catalyst is far superior to other copper salts and chelates except for copper(II) hexafluoroacetylacetonate [$Cu(hfacac)_2$], which exhibits similar efficiency. However, highly nucleophilic vinyl ethers, such as dihydropyran and dihydrofuran cannot be cyclopropanated as they rapidly polymerize on contact with $Cu(OTf)_2$. With these substrates, copper(II) trifluoroacetate or copper(II) hexafluoroacetylacetonate have to be used. The vinylcyclopropanation is stereospecific with *cis*- and *trans*-2-butene. The 7-vinylbicyclo[4.1.0]heptanes formed from cyclohexene are obtained with the same *exo*/*endo* ratio in both the $Cu(OTf)_2$ and $Cu(hfacac)_2$ catalyzed reaction. The preponderance of the *endo*-isomer contrasts with the result from copper-catalyzed cyclopropanations with alkyl diazoacetates.

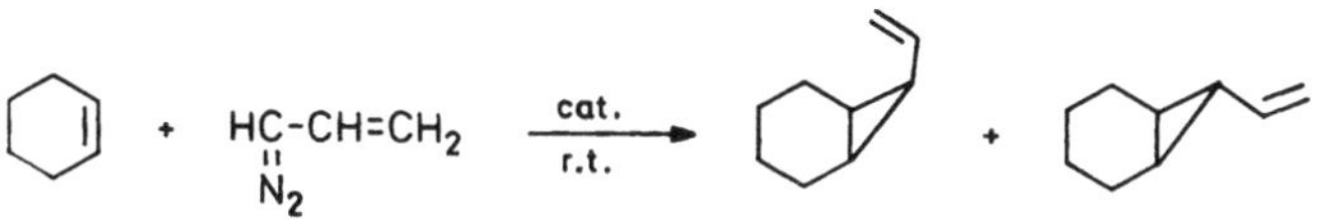

2

cat.: Cu(OTf)$_2$ or Cu(hfacac)$_2$ total yield: 20% *endo* : *exo* 1.2 : 1

However, cyclopropanation of cyclohexene with methyl (α-diazomethyl)acrylate *3* in the presence of copper(II) trifluoroacetate furnished the 7-*exo*-substituted bicyclo-[4.1.0]heptane preferentially [28].

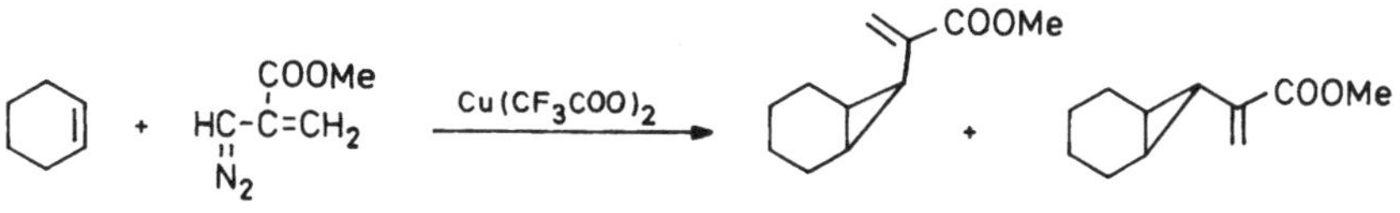

3

total yield: 6% *endo* : *exo* 1 : 2

Palladium-based catalysts also bring about cyclopropanations in high-yield. With palladium acetate/CH_2N_2, styrene[29], unactivated terminal olefins [30], strained olefins [31,32], 1,3-dienes [32], an enamine [33], as well as α,β-unsaturated carbonyl compounds [34,35,36] have been cyclopropanated (Table 1). Contrary to an earlier report, the reaction also works well with cyclohexene if the conditions are chosen appropriately; it seems that the catalyst is rapidly deactivated in the presence of this olefin [32]. Trisubstituted α,β-unsaturated carbonyl compounds were found to be unreactive, and the same is true for the double bonds in diethyl fumarate, maleic anhydride, coumarin and 1,3-dimethyluracil. Whereas the latter two were totally unreactive, [3+2] cycloaddition of diazomethane gave pyrazolines in the former two cases. The last entry of Table 1 shows that an allyl alcohol function can still be cyclopropanated, but methylene insertion into the O—H bond is a competing process.

Instead of $Pd(OAc)_2$, other catalysts such as $PdCl_2$, $PdCl_2 \cdot 2\,PhCN$ and $[(\eta^3\text{-}C_3H_5)PdCl]_2$ can be used without significant loss of activity [32]. Comparison of the $Pd(OAc)_2/CH_2N_2$ reagent with the $CuCl/CH_2N_2$ system reveals some complementary behavior: Cyclopropanation of carbonyl-substituted alkenes works well with palladium, but fails with the copper catalyst. The same was true for cyclopropanation of a β-arylenamine [33], although other enamines and 1,2-enediamines underwent the expected reaction with $CuCl/CH_2N_2$. In intramolecular competition

Table 1. Cyclopropanation of olefins with Pd(II)/diazomethane

Olefin	Conditions[a]	Product	Yield [%]	Ref.
$PhCH{=}CH_2$	A	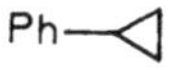	90	29)
n-$C_8H_{17}CH{=}CH_2$	B	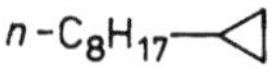	89	30)
$PhOCH_2CH{=}CH_2$	B	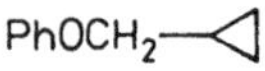	97	30)
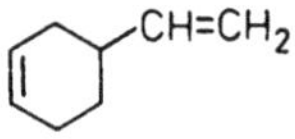	B[b]	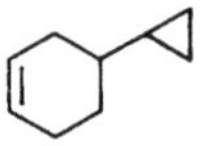	77	30)
	C	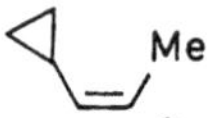	90	32)
$CH_2{=}CH{-}C(Me)_2CH_2CO_2Et$	B	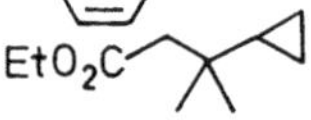	90	32)
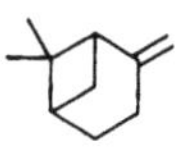	B	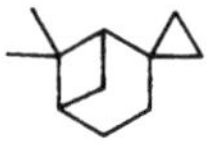	63	32)
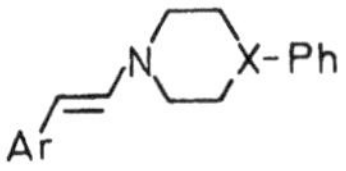 Ar = 3,4-dimethoxyphenyl X = CH, N	D (−15°C → r.t.)	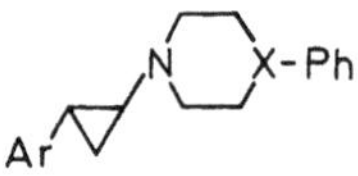	10	33)
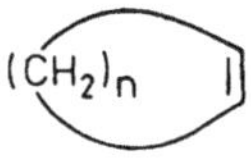	C	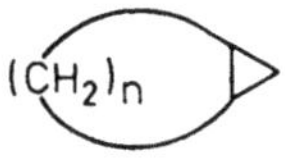	n = 4 : 15–60 n = 5 : 80–82 n = 6 : 93	32)
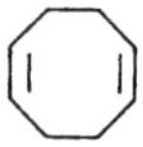	C[c]	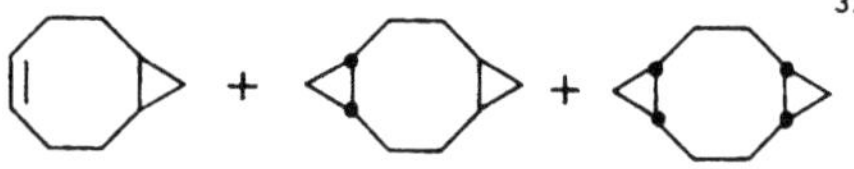 6% 12% 10%		32)
	C[c]	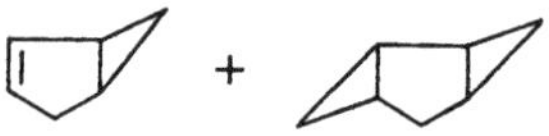 15% 75%		32)
	C, D	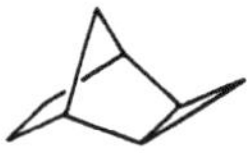	96 (C) 97 (D)	32) 31)
	C, D		94 (C) 63 (D)	32) 31)
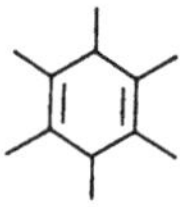	D (−10°C → r.t.)	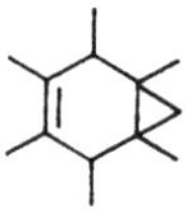	78	31)
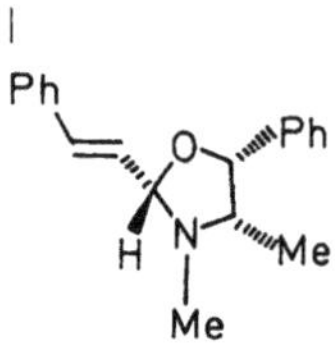	0°C	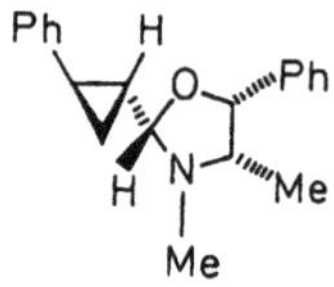	"quant."	37)

Olefin	Conditions[a]	Product	Yield [%]	Ref.
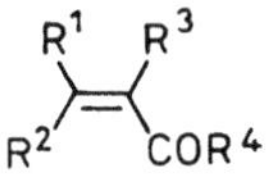	D	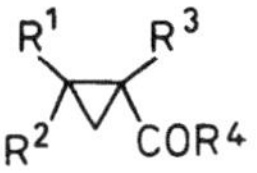		34)

R^1	R^2	R^3	R^4	
Ph	H	H	Ph	98
Ph	H	H	Me	85
Ph	H	H	COOEt	90
H	Ph	H	COOEt	85
Me	H	H	COOMe	89
H	H	Me	COOMe	80

Olefin	Conditions[a]	Product	Yield [%]	Ref.
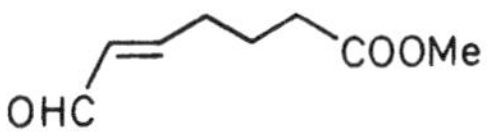	D	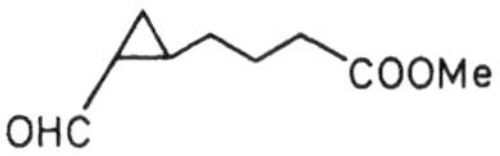	80	36)
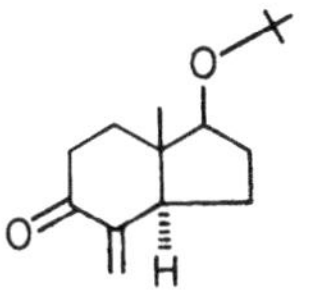	D	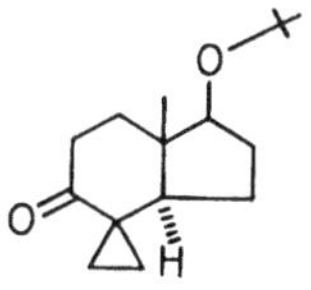	?	34)
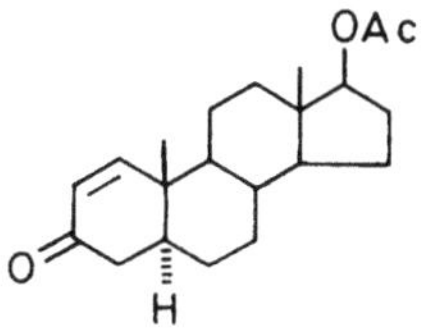	D	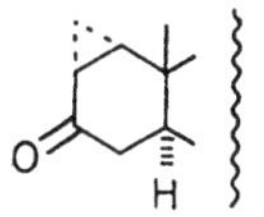	80	34)
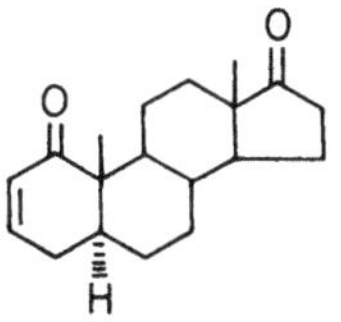	D	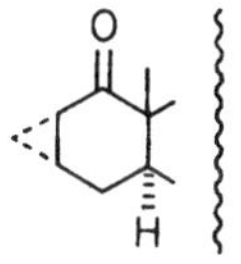	75	34)
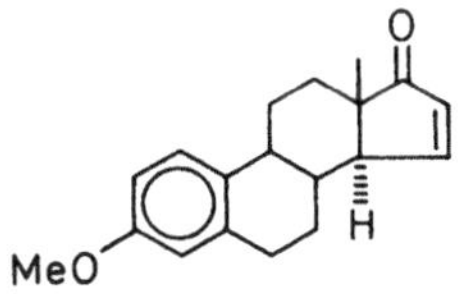	D	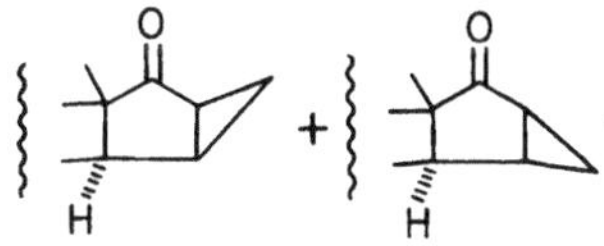	100 (7:3)	34)
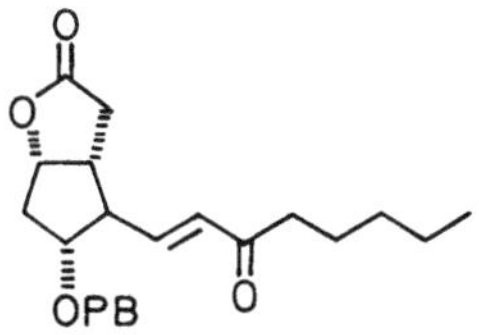		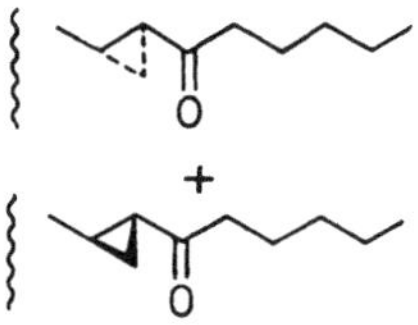	92	35)

PB = 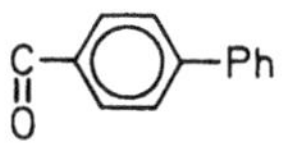

Table 1, continued

Olefin	Conditions[a]	Product	Yield [%]	Ref.
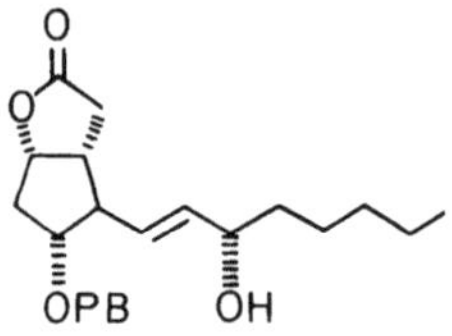	large excess of CH_2N_2	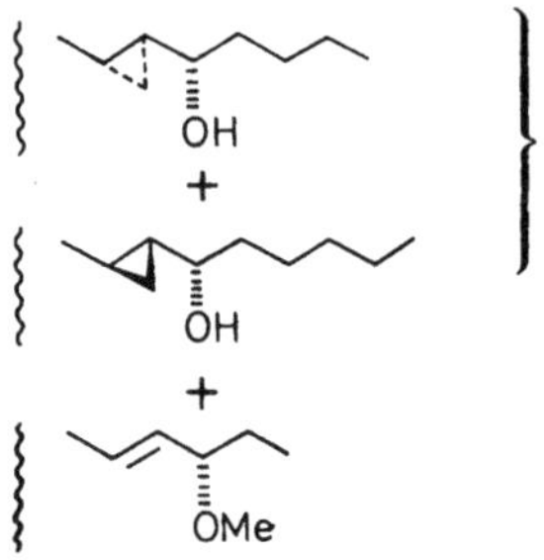	20 20	35)

[a] Excess diazomethane was used in all cases. Catalyst was $Pd(OAc)_2$ if not stated otherwise. A: Gaseous diazomethane added at 0 °C. — B: Catalyst added to the solution of olefin and diazomethane/ether 0 °C. — C: Gaseous diazomethane added at −10 to 0 °C to the solution of olefin in CH_2Cl_2; catalyst $PdCl_2 \cdot 2$ PhCN. — D: Diazomethane/ether added at 0 °C to olefin and catalyst.
[b] With $PdCl_2 \cdot$ PhCN: 3-cyclopropyl-1-cyclohexene (70%); 3-vinylbicyclo[4.1.0]heptane (3%); 3-cyclopropylbicyclo[4.1.0]heptane (7%) [32].
[c] Olefin: CH_2N_2 = 1:2.

of unactivated dienes (e.g. 4-vinylcyclohexene), $Pd(OAc)_2/CH_2N_2$ prefers the less substituted double bond [32], whereas $CuCl/CH_2N_2$ is rather undiscriminating [38]. The cyclopropanation of strained cyclic olefins proceeds in high yield with both catalysts [32,38]; norbornene and norbornadiene yield exclusively *exo*-cyclopropanes in either case. *Exo*-cyclopropanes are also obtained, when the strained olefins *4*–*7* are treated with diazomethane in the presence of bis(μ-chloro-η^3-allylpalladium) [39].

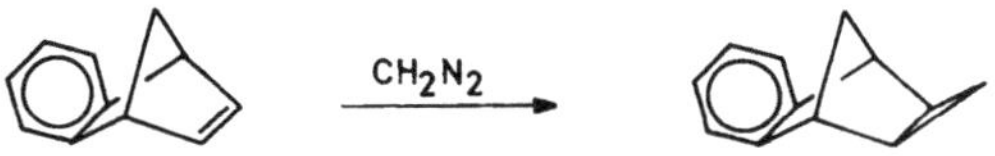

100 %

4

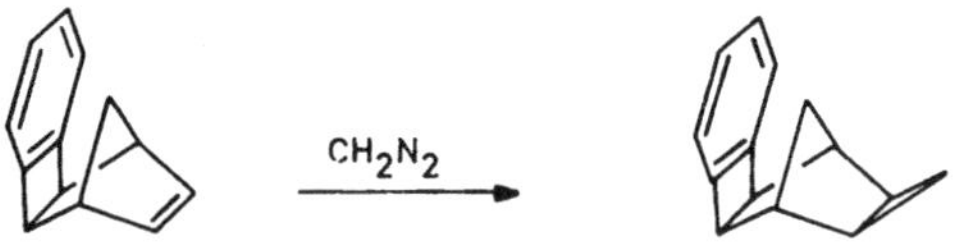

90 %

5

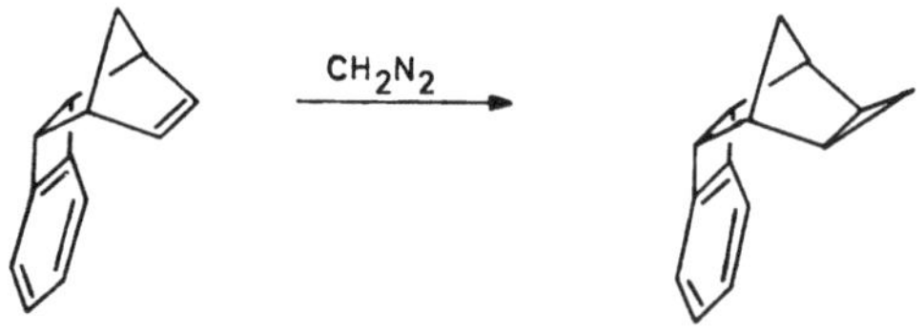

95 %

6

7

Conditions: Excess CH_2N_2, $[(\eta^3-C_3H_5)PdCl]_2$, 0 °C, in ether; $[CH_2N_2]$: [catalyst] = 600:1

Diazomethane is also decomposed by Ni(0) [40–43] and Pd(0) complexes [43]. Electron-poor alkenes such as methyl acrylate are cyclopropanated efficiently with Ni(0) catalysts, whereas with Pd(0) yields were much lower (Scheme 1) [43]. Cyclopropanes derived from styrene, cyclohexene or 1-hexene were formed only in trace yields. In the uncatalyzed reaction between diazomethane and methyl acrylate, methyl 2-pyrazoline-3-carboxylate and methyl crotonate are formed competitively, but the yield of the latter can be largely reduced by adding an appropriate amount of catalyst. It has been verified that cyclopropane formation does not result from metal-catalyzed ring contraction of the 2-pyrazoline. Instead, a nickel(0)-carbene complex is assumed to be involved in the direct cyclopropanation of the olefin. The preference of such an intermediate for an electron-poor alkene is in agreement with the view that nickel carbenoids are nucleophilic [44].

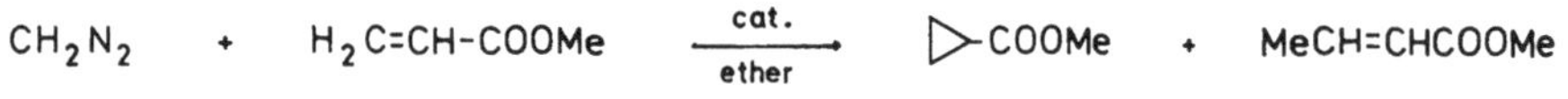

catalyst	temp.	molar ratio cat.: alkene	yield [%]	
"Ni(t-BuN≡C)$_2$"	−78 °C	0.10	66.5	20.4
Ni(PPh$_3$)$_4$	−78 °C	0.025	21.6	33.4
		0.5	77.0	6.7
		1.0	75.4	1.0
[Pd(t-BuN≡C)$_2$]$_2$	−78 °C → r.t.	0.10	0.9	24.4

Scheme 1

As for cyclopropanation of alkenes with aryldiazomethanes, there seems to be only one report of a successful reaction with a group 9 transition metal catalyst: $Rh_2(OAc)_4$ promotes phenylcyclopropane formation with phenyldiazomethane, but satisfactory yields are obtained only with vinyl ethers [45] (Scheme 2). *Cis-* and *trans*-stilbene as well as benzalazine represent by-products of these reactions, and $Rh_2(OAc)_4$ has to be used in an unusually high concentration because the azine inhibits its catalytic activity. With most monosubstituted alkenes of Scheme 2, a preference for the *Z*-cyclopropane is observed; similarly, *syn*-selectivity in cyclopropanation of cyclopentene is found. These selectivities are the exact opposite to those obtained in reactions of ethyl diazoacetate with the same olefins [45]. Furthermore, they are temperature-dependent; for example, the *cis/trans* ratio for 1-ethoxy-2-phenylcyclopropane increases with decreasing temperature.

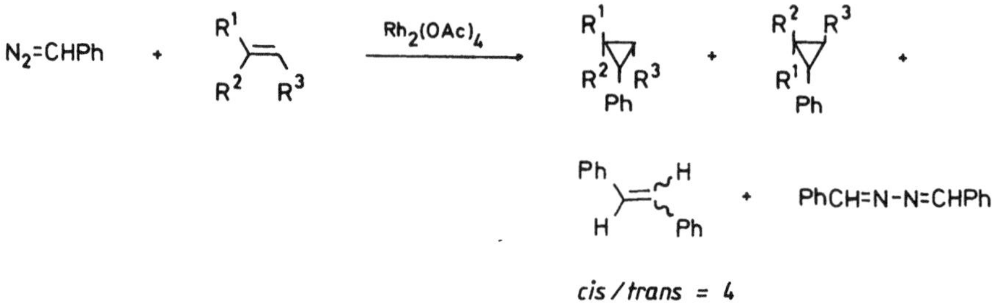

Conditions: ether, 25 °C; $[N_2CHPh]$: [cat.] = 25:1

R^1	R^2	R^3	cyclopropane	
			yield [%]	Z/E ratio
Ph	OMe	H	98	0.67
Me	OMe	H	82	1.2
t-Bu	OMe	H	92	0.15
n-BuO	H	H	92	2.5
EtO	H	H	54	3.0
AcO	H	H	7	3.8
Ph	H	H	38	3.3
Me	Me	Me	23	12
n-Bu	H	H	6	0.90
H	$-(CH_2)_3-$		6	1.6 (*syn/anti*)

Scheme 2

The limitation to electron-rich alkenes in Rh(II)-catalyzed cyclopropanation with phenyldiazomethane leaves untouched the great versatility of zinc halides for this purpose; with this catalyst, efficient and very mild cyclopropanation of 1,3-dienes and unactivated alkenes has been reported [46].

Only one report mentions the cyclopropanation with diazodiphenylmethane in the presence of a group VIII metal catalyst. Remarkably enough, the selectivity of the reaction with 5-methylene-bicyclo[2.2.1]hept-2-ene (**8**) can be reversed completely. With $Rh_2(OAc)_4$ as catalyst, the exocyclic double bond is cyclopropanated exclusively (>100:1), whereas in the presence of bis(benzonitrile) palladium(II) chloride the endocyclic C=C bond is attacked with very high selectivity (>50:1) [47].

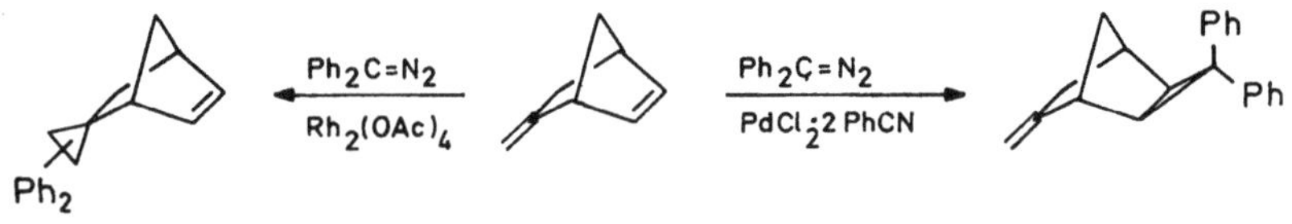

8

Apart from these findings, the limited application of $ZnCl_2$ (cyclopropanation of some cyclic 1,3-dienes, isoprene and ethyl vinyl ether [48,49]) and copper(II) acetylacetonate (cyclopropanation of enamines [50]) still stand alone.

Other than for the reaction of diazomethane with methyl acrylate (see Scheme 1),

Ni(0) and Pd(0) are not efficient in catalyzing the cyclopropanation of diethyl maleate with diazofluorene [43]. In contrast to CH_2N_2, which is decomposed even at −78 °C, diazofluorene forms a thermally stable complex **9** with "Ni(*t*-BuNC)$_2$", in which the diazo group is η^2-coordinated to the metal atom. Only at 100 °C is the *trans*-cyclopropane **10** formed from **9** and diethyl maleate, but as free diazofluorene yields the same cyclopropane (via ring contraction of a primarily formed 1-pyrazoline) at 30 °C, it is likely that the nickel complex releases diazofluorene at 100 °C which then reacts without catalytic assistance. The formation of cyclopropane **10** (21 %) by thermal decomposition of Pd(PPh_3)$_2$(diazofluorene) at 80 °C in the presence of diethyl maleate [43] can be explained in the same way. Other Ni(0) or Pd(0)-diazoalkane complexes, such as Ni(PPh_3)$_2$(diazofluorene), Pd(*t*-BuNC)$_2$ (diazofluorene) and Ni(*t*-BuNC)$_2$ (diazodiphenylmethane) [51] are thermally more labile than **9**, but their aptitude to cyclopropanation reactions has not been tested. The Ni(0)-diazoalkane complexes which are formed in situ from diazocyclopentadiene or tetrachloro-diazocyclopentadiene and Ni(C_2H_4)(PPh_3)$_2$ and which are assumed by analogy with **9** to have the stoichiometry Ni(PPh_3)$_2$(diazoalkane), do not yield cyclopropanation products with diethyl maleate [52].

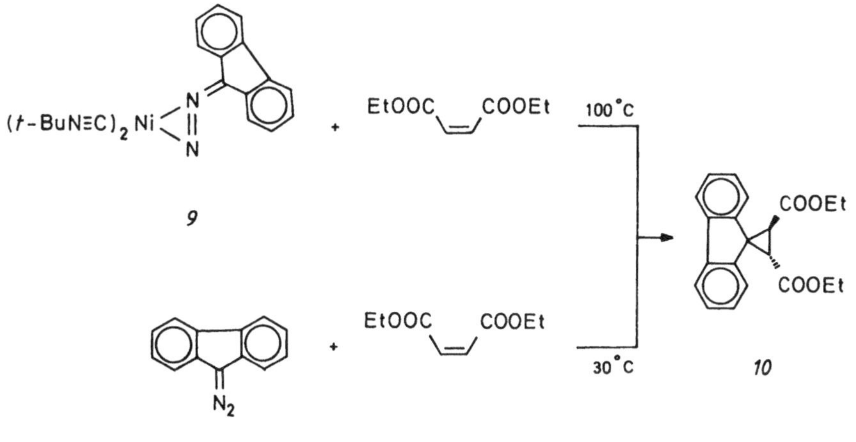

2.2 Cyclopropanation with α-Diazocarbonyl Compounds

2.2.1 The Influence of the Catalyst on Reactivity

Copper powder, copper bronze, Cu_2O, CuO, $CuSO_4$, CuCl and CuBr were the first catalysts which were used routinely for cyclopropanation of olefins as well as of aromatic and heteroaromatic compounds with diazoketones and diazoacetates. Competing insertion of a ketocarbene unit into a C—H bond of the substrate or solvent remained an excpetion [53] in contrast to the much more frequent intramolecular C—H insertion reactions of appropriately substituted α-diazoketones or diazoacetates [12]. Reviews dealing with the cyclopropanation chemistry of diazoacetic esters [6] (including consideration of the efficiency of the copper catalysts mentioned above) and diazomalonic esters [9, 13] as well as with intramolecular cyclopropanation reactions of diazoketones [12] have appeared.

Catalysts more efficient than the catalysts mentioned above are copper(I) trifluoromethansulfonate, copper(II) trifluoromethanesulfonate, copper(II) tetrafluoroborate, copper(II) acetylacetonate and related chelates as well as complexes of copper(I) halides. The first three of these catalysts have the advantage that their very weakly nucleophilic anions do not interfere with the catalytic action of the copper ion. Moreover, $Cu(O_3SCF_3)$, $Cu(BF_4)_2$, copper(I) halide/phosphite or sulfide complexes as well as copper(II) acetylacetonate are readily soluble in many commonly used solvents, thus creating conditions of homogeneous catalysis. The assumption that yields of cyclopropanation under heterogeneous conditions are generally superior to those of homogeneous catalysis [54] can no longer be maintained. A more recent investigation [55] of the reaction between dimethyl diazomalonate and cyclohexene in the presence of copper(II) acetylacetonate, $(MeO)_3P \cdot CuCl$ or $(MeO)_3P \cdot CuI$ has revealed a strong dependence on catalyst concentration for all products (cyclopropane, C—H insertion product, carbene dimer). Under optimum conditions, homogeneous catalysis gave consistently higher yields of cyclopropanation than heterogeneous catalysis. Another example is found with the cyclopropanation of 2-butene by ethyl diazoacetate, which gives a ca. 50% yield with copper(I) trifluoromethanesulfonate [25], but only a 5–10% yield under the heterogeneous conditions of $CuSO_4$ catalysis [56]. Some experiments have shown that even under formally heterogeneous conditions, the active catalyst may exist in solution [14]. This means that in these cases homogeneous catalysis is actually also taking place, but the concentration of the active catalytic species is not at its optimum. An additional advantage of soluble copper catalysts is that a considerable lowering of the decomposition temperature of diazocarbonyl compounds occurs, such that reactions can be performed at room temperature or below. Nevertheless, higher reaction temperatures might be desirable in some cases. For example, cyclopropanation of cyclohexene by dimethyl diazomalonate can be favored over the C—H insertion reactions by raising the temperature [57].

Among the homogeneous copper-based catalysts, trialkyl phosphite-copper(I) iodides have been recommended for cyclopropanation with diazoacetic esters [58] and diazomalonic esters [9], and $(n\text{-}Bu_2S) \cdot CuI$ for cyclopropanation with α-diazoketones [54]. Furthermore, copper(II) acetylacetonate (with somewhat restricted solubility) is a versatile catalyst. A recent comparison of the influence of various copper catalysts on the yields and stereoselectivities of cyclopropanation is given in Table 2 [59]. It can be seen that $Cu(OTf)_2$ generally produces the best results, although its drawback is the ease with which vinyl ethers are polymerized. This problem, in some cases only, can be partly overcome, either by lowering the reaction temperature or by performing the reaction in a co-solvent. $Cu(acac)_2$ is, on average, less reactive than the CuCl-phosphite complexes, and an elevated reaction temperature is required to improve the yields. Alternatively, $Cu(acac)_2$ was found to serve better for cyclopropanation of silyl enol ethers with methyl diazoacetate than $CuCl \cdot P(OMe)_3$, copper bronze and copper(II) hexafluoroacetylacetonate [60].
Thorough investigations with dimethyl diazomalonate and catalysts of the type $(RO)_3P \cdot CuX$ have revealed that the efficiency of competing reaction paths, the *syn/anti* or *E/Z* selectivity in cyclopropane formation as well as the *cis/trans* ratio of carbene dimers depend not only on catalyst concentration and temperature but also on the nature of R [58] and of the halide anion X [57,61]. Furthermore, the cyclopropane yield can be augmented in many cases at the expense of carbene dimer

Table 2. Stereoselectivities for cyclopropane formation from olefins and ethyl diazoacetate with representative copper catalysts[a] (reproduced from reference 59, with the permission of the American Chemical Society)

R	$CuCl \cdot P(O\text{-}i\text{-}Pr)_3$		$CuCl \cdot P(OPh)_3$		$Cu(OTf)_2$		copper bronze		$Cu(acac)_2$	
	yield,[b]		yield,[b]		yield,[b]		yield,[b]		yield,[b]	
	%	t/c^c	%	t/c^c	%	t/c^c	%	t/c^c	%	t/c^c
	Monosubstituted Olefins $H_2C{=}CHR$									
Ph	88	2.8	84	2.5	97	(1.9)[e]	53	(1.9)[e]	71	(2.6)[e]
OEt	61	1.9	64	2.2	55	(2.4)[d]			15	1.6
t-Bu	23	7.3			54	5.5	5	8.1	20	(10.4)[e]
$C(Ph){=}CH_2$	7/28	3.5			14/48	1.8	8/30	3.2	11/42	3.1
$C(Me){=}CH_2$	28/81	3.4	25/73	3.6	19/57	1.8	7/20	4.0	18/55	3.3
	Di- and Trisubstituted Olefins									
2-phenyl-1,3-butadiene (1,2-position)	21/28	1.1			34/48	1.0	22/30	1.2	31/42	1.0
isoprene (1,2-position)	53/81	1.3	48/73	1.2	38/57	1.0	13/20	1.2	37/55	1.3
cyclohexene	28	6.8	25	7.0	80	6.8	23	7.5	18	(6.5)[e]
2,5-dimethyl-2,4-hexadiene	55	2.7	66	2.7	93	2.3	34	2.7	76	(1.8)[e]
1-methoxycyclohexene	54	4.2	35	4.5					44	4.8

[a] Unless indicated otherwise, reactions were performed at 25 °C; [b] For reactions with dienes, yields are presented as (% yield of cyclopropane isomers/(total % yield of cyclopropane products); [c] Precision ±5% of reported value; [d] Reaction performed at 0 °C; [e] Reactions performed at 60 °C.

formation by adding the diazo compound at a controlled rate to the mixture of olefin and catalyst. These few remarks are made to point out that, even with a given alkene, diazo compound and catalytically active metal, the yields of cyclopropanation as well as other carbenoid reactions may quickly become dependent on a hopelessly complicated multi-parameter system. This may not always be so, but these facts have to be kept in mind if the efficiency of a particular catalyst is to be evaluated. It seems, however, that such parameters have been thoroughly optimized in recent studies which compare the efficiency of different catalytic systems.

Table 3. Yields and *trans/cis* ratio in the presence of different catalysts for the following reaction[a, b]:

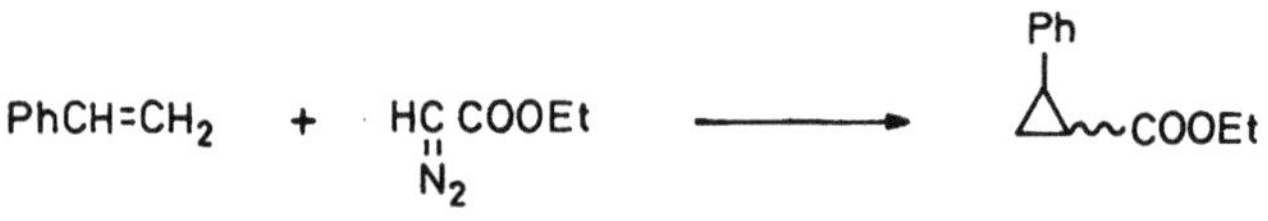

catalyst	yield [%]	*trans/cis*	catalyst	yield [%]	*trans/cis*
$PdCl_2$	70		$RhCl_3 \cdot 3\,H_2O$	7	
$Pd(OAc)_2$	98	2.0	$RhCl(PPh_3)_3$	12	
$PdCl_2 \cdot 2\,PhCN$	65	2.3	$Rh_2(OAc)_4$	92	1.5
$Pd(PPh_3)_4$	57	2.2	$Rh_2(OOC\text{-}t\text{-}C_4H_9)_4$	60	1.5
Pd on C	0		$Rh_2(OOC\text{-}n\text{-}C_6H_{13})_4$	95	1.3
$Cu(acac)_2$	65	2.1	$Rh_2(OOCCF_3)_4$	66	0.9
$Cu(OTf)_2$	80		$Mo_2(OAc)_4$	5	
			$Ru_2(OAc)_4Cl$	38	1.8

[a] From ref. 64; [b] Reaction conditions: 22 °C; molar ratio 3000 (olefin)/1(catalyst)/200(EDA).

Table 4. Efficiency of different catalysts for the following reaction[a, b]:

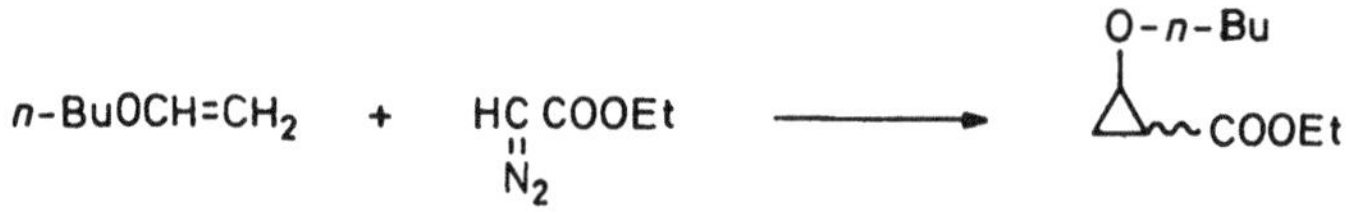

catalyst	yield [%]	catalyst	yield [%]
Cu bronze	95	$Rh_6(CO)_{16}$	86[c]
$CuCl \cdot P\text{-}(O\text{-}i\text{-}Pr)_3$	71	$[Rh(CO)_2Cl]_2$	58[c]
$Cu(acac)_2$[c]	71	$Rh_2(OAc)_4$	86[c]
$Pd(PPh_3)_4$	31	$Ru_3(CO)_{12}$	65
$PdCl_2 \cdot 2\,PhCN$[c]	34	$Re_2(CO)_{10}$	18
$CO_2(CO)_8$	18	$[Ru(CO)_3Cl_2]_2$	polymerization
$Fe(CO)_5$	16	$RuCl_3(CO)(PPh_3)_2$	polymerization
$Mo(CO)_6$	38		

[a] From ref. 66;
[b] Reaction conditions: EDA in cyclohexane was added to a mixture of olefin (fivefold molar excess) and catalyst (0.5 mol%, for copper bronze 11 mol %); reaction temperature 65 °C unless stated otherwise;
[c] At 25 °C; EDA dissolved in diethyl ether.

The dominant role of copper catalysts has been challenged by the introduction of powerful group VIII metal catalysts. From a systematic screening, palladium(II) and rhodium(II) derivatives, especially the respective carboxylates [62) 63) 64)], have emerged as catalysts of choice. In addition, rhodium and ruthenium carbonyl clusters, $Rh_6(CO)_{16}$ [65)] and $Ru_3(CO)_{12}$ [66)], seem to work well. Tables 3 and 4 present a comparison of the efficiency of different catalysts in cyclopropanation reactions with ethyl diazoacetate under standardized conditions.

Representative copper, palladium and rhodium catalysts have been tested for their ability to promote cyclopropanation of olefins with different steric and electronic properties. The results of these studies do not only offer some insight into mechanistic pathways but also pave the way to an appropriate catalyst choice if it comes to questions of regioselectivity and stereoselectivity. These aspects will be dealt with in the following Sections. It has been proposed that basically two mechanisms are operating in transition-metal catalyzed cyclopropanation reaction [64)]. One is a carbenoid pathway with a diazo compound/metal complex or a metal-carbene attacking the olefin. The second possibility is a coordination-type mode, according to which cyclopropanation occurs by intramolecular reaction between a metal-associated diazoalkane or carbene and the olefin coordinated to the same metal atom. Rhodium(II) carboxylates, being dinuclear complexes with only one available coordination site at each metal atom [67)], are expected to react in the carbenoid mode. Alternatively, palladium compounds are known to form stable complexes with olefins [68)], so that in these cases, a coordination mechanism is feasible. Finally, copper catalysts are expected to initiate a carbenoid reaction except in those cases where very weakly coordinating ligands such as TfO^- or BF_4^- are present, which can easily be displaced by an olefin [69)]. The efficiency of $Rh_2(OAc)_4$, $Pd(OAc)_2$ and $Cu(OTf)_2$ to catalyze cyclopropanation of various olefins (Table 5), including dienes with equivalent double bonds, can be summarized as follows [64)]:

- $Rh_2(OAc)_4$ is the most effective and versatile of the three catalysts used. Terminal and non-terminal olefins, strained olefins (norbornene, norbornadiene) and conjugated olefins (styrene) all react in good yield.
- $Pd(OAc)_2$ works well with strained double bonds as well as with styrene and its ring-substituted derivatives. Basic substituents cannot be tolerated, however, as the failures with 4-(dimethylamino)styrene, 4-vinylpyridine and 1-vinylimidazole show. In contrast to $Rh_2(OAc)_4$, $Pd(OAc)_2$ causes preferential cyclopropanation of the terminal or less hindered double bond in intermolecular competition experiments. These facts are in agreement with a mechanism in which olefin coordination to the metal is a determining factor but the reluctance or complete failure of Pd(II)-diene complexes to react with diazoesters sheds some doubt on the hypothesis of Pd-olefin-carbene complexes (see Sect. 11).
- $Cu(OTf)_2$ generally gives yields intermediate between those of the other two catalysts, but with a closer resemblance to rhodium. In competition experiments, the better coordinating norbornene is preferred over styrene, just as in the case with $Pd(OAc)_2$. $Cu(acac)_2$, however, parallels $Rh_2(OAc)_4$ in its preference for styrene. These findings illustrate the variability of copper-promoted cyclopropanations, and it was suggested that in the $Cu(OTf)_2$-catalyzed reactions of diazoesters, basic by-products, which are formed as the reaction proceeds, may gradually suppress

Table 5. Yields of cyclopropanation of various olefins by diazoacetic esters in the presence of $Rh_2(OAc)_4$, $Pd(OAc)_2$ or $Cu(OTf)_2$ [a, b]

olefin	diazo-ester	yield [%] with		
		$Pd(OAc)_2$	$Cu(OTf)_2$	$Rh_2(OAc)_4$
1-hexene	MDA	30	36	86
cis-2-butene	MDA	24		54
trans-2-butene	MDA	21		
cis-3-hexene	EDA	15	15	
cis-2-octene	EDA	5	40	65
trans-2-octene	EDA	2	14	24
trans-4-octene	EDA	12	8	7
2,3-dimethyl-2-butene	EDA	5	30	70
cyclopentene	EDA	60	60	95
cyclohexene	EDA	21	54	88
cycloheptene	EDA	40	30	75
cyclooctene	EDA	20	28	95[c]
norbornene	EDA	87	95	95
indene	EDA	20	55	71
styrene	EDA	98	80	90
α-methylstyrene	EDA	42		
p-methylstyrene	EDA	81		
p-methoxystyrene	EDA	79		
p-chlorostyrene	EDA	86		
p-nitrostyrene	EDA	71		
p-(dimethylamino)styrene	EDA	0		
1,1-diphenylethylene	EDA	0		
trans-1,2-diphenylethylene	EDA	0		
vinylacetate	EDA	5	22	77
ethyl vinyl ether	EDA	42	polymer.	85
dihydropyran	EDA	20	55	71
1,5-hexadiene	EDA	37	60	80
1,3-cyclohexadiene	EDA	18	53	90
1,5-cyclooctadiene	EDA	10	25	64
norbornadiene	EDA	95	47	88
dimethyl maleate	EDA	traces	traces	traces

[a] From ref. 64; [b] Reaction conditions: 22 °C; molar ratio 3000 (olefin)/1(catalyst)/200 (diazo ester); [c] 54% isolated yield reported in reference 70. Reaction conditions: room temp.; molar ratio: 7400 (olefin)/1(catalyst)/1020(diazo ester).

the coordination mechanism and give way to the carbenoid mode as in rhodium catalysis.

The air-stable rhodium carbonyl cluster $Rh_6(CO)_{16}$ displays a similarly high catalytic efficiency as does $Rh_2(OAc)_4$; Table 6 gives some examples. Comparisons of $Rh_6(CO)_{16}$ with $Rh_2(OAc)_4$, $PdCl_2 \cdot 2\,PhCN$ and $CuCl \cdot P(O\text{-}i\text{-}Pr)_3$ are given in Table 7. They show once more that the rhodium catalysts are superior to other transition-metal catalysts. $Rh_2(OAc)_4$ gives consistently the highest yields, whereas the activity of $Rh_6(CO)_{16}$ depends on the olefin used. Because of its insolubility in olefins, it is probably the extent to which an active, homogeneous catalyst is formed, which imposes certain limits on the versatility of $Rh_6(CO)_{16}$. Results comparable to those in Table 7 were also obtained with monosubstituted 1,3-dienes [59,71,72].

Table 6. Cyclopropanation reactions with ethyl diazoacetate using equimolar amounts of alkene and diazo ester[a, b]

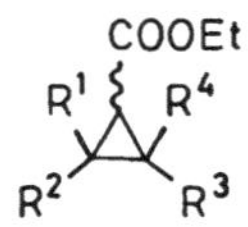

olefin	catalyst	[EDA]/[catalyst]	addition rate[c] [mmol/h]	R^1	R^2	R^3	R^4	isolated yield [%]	E/Z ratio
α-methoxystyrene	$Rh_2(OAc)_4$	100	7.5, 3.8	OMe	Ph	H	H	94	1.0
2-methoxypropene	$Rh_2(OAc)_4$	200	10, 5.0	OMe	Me	H	H	65	2.3
	$Rh_6(CO)_{16}$	500	5.0, 2.0					63	2.0
	$CuCl \cdot P(O\text{-}i\text{-}Pr)_3$	200	0.25					40	2.5
ethyl vinyl ether	$Rh_2(OAc)_4$	300	6.7, 3.5	OEt	H	H	H	75	1.6
1-methoxycyclohexene	$Rh_2(OAc)_4$	2000	0.5, 0.25				H	80	2.4 (*anti/syn*)
cyclohexene	$Rh_2(OAc)_4$	200	10, 5.0	OMe	$-(CH_2)_4-$		H	80	4.5 (*anti/syn*)
	$Rh_6(CO)_{16}$[d]	1000	5.0, 2.0	H	$-(CH_2)_4-$			43	3.0 (*anti/syn*)
	$CuCl \cdot P(O\text{-}i\text{-}Pr)_3$		0.25					40	12 (*anti/syn*)
2,5-dimethyl-	$Rh_2(OAc)_4$[e]	1100	0.60	Me	Me	H	$-CH{=}C(Me)_2$	58	2.2
2,4-hexadiene	$Rh_6(CO)_{16}$	200	0.33					50	2.3

[a] From Ref. [73]; [b] Reaction conditions: 25 °C, components dissoled in anhydrous ether; [c] Addition rate for EDA solution. The first half was added at the faster rate; [d] In refluxing cyclohexane; [e] No additional solvent; [f] In toluene at 60 °C.

Table 7. Yields of cyclopropanation and stereoselectivities with ethyl diazoacetate in the presence of different catalysts[a, b, c]

olefin	$Rh_2(OAc)_4$		$Rh_6(CO)_{16}$		$CuCl \cdot P(O\text{-}i\text{-}Pr)_3$		$PdCl_2 \cdot 2\,PhCN$	
	yield [%]	*E*/*Z*	yield [%]	*E*/*Z*	yield [%]	*E*/*Z*	yield [%]	*E*/*Z*
allyl bromide	55	1.1	30	1.1	7	1.8		
allyl chloride	90	1.2	24	1.2	50	1.8		
styrene	93	1.6	86	1.7	88	2.8	52	1.6
ethyl vinyl ether	88	1.7	62	1.7	61	1.9	43	1.5
n-butyl vinyl ether	84	1.7	69	1.8	51	2.0	34	1.6
3,3-dimethyl-1-butene	87	4.2	42	4.5	23	7.3	34	2.5
2-methoxypropene	78	1.0	72	1.0	67	1.1	66	0.93
3,3-dimethyl-2-methoxy-1-butene	70	0.71	83	0.72	7	1.2	28	0.61
cyclohexene	90	3.8	88	3.9	28	6.8	31	2.2
3,4-dihydropyran	91	6.5	82	6.8	18	6.3	41	3.8
2,5-dimethyl-2,4-hexadiene	81	1.8	87	1.9	55	2.7	20	2.3
1-methoxycyclohexene	78	2.5	59	2.6	54	4.2	39	1.5

[a] From Ref. [59]; [b] Reaction conditions: 25 °C; EDA (2.0 mmol) added over a 6–8 h period to the mixture of alkene (20 mmol) and catalyst (0.01 mmol); [c] Yields determined by GC, precision ±5%.

The exceptionally high activity of $Rh_6(CO)_{16}$ is exemplified by the 90% yield of ethyl chrysanthemate **11** from 2,5-dimethyl-2,4-hexadiene, even if the ratio EDA/catalyst was as high as 2000 [65]. When the reactions are carried out in a CO atmosphere, the catalyst can be recovered quantitatively.

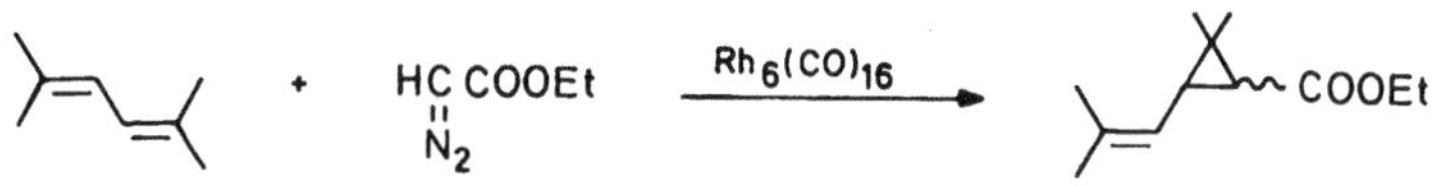

11

(>90%)

Conditions: 60 °C; EDA added over 6 h to a sevenfold molar excess of olefin

The common by-products obtained in the transition-metal catalyzed reactions are the formal carbene dimers, diethyl maleate and diethyl fumarate. In accordance with the assumption that they owe their formation to the competition of olefin and excess diazo ester for an intermediate metal carbene, they can be widely suppressed by keeping the actual concentration of diazo compound as low as possible. Usually, one attempts to verify this condition by slow addition of the diazo compound to an excess (usually five- to tenfold) of olefin. This means that the addition rate will be crucial for the yields of cyclopropanes and carbene dimers. For example, $Rh_6(CO)_{16}$-catalyzed cyclopropanation of *n*-butyl vinyl ether with ethyl diazoacetate proceeds in 69% yield when EDA is added during 30 minutes, but it increases to 87% for a 6 h period. For styrene, the same differences were observed [65].

Whereas control of the rate of addition of the diazoester generally meets with increased yields when $Rh_2(OAc)_4$, $Rh_6(CO)_{16}$ and $CuCl \cdot P(OR)_3$ are used, it has no effect on cyclopropane yields in the case of $PdCl_2 \cdot 2\,PhCN$ [59].

With the rhodium and copper catalysts, even the combination of equimolar amounts of olefin and diazoester will allow high yields of cyclopropanes if the addition rate is controlled meticulously (see Table 6 for examples). This circumstance is particularly useful for cyclopropanation of olefins which are in short supply. In combination with $Rh_6(CO)_{16}$, the easy recovery of the unchanged catalyst (by diluting the mixture with hexane and separating the precipitated catalyst from the liquid [65]) may render such a procedure particularly attractive from an economical point of view.

All reactions listed in Tables 5–7 were carried out under a nitrogen atmosphere, but with the rhodium or palladium catalysts no noticeable or only minor reduction in cyclopropane yields was observed when air was present. In contrast, air clearly had a yield-diminishing effect in the $CuCl \cdot P(O\text{-}i\text{-}Pr)_3$-catalyzed reactions, especially with cyclohexene and 3,4-dihydropyran. Cyclohexene was oxidized to 2-cyclohexen-1-one, and 3,4-dihydropyran gave 5,6-dihydro-4-pyrone and 5,6-dihydro-2-pyrone, albeit in yields below 8% [59].

In other cases, oxidation of the rhodium or palladium ketocarbenoid to a 1,2-dicarbonyl compound is well established: The $Rh_2(OAc)_4$-catalyzed decomposition of 4-diazo-1,2,5,6-tetramethyltricyclo[3.1.0.0^{2,6}]hexan-3-one (**12**) in the presence of 2,3-dimethyl-2-butene and without exclusion of air does not lead to the formal [3 + 2] cycloaddition product of the rearranged ketocarbene **14** to the olefin, as is the case with $AgClO_4$ as catalyst. Instead, compound **15** is isolated in low yield, probably arising from the reaction sequence shown below, which includes air oxidation of ketocarbene **14** to the 1,2-benzoquinone [74].

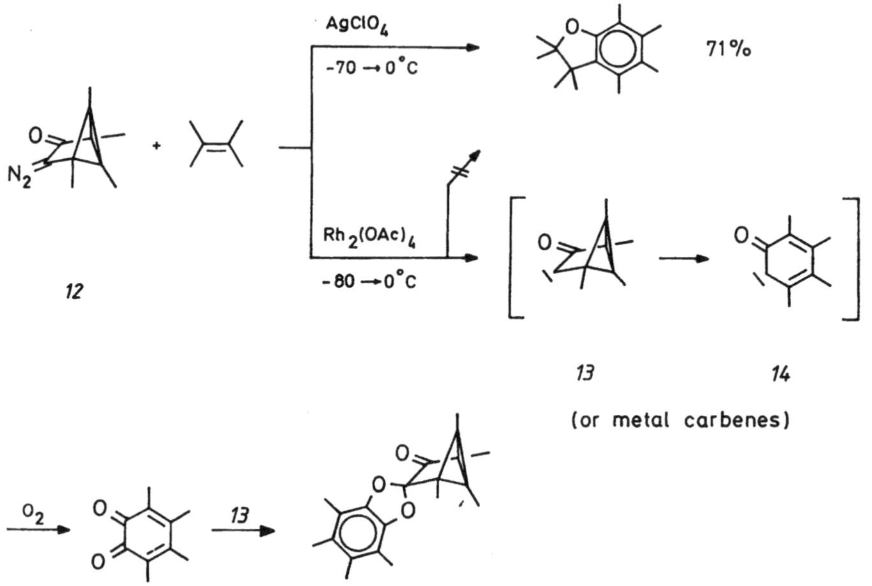

12 13 14 15 (15%)

Azibenzil, in the presence of O_2 and $Pd(OAc)_2$ or $PdCl_2 \cdot 2\,PhCN$, forms an intermediate metal-oxygen-carbene complex which is able to epoxidize aliphatic and alicyclic olefins; azibenzil itself is transformed into benzil [75].

Diazo ester/rhodium(II) carboxylate combinations other than EDA/$Rh_2(OAc)_4$ have been tested [63, 64]. It turned out that the solubility of the rhodium(II) carboxylate greatly influenced the efficiency of cyclopropanation. For the reaction of monoolefins with ethyl diazoacetate, markedly higher yields than with Rh(II) acetate were obtained with the better soluble rhodium(II) butanoate and rhodium(II) pivalate, the latter one being soluble even in pentane. However, only poor yields resulted from the use of rhodium(II) trifluoroacetate, even though this compound is readily soluble. $Rh_2(CF_3COO)_4$, in contrast to the other rhodium(II) carboxylates, is able to form 1:1 complexes with olefins [76], particularly with electron-rich ones; thus, competition of olefin and diazo compound for the only available coordination site at the metal atom could be responsible for the reduced catalytic action of $Rh_2(CF_3COO)_4$ (as will be seen in Section 4.1, this complex is an excellent catalyst for cyclopropanation of aromatic substrates). The diazoester substituent also has some influence on the yields. Increasing yields were obtained in the series methyl ester, ethyl ester, *n*-butyl ester; an explanation for this may be a higher degree of stabilization and/or solubility of an intermediate metal carbene [63].

For the synthesis of permethric acid esters **16** from 1,1-dichloro-4-methyl-1,3-pentadiene and of chrysanthemic acid esters from 2,5-dimethyl-2,4-hexadienes, it seems that the yields are less sensitive to the choice of the catalyst [72,77]. It is evident, however, that $Rh_2(OOCCF_3)_4$ is again less efficient than other rhodium acetates. The influence of the alkyl group of the diazoacetate on the yields is only marginal for the chrysanthemic acid esters, but the yield of permethric acid esters *16* varies in a catalyst-dependent non-predictable way when methyl, ethyl, *n*-butyl or *t*-butyl diazoacetate are used [77].

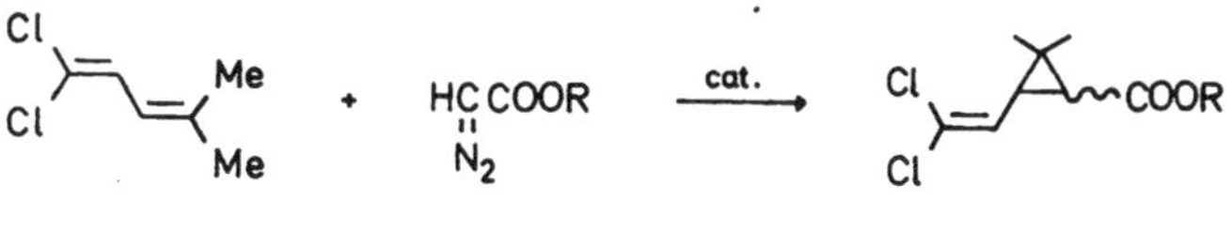

16

cat.	R=Et		R=*n*-Bu	
	yield [%]	ratio *Z*/*E*	yield [%]	ratio *Z*/*E*
$Rh_2(OAc)_4$	54	1	56	1.29
$Rh_2(OOCCF_3)_4$	29	0.93	45	0.8
$Rh_2(OOCC_6F_5)_4$ [a]	64	0.94	56	0.87
$Rh_2(OOCCMe_3)_4$	56	1.15	32	1.29

[a]) Rh(II) pentafluorobenzoate
Conditions: 22 °C; molar ratio 800 (diene)/200 (diazo ester)/1 (catalyst)

Scheme 3

Rhodium(II) pivalate has also been recommended for the cyclopropanation of vinyl halides with ethyl diazoacetate [78]. As Table 8 shows, yields with this catalyst are far higher and reaction conditions milder than with copper. Failures are noted,

Table 8. Cyclopropanation of vinyl halides with ethyl diazoacetate in the presence of rhodium(II) pivalate (Rhpiv) or copper

halide	catalyst	solvent	temp. [°C]	yield of cyclopropane [%]	molar ratios olefin/EDA/Rhpiv
	Rh piv	CH_2Cl_2	25	32	1000/118/0.33
	Cu	$ClCH_2CH_2Cl$	80	1.4	
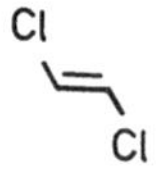	Rh piv	CH_2Cl_2	25	1.5	1000/118/0.33
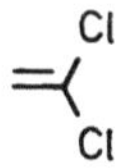	Rh piv	—	25	76	360/76/0.33
	Cu	$ClCH_2CH_2Cl$	80	20	
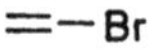	Rh piv	$ClCH_2CH_2Cl$	25	62	374/77/0.10
	Cu		100	9	

however, with *trans*-1,2-dichloroethylene; trichloroethylene and tetrachloroethylene did not react at all.

2.2.2 Regioselectivity

The catalytic cyclopropanation of 1,3-dienes leads exclusively or nearly so to mono-cyclopropanation products, as long as no excess of diazocarbonyl compound is applied. The regioselectivity has been tested for representative rhodium, copper and palladium catalysts [59,71,72], and the results are displayed in Table 9.

The following observations were made:

a) The observed regioselectivity depends on the metal atom, but is virtually unaffected by its ligands or its original oxidation state. This is a remarkable result for the copper catalysts inasmuch as a decisive influence of their nature on the extent of intermolecular competition between 2,3-dimethyl-2-butene and 1-hexene was found for cyclopropanation with ethyl diazoacetate as well as with diazomethane [25].
b) In 2-substituted 1,3-dienes, the elctron-richer double bond is preferred with the rhodium and copper catalysts, an exception being 2-*t*-butyl-1,3-butadiene. With palladium, there is a tendency for the unsubstituted double bond of the diene to react, in agreement with the assumption that palladium-based catalysts act via a coordination mechanism (see Sect. 2.2.1).
c) In 1-substituted butadienes, the less-hindered (unsubstituted) double bond is generally preferred, no matter how nucleophilic the other double bond is. This seems to be related more to the general pattern of 1,2-disubstitution than to the *E* or *Z* configuration at the double bond, as no difference in regioselectivity was found between *E* and *Z* isomers of 1-phenylbutadiene and 1-chlorobutadiene, respectively [71]. In contrast to its behavior towards 2-substituted 1,3-dienes, palladium seems to tolerate the *E*-substituted double bond more than the other catalysts. This aspect still awaits an explanation, especially since it is not a singular case: Contrary to $Rh_2(OAc)_4$ and $Cu(OTf)_2$, $Pd(OAc)_2$ does not discriminate between the *Z*- and *E*-double bond of 2,4-hexadiene and largely prefers the *E*-double bond of *Z,E*-1,5-cyclodecadiene in cyclopropanations with ethyl diazoacetate (Scheme 4) [72].
d) The regioselectivities for $Rh_6(CO)_{16}$ and $CuCl \cdot P(O\text{-}i\text{-}Pr)_3$ are consistently higher than those for $Rh_2(OAc)_4$, whereas $PdCl_2 \cdot 2\,PhCN$ exhibits lower (including reversed) regioselectivity. As the examples of Table 9 indicate, the preferred site of cyclopropanation results from a delicate balance of steric and electronic factors.

Several other examples of regioselective cyclopropanation of 1- and 2-substituted butadienes in the presence of copper catalysts are known (Scheme 5). 2-Trimethyl-siloxy-1,3-butadiene parallels the behavior of other 2-substituted butadienes (see Table 9) in that the electron-rich double bond is cyclopropanated [60]. With the 1-methoxy-, acetoxy- or trimethylsilyloxy-substituted butadienes **17**, **18** and **19**, both double bonds are cyclopropanated, thus giving rise to sometimes unseparable mixtures of regio- and stereoisomers [79]. Perhaps, the yields of separated and isolated regioisomers in some cases do not reflect the true regioselectivity as considerable

Table 9. Regioselectivities in the cyclopropanation of 1,3-dienes with EDA in the presence of various catalysts[a]. The yields [%] of cyclopropanation at each of the double bonds are given[a]

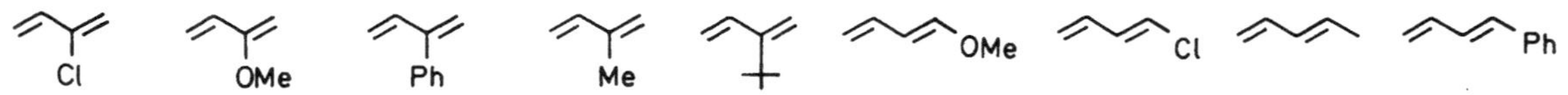

catalyst[b]	yields [%][c]																	
	diene 1		diene 2		diene 3		diene 4		diene 5		diene 6		diene 7		diene 8		diene 9	
$Rh_2(OAc)_4$	22	54	0	88	25	56	36	57	45	25	80	10	73	0	54	7	82	2
$Rh_6(CO)_{16}$	24	56	0	63	20	51	27	43	5	4	58	5	21	0	64	8	39	1
CuCl · P(O-*i*-Pr)$_3$	5	28	0	38	7	21	28	53	8	12	48	9	17	0	21	2	55	1
$Cu(OTf)_2$[d]							25	49										
$PdCl_2$ · 2 PhCN	<8%		0	18	5	5	16	8	4	2	16	6	<8%		24	7	12.5	0.5
$Pd(OAc)_2$[d]							37	11										

catalyst[b]	yields [%][c]			
	diene 1		diene 2	
CuCl · P(O-*i*-Pr)$_3$	7	21	28	53
CuCl · P(OPh)$_3$			25	48
$Cu(OTf)_2$[c]	14	34	19	38
copper bronze	8	22	7	13
$Cu(acac)_2$	11	31	18	37

[a] From Refs. [59] and [72]; [b] 25 °C; EDA (2 mmol) added over 6–8 h to diene (10 mmol) and catalyst (0.02 mmol); [c] Yields determined by GC; [d] 22 °C; EDA (5 mmol) added over 4 h to diene (20 mmol) and catalyst (0.025 mmol).

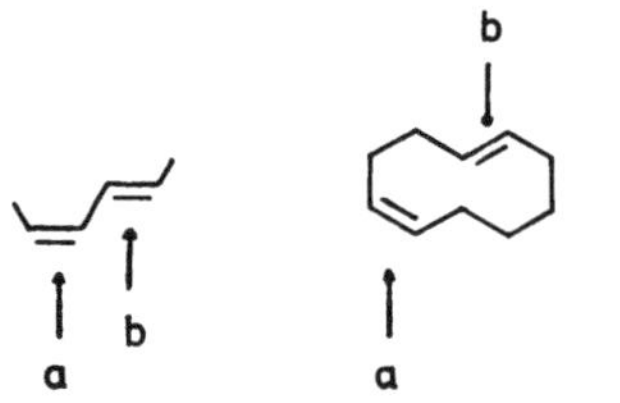

catalyst	yield [%]		yield [%]	
	a	b	a	b
$Rh_2(OAc)_4$	66	21	25	25
$Cu(OTf)_2$	66	21	11	11
$Pd(OAc)_2$	30	30	1	40

Scheme 4. Yields of intramolecular competition between *E* and *Z* double bonds upon cyclopropanation with EDA in the presence of different catalysts. Conditions: 22 °C; molar ratio 800 (diene)/200 (EDA)/1 (catalyst).

losses in the material balance occurred during work-up. From the given values, it seems however, that ethyl diazoacetate and diazoacetone react with similar regioselectivity. For α-terpinene **20**, the carbenoid cyclopropanation was expected to occur preferentially at the less hindered double bond, and this was indeed the case in the only isolated product [80].

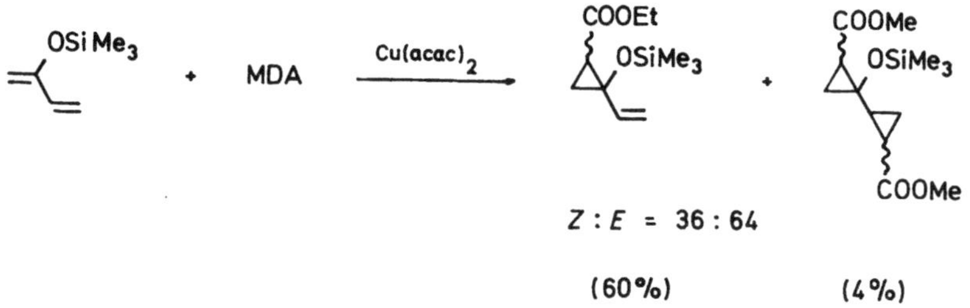

Conditions: 90–100 °C; equimolar amounts of olefin and MDA; benzene solution of MDA slowly added to the olefin

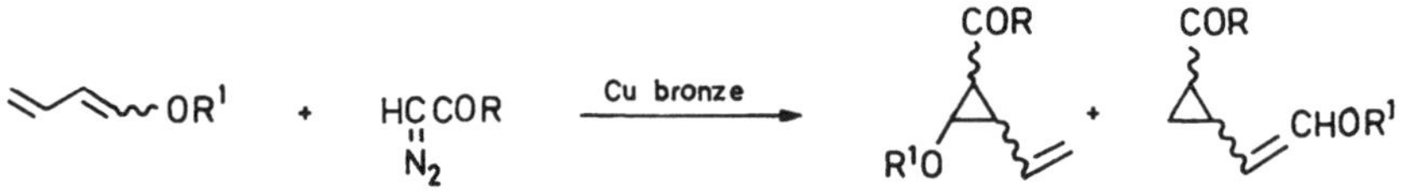

17

R^1 = Me	R = OEt	8%	34%
	Me	13% [a]	33% [a]
R^1 = $SiMe_3$	R = OEt	total: 62% [b]	
	Me	total: 44% [b]	

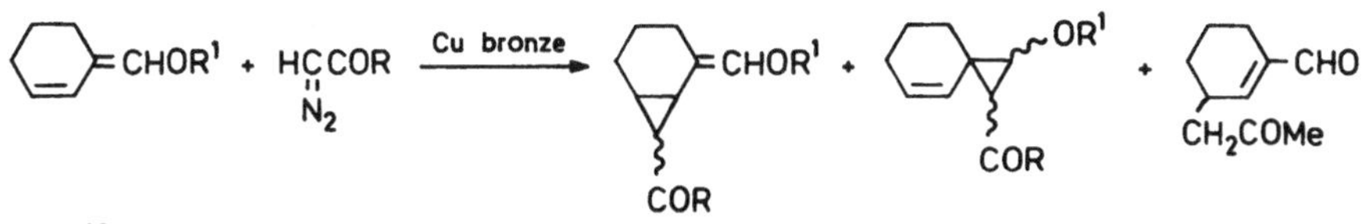

18

R^1 = Ac	R = OEt	28% [a]		8% [a]	–
	Me	19% [a]		6% [a]	–
R^1 = $SiMe_3$	R = OEt	2.7	:	1	–
	Me	2.3	:	1.8 :	1

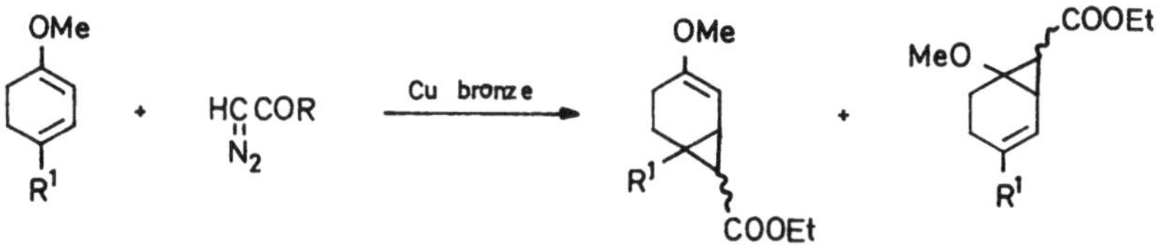

19

R = OEt	R^1 = H	18%	43%
	Me	5%	22%
R = Me	R^1 = H	total: 66% [b]	

Conditions for reactions of *17–19*: 80 °C, cyclohexane (for EDA),
70 °C, cyclohexane (for diazoacetone)

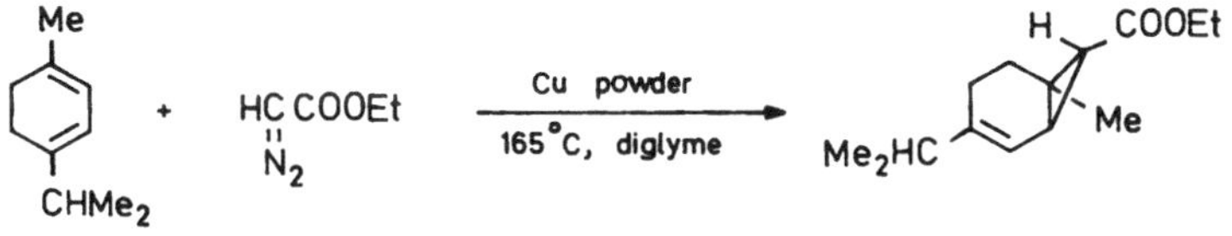

20 (58%)

[a] Regioisomers not separated

[b] Mixture not separated, no stereochemical assignment

Scheme 5

1,3,5-Cycloheptatriene, as an example of a conjugated triene, is mainly cyclopropanated at an outer double bond (Scheme 6). This is true for $Rh_2(OAc)_4$, $Cu(OTf)_2$ and $Pd(OAc)_2$, but the highest yield is obtained again with the rhodium catalyst [72]. Twofold cyclopropanation occurs to only a minor extent, as long as an excess of olefin is applied. With equal amounts of diazo ester and cycloheptatriene, double cyclopropanation increases and even traces of the triply cyclopropanated triene are found with $Rh_2(OAc)_4$ and $Cu(OTf)_2$. This behavior essentially parallels the earlier results of CuBr-catalyzed cyclopropanation of cycloheptatriene with ethyl diazoacetate[81].

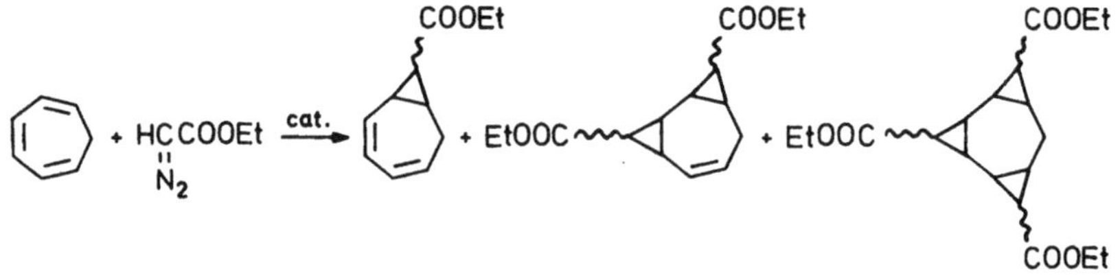

catalyst	yield [%][a]		
$Rh_2(OAc)_4$	90(41)	8(20)	0(5)
$Cu(OTf)_2$	78(40)	5(20)	0(2)
$Pd(OAc)_2$	22	0	0

[a] Conditions: 22 °C; molar ratio 800 (triene)/200 (EDA)/1 (catalyst); values in parentheses are yields obtained from equimolar amounts of cycloheptatriene and EDA

Scheme 6

In molecules containing a 1,3-diene unit and an isolated double bond, the diene is cyclopropanated preferentially (Scheme 7)[72,82]. What has been said about the influence of steric and electronic factors as well as the nature of the catalyst (see above), can also be applied to explain the product distribution in these cases. The inertness of a trisubstituted double bond and the low reactivity of an *E*-disubstituted olefinic bond are quite obvious in these intramolecular competitions.

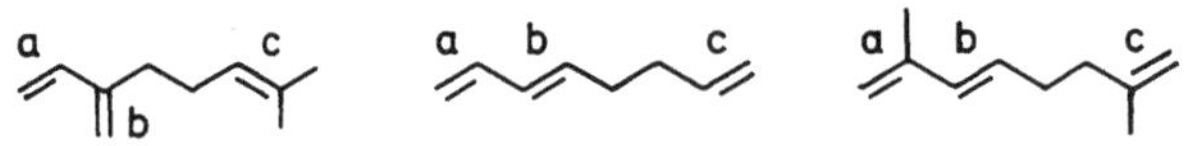

21

catalyst	yield [%]								
	a	b	c	a	b	c	a	b	c
$Rh_2(OAc)_4$	39	48	0	87	10	traces	38	0	28
$Cu(OTf)_2$	31	41	0	56	8	traces	20	0	13
$CuSO_4$	30	20	0						
$Pd(OAc)_2$	25	8	0	20	7	traces	8	0	5

Conditions: As in Scheme 6 for $Rh_2(OAc)_4$, $Cu(OTf)_2$. $Pd(OAc)_2$. $CuSO_4$: room temp., no solvent

Scheme 7

In view of our earlier statement, that the regioselectivity remains unaffected by the ligands of the catalyzing metal, the reversal of regioselectivity for myrcene (**21**), when $Cu(OTf)_2$ is replaced by $CuSO_4$, comes as a surprise. As the two observations were made in different studies, however, this difference should not be overestimated. Standardized conditions would be necessary to clarify this point.

Only a few results are available concerning competitive cyclopropanation of non-conjugated dienes. The case of 1,4-hexadiene [72] (mixture of *Z* and *E* isomers) illustrates the reactivity difference between a monosubstituted and a 1,2-disubstituted double bond, whereas in limonene (**24**) [47], a 1,1-disubstituted and a trisubstituted double bond compete for the carbenoid derived from ethyl diazoacetate. In both cases, the less substituted double bond reacts preferentially (Scheme 8).

22 *23*

catalyst[a]	*22* + *23* [%]	*22*:*23*
$Rh_2(OAc)_4$	87	2.9
$Cu(OTf)_2$	56	1.7
$Pd(OAc)_2$	53	5.6

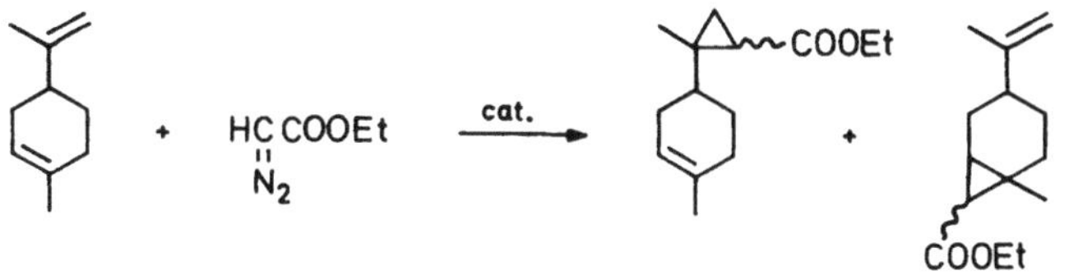

24 *25* *26*

catalyst[b]	*25* + *26* [%]	*25*:*26*
$Rh_2(OAc)_4$	98	3.4
$Cu(OTf)_2$	59	3.9
$PdCl_2 \cdot 2\,PhCN$	32	3.6

[a] Conditions: see Scheme 6.
[b] Conditions: room temp.; molar ratio 1000 (olefin)/100 (EDA)/1–2 (catalyst).

Scheme 8

It can be seen that for limonene the regioselectivities are virtually independent of the catalyst, and this is also true when $Rh_6(CO)_{16}$, Cu bronze, $CuCl \cdot P(O\text{-}i\text{-}Pr)_3$ or $Pd(CF_3COO)_2$ are taken as catalysts.

The preference for the less substituted double bond also determines the outcome of the copper-catalyzed cyclopropanation of isotetraline with dimethyl diazomalonate which gives **27** and its dehydrogenated relative **28** [83]; the same behavior of the carbenoid derived from ethyl diazoacetate has been reported [84].

In contrast to the behavior of 1,4-hexadiene and limonene, the regioselectivity of the norbornene derivatives **29** and **32** strongly depends on the catalyst. A preference for the exocyclic double bond is found with $Rh_2(OAc)_4$, but the "1,2-disubstituted" endocyclic C=C bond is distinctly favored with $PdCl_2 \cdot 2\ PhCN$ [47].

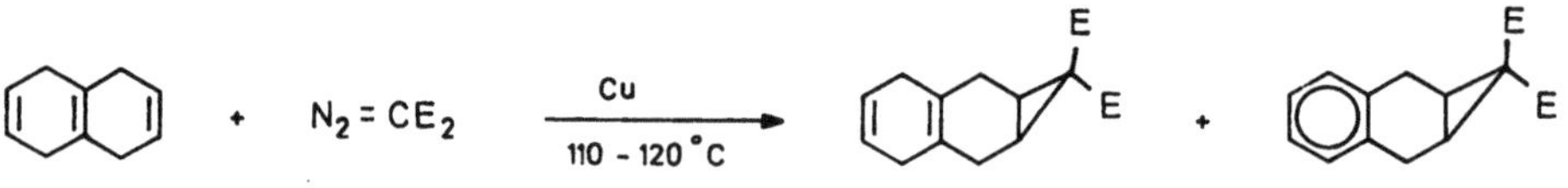

27 (14%) 28 (7%)

E = COOMe

$Cu(OTf)_2$ is an intermediate case (Scheme 9). These findings parallel the catalyst's control over the regioselectivity of cyclopropanation with diazodiphenylmethane [47] (see Sect. 2.1).

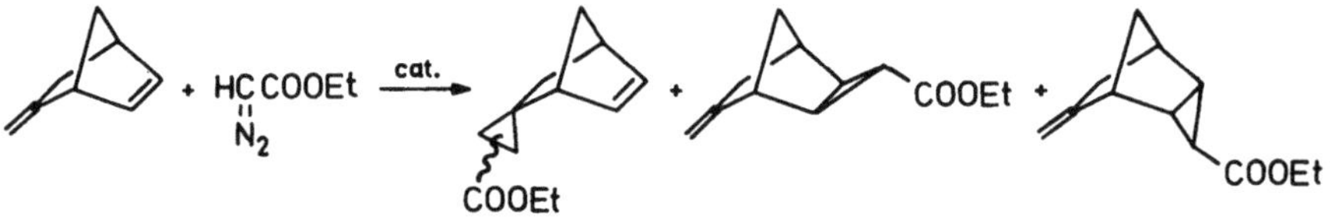

29 30 exo-31 endo-31

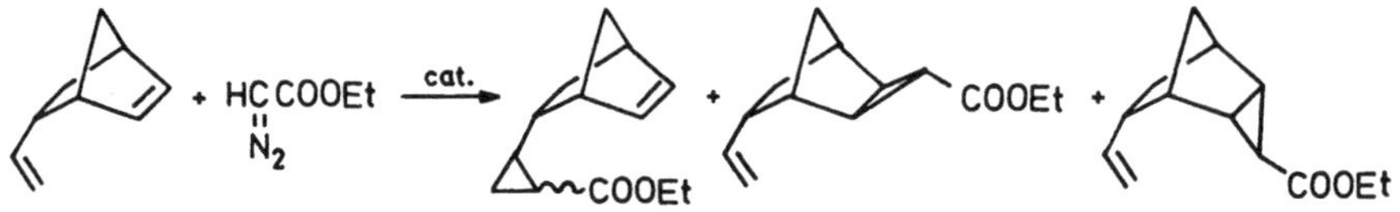

32 33 exo-34 endo-34

catalyst	*30* + *31* [%]	*30*:*31*	exo-*31*: endo-*31*	*33* + *34* [%]	*33*:*34*
$Rh_2(OAc)_4$	93	2.3	2.6	not given	1.2
$Cu(OTf)_2$	53	1.0	1.3	not given	0.30
$PdCl_2 \cdot 2\ PhCN$	82	0.16	1.2	not given	0.20

Scheme 9

As an explanation, if was suggested that the degree of charge development in the transition state determines the preferred site of cyclopropanation: A transition state with little charge development should prefer the endocyclic double bond (Pd catalysis), whereas one with much charge development should favor the exocyclic bond (Rh catalysis).

2.2.3 Stereoselectivity

Cyclopropanation of C=C bonds by carbenoids derived from diazoesters usually occurs stereospecifically with respect to the configuration of the olefin. This has been confirmed for cyclopropanation with copper [25,57,60,85], palladium [86], and rhodium catalysts [59,87]. However, cyclopropanation of *cis*-D_2-styrene with ethyl diazoacetate in the presence of a (1,2-dioximato)cobalt(II) complex occurs with considerable geometrical isomerization [88]. Furthermore, CuCl-catalyzed cyclopropanation of *cis*-2-butene with ω-diazoacetophenone gives a mixture of the *cis*- and *trans*-1,2-dimethylcyclopropanes [89].

For many catalytic cyclopropanations, the stereoselectivity describing the stereochemical relation between substituents at the carbenoid and those at the double bond is not very pronounced. *E/Z* or *syn/anti* ratios of ca. 1–3 in favor of the less congested isomer may be considered normal (for examples see Tables 6 and 7). The stereochemical outcome can be expected to be governed by the nature of the olefin, the diazo compound and the catalyst.

For cyclopropanations with ethyl diazoacetate, a rather weak influence of the olefin structure has been noted [59,60] (Table 7). The preference for the sterically less crowded cyclopropane is more marked for 1,2-disubstituted than for 1,1-disubstituted olefins. The influence of steric factors becomes obvious from the fact that the ratio *Z*-**36**/*E*-**36**, obtained upon cyclopropanation of silyl enol ethers **35**, parallels Knorr's [90] empirical substituent parameter λ^d of the group R [60]. These *Z/E* ratios, however, do not represent the thermodynamic equilibrium of both diastereomers.

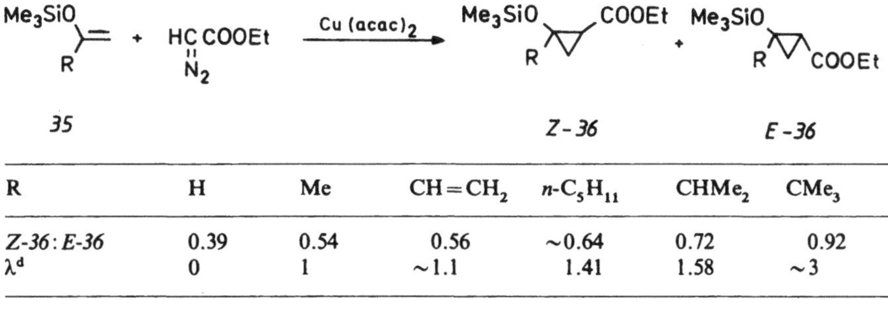

35   Z-36   E-36

R	H	Me	$CH=CH_2$	n-C_5H_{11}	$CHMe_2$	CMe_3	Ph
Z-36:*E-36*	0.39	0.54	0.56	~0.64	0.72	0.92	0.92
λ^d	0	1	~1.1	1.41	1.58	~3	3

A striking example for the preferred formation of the thermodynamically less stable cyclopropane is furnished by the homoallylic halides **37**, which are cyclopropanated with high *cis*-selectivity in the presence of copper chelate **38** [91]. The cyclopropane can easily be converted into *cis*-permethric acid. In contrast, the direct synthesis of permethric esters by cyclopropanation of 1,1-dichloro-4-methyl-1,3-pentadiene using the same catalyst produces the *trans*-permethric ester (*trans*-**39**) preferentially; in a similar fashion, mainly *trans*-chrysanthemic ester (*trans*-**40**) was obtained when starting with 2,5-dimethyl-2,4-hexadiene [92].

A similar, although less marked difference characterizes the cyclopropanation of olefins **41** and **42**. In the presence of either copper or copper complexes whose chelating ligands contain an azomethine moiety derived from an α-amino acid, no stereoselectivity was observed with diene **41**, whereas the cyclopropanes derived from **42** occur with *cis/trans* ratios of 57:43 to 69:31, depending on the catalyst [93].

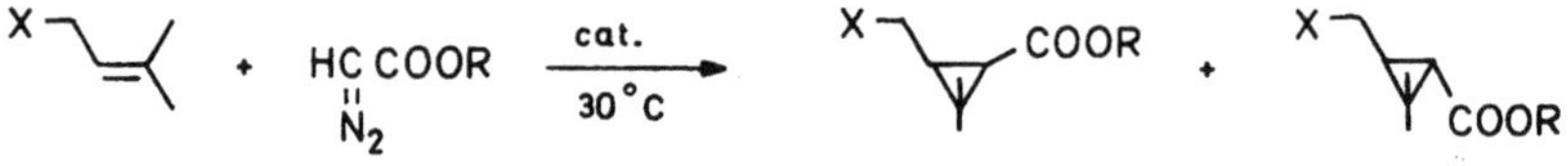

37

X	R	yield [%]	*cis*: *trans*
CCl_3	Et	59	85:15
	l-menthyl	54	85:15
$CHCl_2$	Et	71	88:12
	l-menthyl	57	86:14
CH_2Cl	Et	73	84:16
	l-menthyl	59	83:17

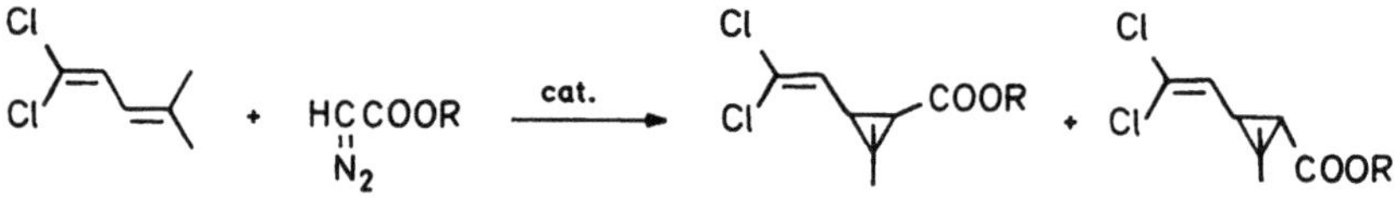

cis - *39* *trans* - *39*

R = *l*-menthyl

total yield : 52 %

cis : *trans* = 36 : 64

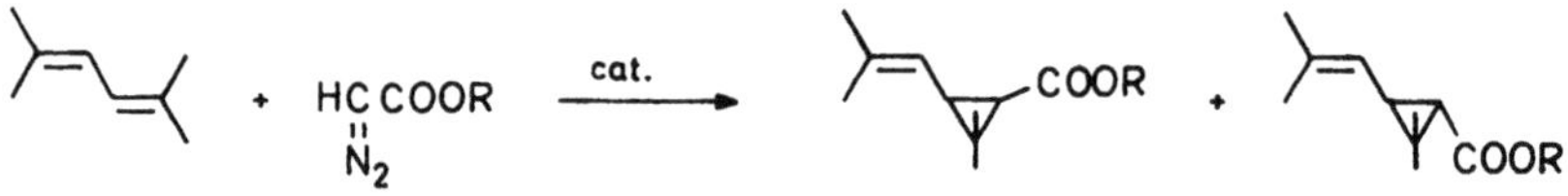

cis - *40* *trans* - *40*

R = *l*-menthyl

total yield : 72 %

cis : *trans* = 7 : 93

cat.:

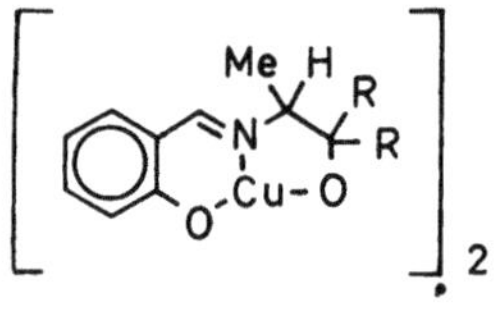

R =

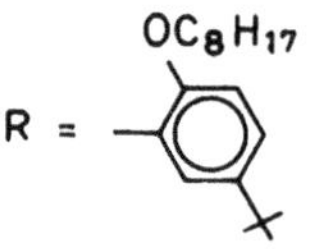

38

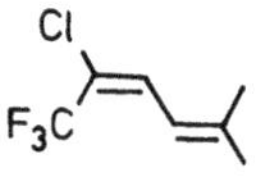

41

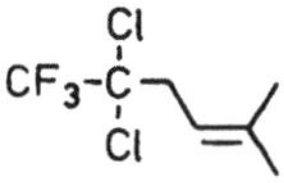

42

The reason for the different behavior of dienes like **41** and monoenes **37** or **42** is not yet established. It is hard to believe that simple steric factors should make up for the different orientation of the olefin that approaches a metal carbene intermediate. More likely is stereochemical control by an ylide-type interaction between the halogen atom of the (sterically more flexible) monoenes **37** or **42** and the electrophilic metal carbene.

Diastereofacial differentiation occurs upon cyclopropanation of the substituted cyclohexene **43** with methyl diazoacetate. Only the two stereoisomers *endo*-**44** and *exo*-**44** were found, both with a 5-*anti* methyl group [60]. In contrast, the ring substituents in 1-trimethylsiloxy-cyclohexenes **45** and **46** are not efficient for such a differentiation, so that the four possible diastereomers are actually formed.

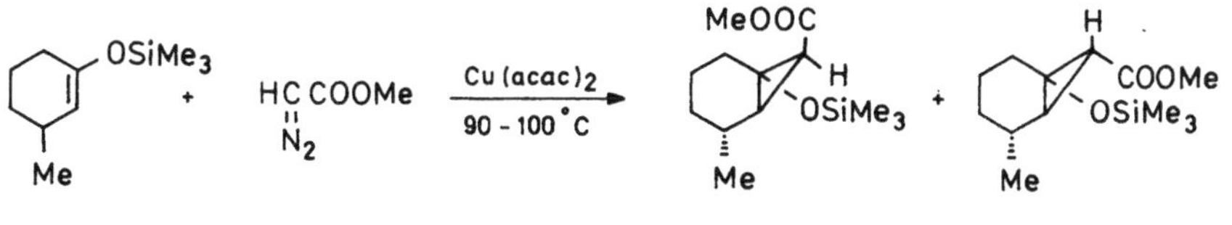

43 *endo - 44* *exo - 44*

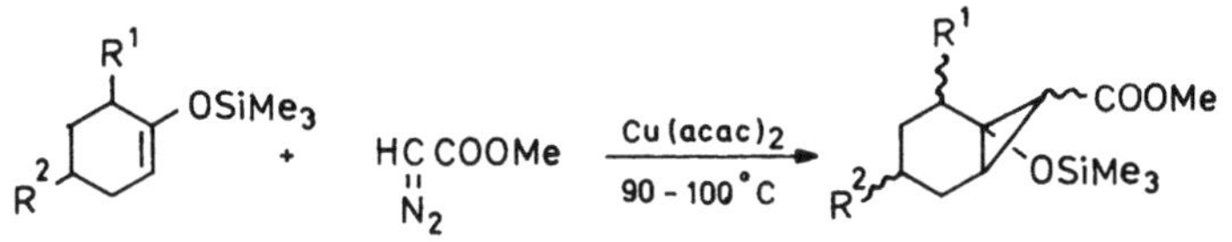

45: R^1 = Me, R^2 = H — 4 stereoisomers: 30 : 30 : 25 : 15

46: R^1 = H, R^2 = *t*-Bu — 25 : 25 : 25 : 25

The influence on stereoselectivity which is exerted by the diazo compound can best be seen when comparing the results from cyclopropanation reactions with diazo esters having different ester residues. As Scheme 3 shows, switching from ethyl diazoacetate to *n*-butyl diazoacetate has no remarkable consequence for the *Z*/*E* ratio of the permethric esters formed [72]. With the different rhodium catalysts mentioned there, not even a general correlation between *Z*/*E* ratio and steric bulk of the ester residue can be established. If, however, the size of the ester group is increased more drastically, a distinct preference for the sterically less crowded cyclopropane results. For example, the following ratios of chrysanthemic esters (*trans*-**40**:*cis*-**40**) have been obtained upon cyclopropanation of 2,5-dimethyl-2,4-hexadiene with alkyl diazoacetates in the presence of catalyst **38**: ethyl, 51:49; *t*-butyl, 75:25; 1-adamantyl, 84:16; 2,3,4-trimethyl-3-pentyl, 92:8; *l*-menthyl, 93:7 [92]. A surprising case, which does not fit into this picture, is provided once again by the cyclopropanation reaction of homoallylic halides **37**: The same *Z*/*E* ratios were obtained with ethyl diazoacetate and *l*-menthyl diazoacetate [91]. This implies that the remarkable *Z* selectivity is solely governed by the respective chloroethyl substituent, and it supports the opinion expressed above that electronic rather than steric factors account for the unusual *cis* stereoselectivity.

In order to elaborate the contribution of the catalyst to the stereoselectivity of cyclopropanation reactions, extensive comparisons of different catalysts have been carried out. For the cyclopropanation of various monoolefins and 1,3-dienes with ethyl diazoacetate, $Rh_2(OAc)_4$, $Rh_6(CO)_{16}$, CuCl · P(O-*i*-Pr)$_3$ and $PdCl_2$ · 2 PhCN have been compared in two sets of experiments (Tables 6 and 7) and CuCl · P(O-*i*-Pr)$_3$, CuCl · P(OPh)$_3$, Cu(OTf)$_2$, copper bronze and Cu(acac)$_2$ in another (Table 2) [59]. Furthermore, the stereoselectivity of cyclopropanation of ring-substituted styrenes with ethyl diazoacetate in the presence of $CuSO_4$, $CuCl_2$, CuCl, copper(II) stearate and copper(I) stearate has been compared [94]. From these investigations, it can be concluded that, at least for the catalysts under investigation, *E*/*Z* selectivities are rather independent of the initial oxidation state of the catalyzing metal and whether the catalyst is homogeneous or heterogeneous. In addition, relatively little dependence on the ligands attached to the metal is observed. There is, however, a distinct influence of the metal itself, and the following sequence concerning the *trans*/*cis* ratio is usually observed: Cu > Rh > Pd, i.e. Cu shows the largest preference for the sterically less crowded cyclopropane (It will be noted, however, that for cyclopropanation of styrene, consistently higher *trans*/*cis* selectivity for Pd catalysis than for Rh catalysis is given in Table 3, contrary to the values for the same olefin in Table 7).

Decreasing *trans*/*cis* ratios for the cyclopropanes derived from styrene and ethyl diazoacetate were also found with chelate complexes of Cu, Pd and Co, in that sequence [95]. Among the copper catalysts, Cu(OTf)$_2$ represents a particular case, since its *trans*/*cis* selectivity is lower than for the other four copper catalysts in Table 2, when styrene and 1,3-dienes are cyclopropanated; cyclopropanation of non-conjugated olefins seems not to be affected by this anomaly. Based on styrene as substrate, the *trans*/*cis* selectivity resulting from the use of Cu(OTf)$_2$ corresponds to the value found for the purely thermal reaction [94] and is close to that for other simple salts ($CuSO_4$, CuCl) and copper metal. Higher *trans*/*cis* ratios are brought about by catalysts with more bulky ligands, but the changes are rather small.

It has been pointed out earlier that the *anti*/*syn* ratio of ethyl bicyclo[4.1.0]heptane-7-carboxylate, which arises from cyclohexene and ethyl diazoacetate, in the presence of CuI · P(OMe)$_3$ depends on the concentration of the catalyst [57]. Doyle reported, however, that for most combinations of alkene and catalyst (see Tables 2 and 7) neither concentration of the catalyst (0.5–4.0 mol-%) nor the rate of addition of the diazo ester nor the molar ratio of olefin to diazo ester affected the stereoselectivity. Thus, cyclopropanation of cyclohexene in the presence of copper catalysts seems to be a particular case, and it has been stated that the most appreciable variations of the *anti*/*syn* ratio occur in the presence of air, when allylic oxidation of cyclohexene becomes a competing process [59]. As the yields for cyclohexene cyclopropanation with copper catalysts [except Cu(OTf)$_2$] are low (Table 2), such variations in stereoselectivity are not very significant in terms of absolute yields anyway.

Some care must be taken in drawing conclusions from the *E*/*Z* or *syn*/*anti* selectivity of a given catalyst/alkene combination. The intrinsic stereoselectivity may be altered in some cases by subsequent isomerizations initiated by the catalyst. For example, epimerization of disubstituted vinylcyclopropanes is effectively catalyzed by palladium compounds; the *cis* → *trans* rearrangement of ethyl chrysanthemate or of chrysanthemic acid occurs already at room temperature in the presence of $PdCl_2 \cdot L_2$ · (L = MeCN, EtCN, PhCN) [96]. Oxycyclopropane carboxylic esters undergo metal-catalyzed structural isomerizations to vinyl ethers, in which the *Z*-isomer generally reacts faster than the *E*-isomer [97]. As temperatures of 100–180 °C are usually needed for these transformations, this possibility pertains to a consideration of *E*/*Z* selectivity only for some copper-catalyzed cyclopropanation reactions, which need be carried out in that temperature range, but not for Rh- or Pd-catalyzed reactions at room temperature. Furthermore, the Rh(I)-catalyzed epimerization as well as structural rearrangements of vinylcyclopropanes occurs at relatively low (ca. 35–80 °C) temperatures [27]. Rh(I) compounds are, however, quite uncommon cyclopropanation catalysts, and the more often used Rh(II) derivatives are not reactive in this sense.

The search for catalysts which are able to reverse the ratio of cyclopropane diastereomers in favor of the thermodynamically less stable isomer has met with only moderate success to date. Rh(II) pivalate and some ring-substituted Rh(II) benzoates induce *cis*-selectivity in the production of permethric acid esters [77,98,99] contrary to rhodium(II) acetate, which gives a 1:1 mixture [74,77,98], and some copper catalysts [98] (Scheme 10).

The change in selectivity is not credited to the catalyst alone: In general, the bulkier the alkyl residue of the diazoacetate is, the more of the *cis*-permethric acid ester results [77]. Alternatively, cyclopropanation of 2,5-dimethyl-2,4-hexadiene instead of 1,1-dichloro-4-methyl-1,3-pentadiene leads to a preference for the thermodynamically favored *trans*-chrysanthemic acid ester for most catalyst/alkyl diazoacetate combinations [77]. The reasons for these discrepancies are not yet clear, the interplay between steric, electronic and lipophilic factors is considered to determine the stereochemical outcome of an individual reaction [77]. This seems to be true also for the cyclopropanation of isoprene with different combinations of alkyl diazoacetates and rhodium catalysts [77].

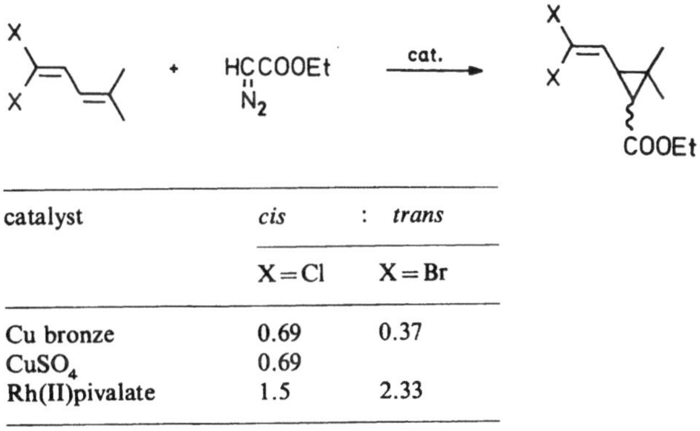

catalyst	*cis* :	*trans*
	X=Cl	X=Br
Cu bronze	0.69	0.37
$CuSO_4$	0.69	
Rh(II)pivalate	1.5	2.33

Conditions: 20 °C; 1,2-dichloroethane; molar ratio 600 (diene)/150 (EDA)/1 (catalyst).

Scheme 10

Rh(II) pivalate is, however, still not efficient in producing more of the *syn* than of the *anti* isomer of ethyl bicyclo[4.1.0]heptane-7-carboxylate from cyclohexene and ethyl diazoacetate [87,98]. It needs a rhodium(III) porphyrin **47** to be successful in this case

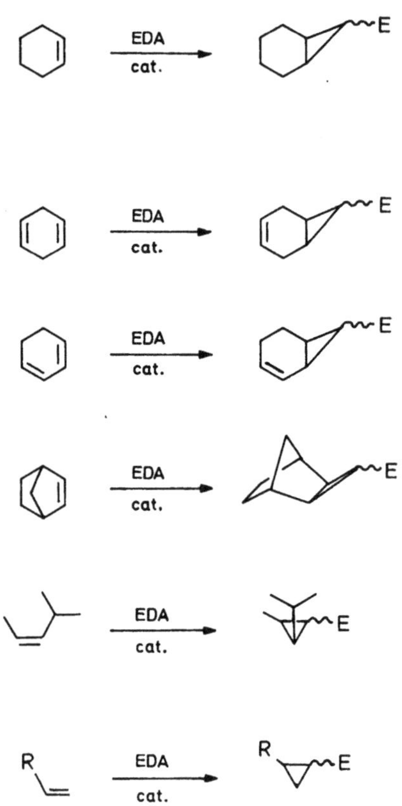

Catalyst	*syn : anti*
$CuSO_4$	0.16
Rhpiv	0.32
47a	0.74
47b	0.83
47c	1.17
Rhpiv	0.51
47a	1.4
47c	3.3
Rhpiv	0.43
47a	0.87
47c	2.16
Rhpiv	0.44
47a	1.28
47b	1.52
47c	2.14
Rhpiv	2.2
47a	4.9
47c	6.52

	Z : E R = *n*-Bu	R = Ph
Rhpiv	0.67	0.67
47a	0.73	0.88
47c	0.87	0.98

Rhpiv = Rh(II) pivalate; E = COOEt
Conditions: 60 °C; 1.2-dichloroethane

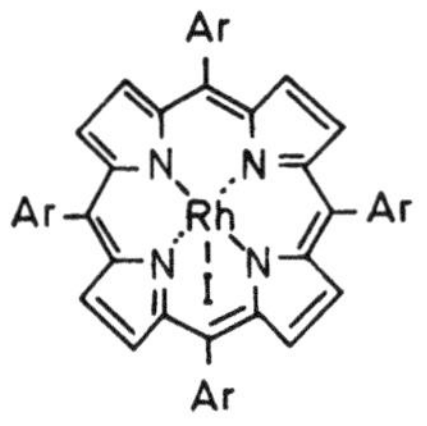

47a Ar = Ph
47b Ar = o-tolyl
47c Ar = mesityl

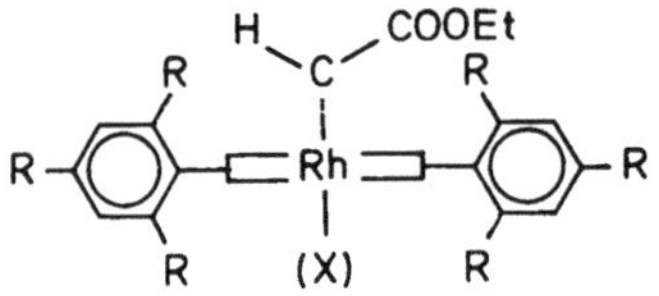

48

Scheme 11

and with several other *cis*-1,2-disubstituted alkenes (Scheme 11)[87,100]. Even with norbornene, the *exo-syn*-cyclopropane is formed preferentially. As the examples of 1-hexene and styrene show, even **47** brings no significant improvement compared to Rh(II) pivalate. *Trans*-1,2-disubstituted alkenes are cyclopropanated in only low yields in the presence of rhodium(III) porphyrins.

The highest *syn/anti* selectivity is always obtained with the iodorhodium(III) *meso*-tetramesitylporphyrin **47c** in which the preferred orientation of the aryl groups is perpendicular to the more or less planar macrocycle. Even though the transition state geometry leading to a cyclopropane from an alkene and an alleged metal carbene **48** is not known, it seems reasonable to assume that all bulky groups (ester group of the carbenoid and substituents on the olefin) prefer to point away from the ligand sphere of the metal, this tendency being increased with R=Me instead of H.

Metal complexes of tetra-4-*tert*-butylphthalocyanine [PcM, M = Mn(III)OAc, Cu(II), Co(II), Ni(II), Fe(II) · $(C_5H_5N)_2$, Rh(III)Cl] have also been tested for their stereoselective potential in the cyclopropanation of styrene with ethyl diazoacetate[101]. The Co(II) and Rh(I) complexes, already highly active at room temperature, produced the 2-phenylcyclopropanecarboxylic esters in a $E\!:\!Z$ isomer ratio of 1.0–1.2 which compares well with the value obtained with the rhodium(III) porphyrin **47a** (1.2). In the other cases, $E\!:\!Z$ ratios of 2.0–2.2 were observed, except for M = Fe(II) · $(C_5H_5N)_2$ where it was (3.0); the $E\!:\!Z$ ratio of the purely thermal reaction was 2.0.

On the whole, it is recognized that the ligands of the catalyst metal play a rather modest role in determining the stereoselectivity of a cyclopropanation reaction. This is certainly surprising, as metal carbenes are assumed to be the reactive intermediates. Notwithstanding this low degree of diastereoselectivity, metal complexes with chiral ligands give rise to enantioselective reactions with a ligand-dependent degree of optical induction. This aspect is dealt with in more detail in Sect. 2.1.4.

2.3 Special Aspects of Reactions between Functionalized Olefins and Ketocarbenoids

2.3.1 Enol Ethers, Enol Acetates and Silyl Enol Ethers

The carbenoid reaction between alkyl diazoacetates and enol ethers, enol acetates and silyl enol ethers furnishes β-oxycyclopropane carboxylates[59,64] (see Tables 2, 4, 5, 6, 7 and Scheme 5). The recently recognized synthetic versatility of these donor/acceptor-substituted cyclopropanes[102,103] (precursors of 1,4-dicarbonyl and β, γ-unsaturated carbonyl compounds, 4-oxocarboxylic acids and esters, among others) gave rise to the synthesis of a large number of such systems with a broad variation of substituents: β-acetoxycyclopropanecarboxylates[79,104], β-alkoxy- or β-aryloxysubstituted cyclopropanecarboxylates[79,102,105–113], 2-alkoxy-1-methyl-1-cyclopropanecarboxylates[114], β-trimethylsilyloxycyclopropanecarboxylates[60,79,115–119]. Although rhodium and (to a lesser extent) palladium catalysts have proven to be suited for cyclopropanation of electron-rich alkenes[59,60,64,97], the majority of transformations are still carried out in the presence of copper catalysts such as copper bronze, copper powder and copper(II)acetylacetonate; this is justified by the generally good yields of cyclopropanes obtained.

Diverging results have been reported for the carbenoid reaction between alkyl diazoacetates and silyl enol ethers **49a–c**. Whereas Reissig and coworkers [60] observed successful cyclopropanation with methyl diazoacetate/Cu(acac)$_2$, Le Goaller and Pierre, in a note without experimental details [118], reported the isolation of 4-oxo-carboxylic esters for the copper-catalyzed decomposition of ethyl diazoacetate. According to this communication, both cyclopropane and ring-opened γ-keto ester are obtained from **49c** but the cyclopropane suffers ring-opening under the reaction conditions.

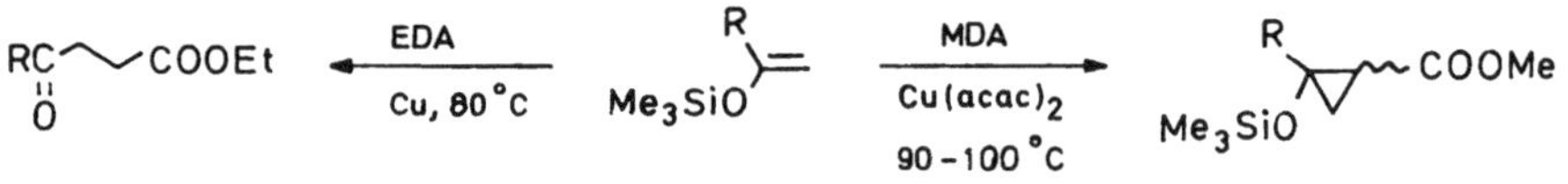

49a: R = Ph
49b: R = *t*-Bu

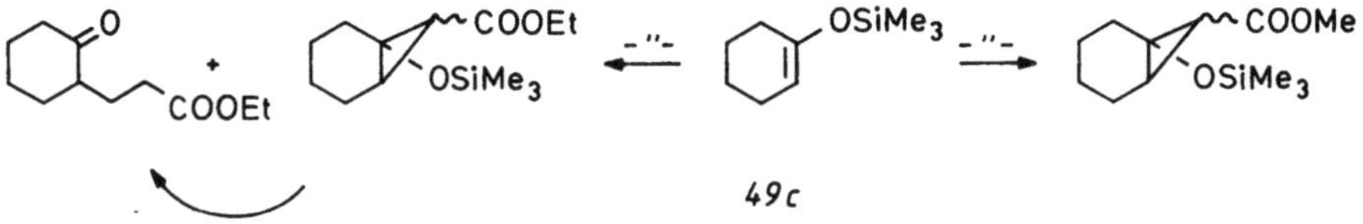

49c

It is not known whether or not this transformation is catalyzed by the transition metal. However, the metal-catalyzed ring-opening reaction of β-alkoxycyclopropane carboxylates yielding vinyl ethers (e.g. **50** → **51** and **52** → **53**) is well documented [97, 120]. Several catalysts are suited [$PtCl_2 \cdot 2\,PhCN$, $Rh_2(OAc)_4$, $[Rh(CO)_2Cl]_2$, $[Ru(CO)_3Cl_2]_2$, Cu bronze, CuCl], but with all of them, reaction temperatures higher than those needed for the carbenoid cyclopropanation reaction are required.

50 *(E/Z)-51*

Cu bronze / 160 °C / 6h: "nearly quantitatively"
CuCl / 160 °C / 2h: 70%

52 *53*

Cu bronze / 180 °C / 4h: 80%

The copper-catalyzed decomposition of ethyl diazoacetate in the presence of a ketene acetal also leads to the corresponding cyclopropane in addition to larger quantities of diethyl maleate and diethyl fumarate [102,121]. If β-unsubstituted or -monosubstituted ketene acetals are used, 4,4-dialkoxy-3-butenoates **55** are formed as by-products (Scheme 12). It was shown that they arise partly from thermal isomerization of cyclopropanes **54**, the 2,2-diethoxycyclopropanecarboxylates being less stable than the 2,2-dimethoxy derivatives. This may explain earlier contradictory results according to which only the cyclopropane was isolated from the reaction of ketene dimethyl acetal with ethyl diazoacetate [122], but exclusively butenoate was obtained after work-up of the reaction between the same diazoester and ketene diethyl acetal [123]. Partly, butenoates **55** are also generated in competition with cyclopropanes rather than by subsequent isomerization of the latter [121]. From the $Rh_2(OAc)_4$-catalyzed reaction between ketene *o*-xylyleneacetal and methyl diazoacetate or diazoacetonitrile, only the cyclopropanation product was obtained (67 and 64% yield, resp. [124]).

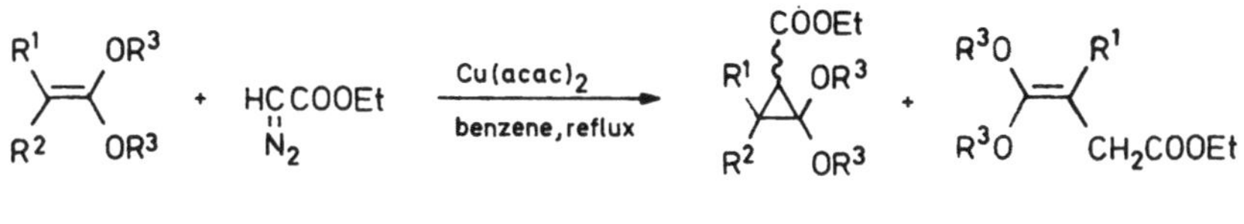

54 55

R¹	R²	R³	yield [%]ᵃ	
H	H	Me	62(50)	5(13)
H	H	Et	62(40)	8(20)
Me	H	Me	53(44)	3(6)
Me	H	Et	63(48)	4(8)
Me	Me	Me	61(25)	—

ᵃ In parentheses yields obtained with Cu powder/benzene/95 °C.

Scheme 12

A different reaction mode emerges when certain α-diazoketones are combined with enol ethers, as dihydrofurans rather than cyclopropanes are isolated. Wenkert and Alonso have used ethyl diazopyruvate **56** and a variety of enol ethers for this transformation (Scheme 13). The dihydrofurans so obtained can be further transformed, e.g. into furoic esters and furanones, and a number of natural products containing a furan ring have been synthesized by taking advantage of this carbenoid reaction [103]. With unactivated olefins such as styrenes [125], indene [125], 4-*t*-butyl-1-methylenecyclohexene [125], cyclohexene [113,126] and cyclohexadiene [28], alkyl diazopyruvates behave like simple diazoketones yielding cyclopropanes. Both cyclopropanation and dihydrofuran formation occur in the reaction between methyl diazopyruvate and *cis*-2-butene; it is, however, not known, whether in that case the heterocycle is a primary product or arises from rearrangement of the primarily formed, but not isolated, all-*cis*-cyclopropane derivative [127].

Dihydrofurans were also isolated from the $Cu(acac)_2$-catalyzed reaction of **56** with

(2-propenyl)-cyclopropane and 1,1-dicyclopropylethylene [128], the electron-donating ability of the cyclopropane ring being responsible for the analogous reactivity of these olefins and enol ethers.

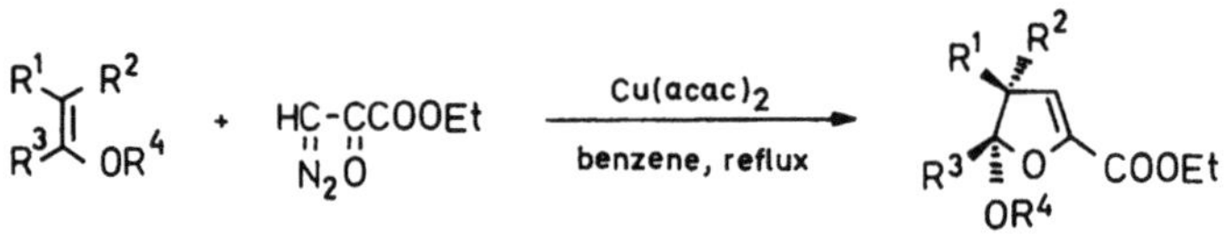

Ref. 122

56

R^1	R^2	R^3	R^4	yield [%]
H	H	H	*n*-Bu	62
Et	H	H	Me (*E*/*Z*)	46
Me	Me	H	Et	79
$-(CH_2)_5-$		H	Me	50
Ph	H	H	Me (*E*/*Z*)	36
H	H	Ph	Me	64

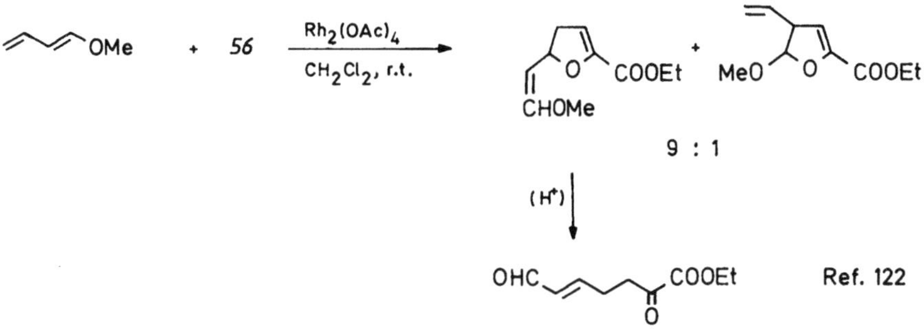

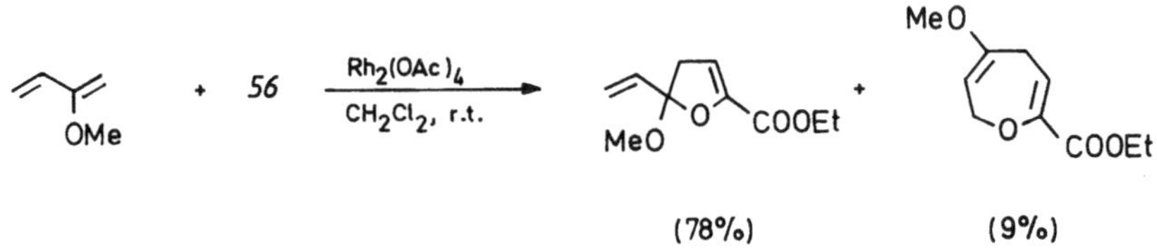

Ref. 122

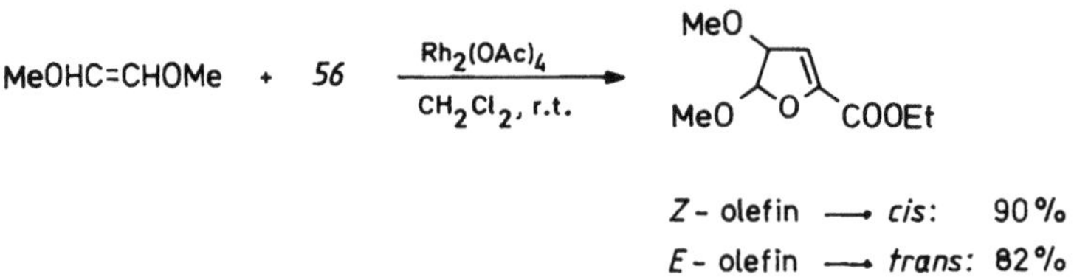

Ref. 122

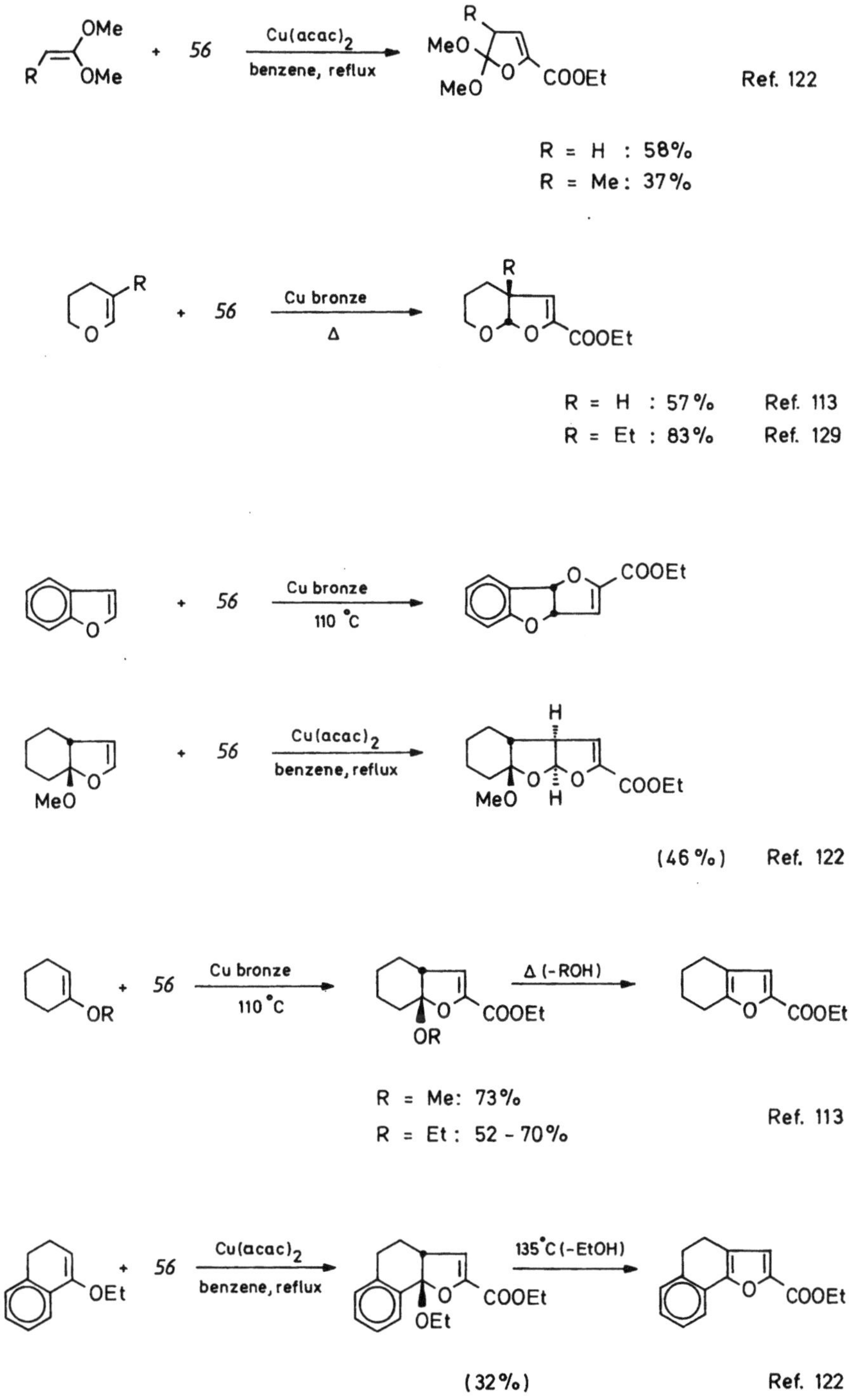
OMe
OMe
R
+ 56
Cu(acac)2
benzene, reflux
R
MeO
MeO
O
COOEt
Ref. 122
R = H : 58%
R = Me: 37%
R
O
+ 56
Cu bronze
Δ
R
O
O
COOEt
R = H : 57%
Ref. 113
R = Et : 83%
Ref. 129
O
+ 56
Cu bronze
110 °C
O
COOEt
O
O
MeO
+ 56
Cu(acac)2
benzene, reflux
H
O
O
MeO
H
COOEt
(46%)
Ref. 122
OR
+ 56
Cu bronze
110 °C
O
COOEt
OR
Δ (-ROH)
O
COOEt
R = Me: 73%
R = Et : 52 - 70%
Ref. 113
OEt
+ 56
Cu(acac)2
benzene, reflux
O
COOEt
OEt
135°C (-EtOH)
O
COOEt
(32%)
Ref. 122

Scheme 13

The regioselectivity of the formal [3 + 2] cycloaddition reactions in Scheme 13 corresponds to the polarization of the olefinic bond and to partial charges in the ketocarbenoid. The only exception occurred with benzofuran, in agreement with the behavior of the carbenoid derived from ethyl 2-diazo-3-oxobutyrate [130]. As several examples in Scheme 13 show, the cycloaddition is sterospecific, i.e. the configuration about the enol ether double bond is retained in the dihydrofuran. The observation that in the case of 1-methoxy-1,3-butadiene highly selective cycloaddition of the unsubstituted double bond occurs, whereas for 2-methoxy-1,3-butadiene cycloaddition takes place at the more electron-rich double bond, finds a close analogy in the cyclopropanation of these same dienes with ethyl diazoacetate (see Table 9).

2-Diazo-1,3-dicarbonyl compounds such as alkyl 2-diazo-3-oxobutyrates **57a, b** and 3-diazo-2,4-pentanedione **57c** behave like the diazopyruvate **56**, as far as their carbenoid cycloaddition behavior is concerned [114,130].

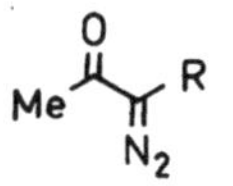

57a: R = COOMe
57b: R = COOEt
57c: R = COMe

For some of the reactions depicted in Scheme 14, copper(II) hexafluoroacetyl-acetonate [Cu(hfacac)$_2$] proved to be a better catalyst than other chelates; neither cyclopropanes nor allylic insertion products were found, and the yield of dihydrofuran was not affected by temperature in the range 70–132 °C [130]. However, in phenylogous vinyl ethers such as 4-methoxystyrene or *trans*-anethole, cyclopropanes (*58a, b*) rather than dihydrofurans resulted [131].

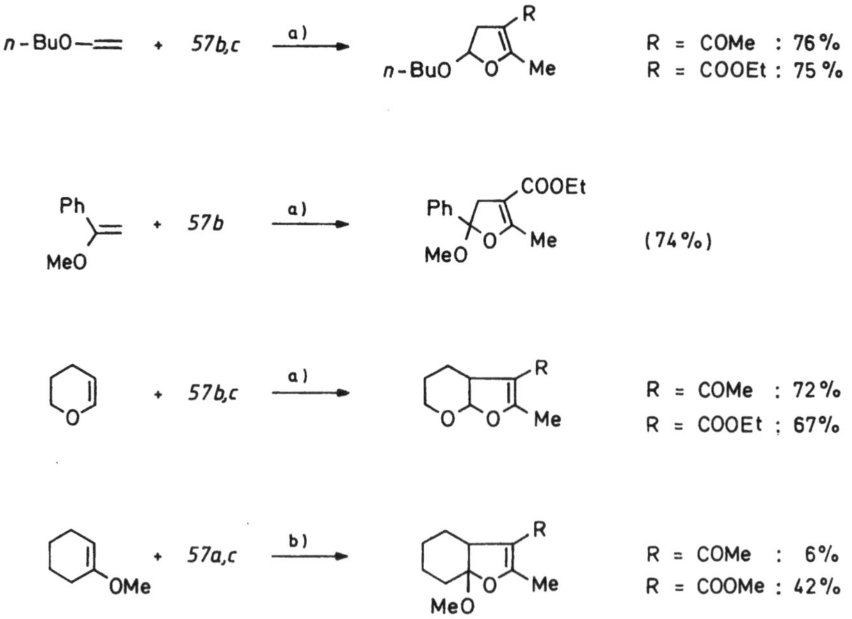

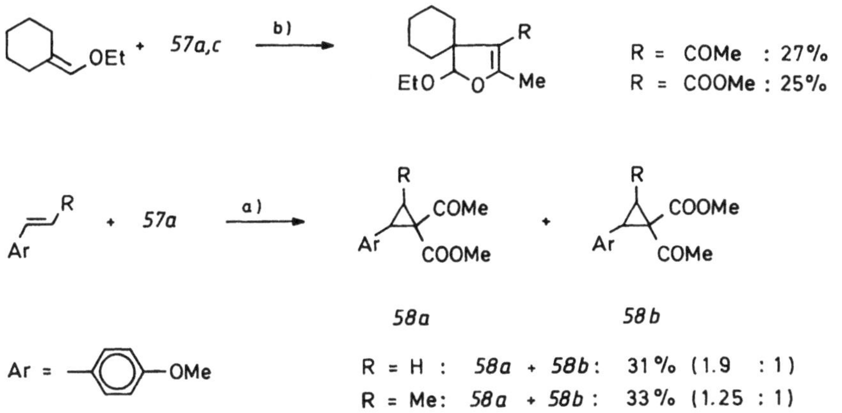

Conditions: a) $Cu(hfacac)_2$, fluorobenzene, reflux (ref. [130]). b) For decomposition of *57c*: $Cu(acac)_2$; reflux; for decomposition of *57a*: CuI · $P(OMe)_3$, 80 °C (ref. [114]).

Scheme 14

From the copper-catalyzed reaction of methyl 2-diazo-3-oxobutyrate **57a** with Z-β-methoxystyrene, dihydrofuran **59** (formed with retention of olefin configuration) and butadienol **60** result [130]. Such an acyclic by-product also occurs when benzofuran is the cycloaddition partner. In that case, however, regioisomers **61** and **62**, arising from the connection of the former diazo carbon with either the 2- or 3-position of the heterocycle, are obtained; similarly, two isomeric dihydrofurans **63** and **64** are formed under $Cu(hfacac)_2$ catalysis [130].

In mechanistic terms, the dipolar intermediate **65a** has been proposed as precursor to both dihydrofuran **66** and acyclic products **67** (Scheme 15) [130]; wheter or not a copper-containing species would be closer to reality than **65a**, is not known. 1,3-Ring closure of **65a** leading to cyclopropane **68** must be inferior to 1,5-cyclization yielding dihydrofuran **66**. It is assumed that the latter is formed stereospecifically, although this hypothesis rests upon weak ground: *cis*-dihydrofuran **59**, so far the only one to be obtained from an acyclic 1,2-disubstituted enol ether and a 2-diazo-1,3-dicarbonyl compound, is isolated in only 21% yield (but note the high-yield stereospecific formation of dihydrofurans from ethyl diazopyruvate and *cis*- or *trans*-1,2-dimethoxyethylene, respectively (Scheme 13)). Provided that the assumption is nevertheless correct, stereospecific stepwise formation of the dihydrofuran implies that ring closure of **65a** is distinctly faster than C/C rotation about the former enol ether double bond in that intermediate (**65a** ⇌ **65b**). Other pathways leading to

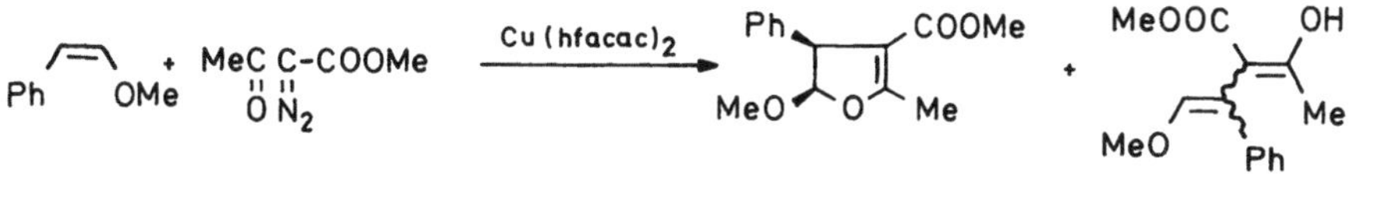

57a *59* *60*

(21%) (9%)

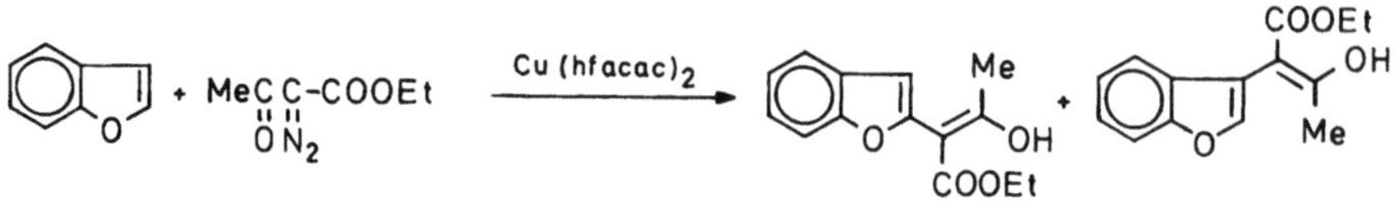

57b 61 62

(36 %) (6 %)

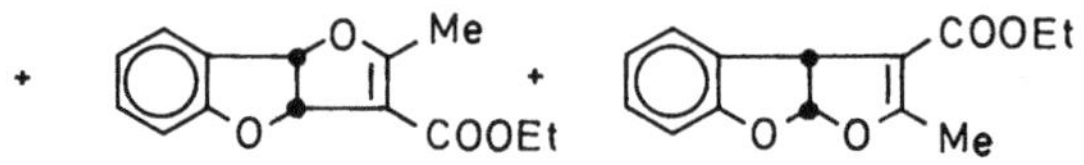

63 64

(16 %) (4 %)

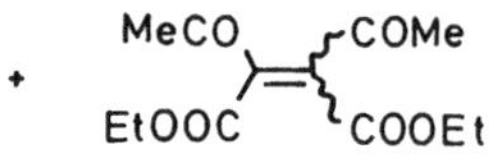

(3 %)

the dihydrofuran can be excluded with some certainty. Thermal or copper-catalyzed ring expansion of a primarily formed cyclpropane ketoester is ruled out by control experiments with cyclopropanes **58a, b,** which do not rearrange to dihydrofurans under authentic reaction conditions (heating in fluorobenzene with or without the copper catalyst). Their pyrolysis at elevated temperature leads to stereoisomeric dihydrofurans, in which the stereochemical information furnished by the olefin has been lost. Furthermore, it has been established that the dihydrofurans do not result from isomerization of the β,γ-unsaturated carbonyl compounds **67** or their tautomers.

A dipolar intermediate **65** can be stabilized, of course, by appropriate substituents. This is the case when 2-diazo-1,3-dicarbonyl compounds **57a–c** are the precursors of the ketocarbenoid; R^1 being a carbonyl unit, the negative charge is effectively delocalized. When a diazopyruvate is used, stabilization of the negative charge in **65a, b** (COOR instead of Me) by the electron-withdrawing properties of the additional ester group is possible. On the other hand, a more efficient stabilization of the positive charge in **65** would result from replacing H by a second OR^3 function; i.e. in the case of a ketene acetal. Hence, the independent formation of both cyclopropanes and 3-butenoates from ketene acetals (having a β-hydrogen) and ethyl diazoacetate (see Scheme 12) may also be attributed to a common dipolar intermediate such as **69** (Scheme 16) [121].

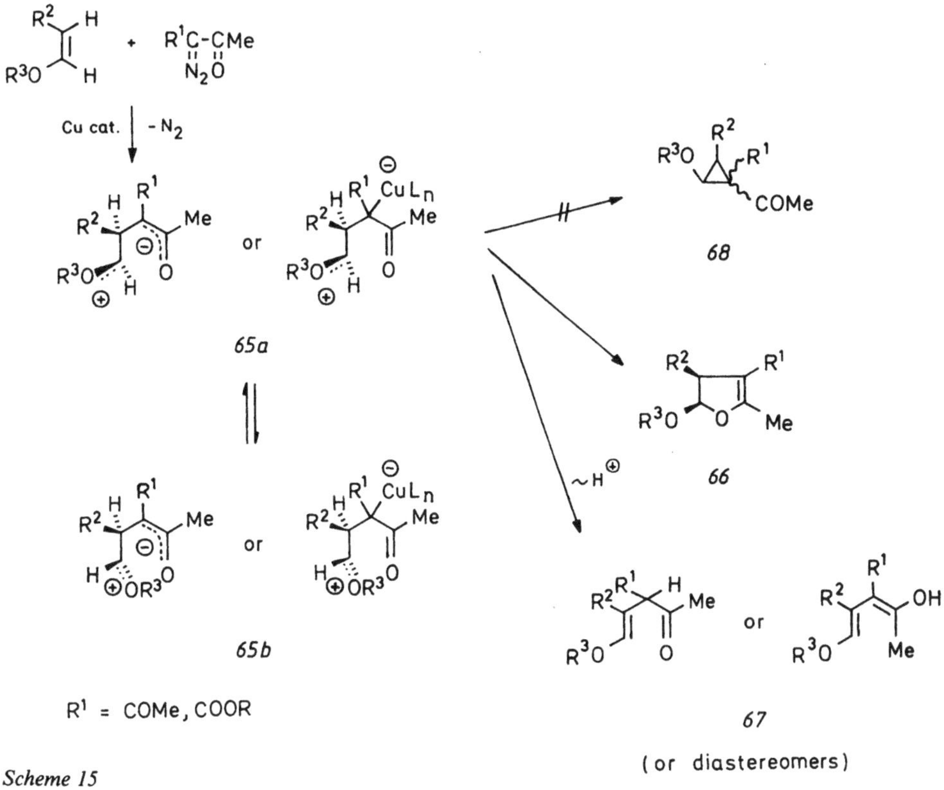

Scheme 15

Participation of **69** in the reaction scheme would also explain why cyclopropanes are obtained from diazoacetic esters, but dihydrofurans from diazoketones [121]. In the latter case, the enolate oxygen in **69** is more nucleophilic, thus favoring 1,5- over 1,3-ring closure.

Diazomalonic esters, in their behavior towards enol ethers, fit neither into the general reactivity pattern of 2-diazo-1,3-dicarbonyl compounds nor into that of alkyl diazoacetates. With the enol ethers in Scheme 17, no dihydrofurans are obtained as was the case with 2-diazo-1,3-dicarbonyl compounds. Rather, copper-induced cyclopropanation yielding **70** occurs with ethoxymethylene cyclohexane [114]. However,

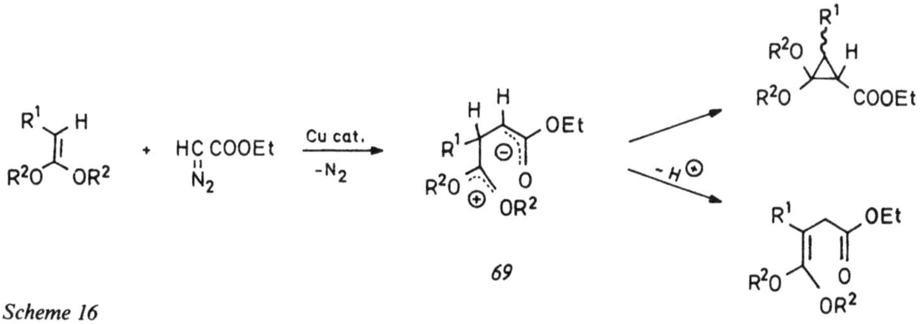

Scheme 16

an enol ether **72** instead of a cyclopropane **71** was isolated from the reaction with 1-methoxycyclohexene [114] or its 5-methyl derivative. The failure to detect **71** could be due to isomerization **71** → **72**, either by copper catalysis or purely thermally by a 1,5-hydrogen shift (having precedent with the cyclopropane derived from a methylene cyclohexane and ethyl diazopyruvate [125]. This would imply, however, different thermal stability of cyclopropanes **70** and **71**, but reasons for such a difference are not obvious. Having recourse to the above-mentioned dipolar intermediate in the ketocarbenoid addition, one notes that intermediate **73** has no other choice but 1,3-ring closure yielding **70**, whereas **74** can dispose of the positive charge by proton elimination leading to **72** (this argument still leaves open the quesion of regiospecific double bond formation, however).

Although non-stereospecific [3 + 2] cycloaddition of ketocarbenes is well established and despite their classification as 2π-rather than 4-π-components, hence not belonging to those 1,3-dipoles which are expected to cycloadd stereospecifically [132],

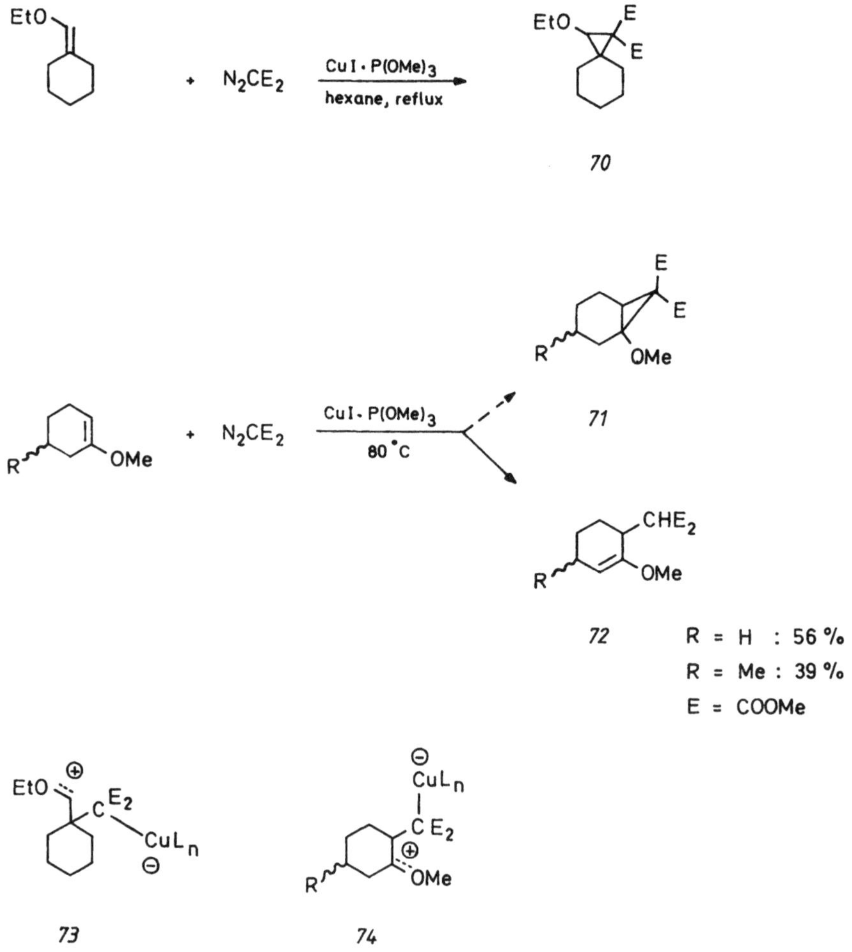

Scheme 17

the mechanistic hypothesis that dipolar intermediates are involved in their reaction with enol ethers has to be confirmed. For this purpose, Alonso made use of the fact that 2-methoxy-3,4-dihydro-2*H*-pyrans suffer ring opening when a positive charge at C–6 is developed. The dipolar intermediate **75** resulting from the carbenoid addition of dimethyl diazomalonate to such a pyran [133] would thus undergo fragmentation to **76** from which final products could arise.

With enol ether **77a**, however, adduct **78a** with intact ring structure was isolated in modest yield, the remainder being polymeric material. The formation of **78a** is readily explained by proton loss from **75a**, as discussed above. More rewarding was the use of enol ether **77b** [133]. Besides **78b**, two epimeric cyclopentanes resulted which are conveniently understood as arising from 1,5-ring closure of the intermediate **76b**. Obviously, fragmentation of **75b** is connected to a better charge stabilization than given in **75a**. Additional experiments support this mechanism. Firstly, no isomerization **78** → **79** occurs under the reaction conditions, i.e. the cyclopentanes are independently formed products. Secondly, the ratio **78b**:**79a, b** alters from 6:1 in cyclohexane to 1:7 in dimethoxyethane, thus underlining considerable charge separation in the intermediates leading to the cyclopentanes. Ethyl diazoacetate gives only stereoisomeric cyclopropanes with enol ether **77b**. They rearrange to an enol ether analogous to **78b** (CH_2CO_2Et instead of CHE_2) in the presence of copper catalysts at temperatures much higher than needed for their formation. This result renders very unlikely the possible genesis of **79a, b** from an unstable oxycyclopropane diester in the diazomalonate reaction.

The possibility that both enol ethers **78** and cyclopentanes **79** owe their formation to the common intermediate **75**, is not excluded by this result, but of course, further mechanistic alternatives can be envisaged.

Whereas metal-catalyzed decomposition of simple diazoketones in the presence of ketene acetals yields dihydrofurans [121,124,134], cyclopropanes usually result from reaction with enol ethers, enol acetates and silyl enol ethers, just as with unactivated alkenes [13]. 1-Acyl-2-alkoxycyclopropanes were thus obtained by copper-catalyzed reactions between diazoacetone and enol ethers [79,105,135], enol acetates [79,135] and

75　　76

a: R = H
b: R = Me

77a　　78a

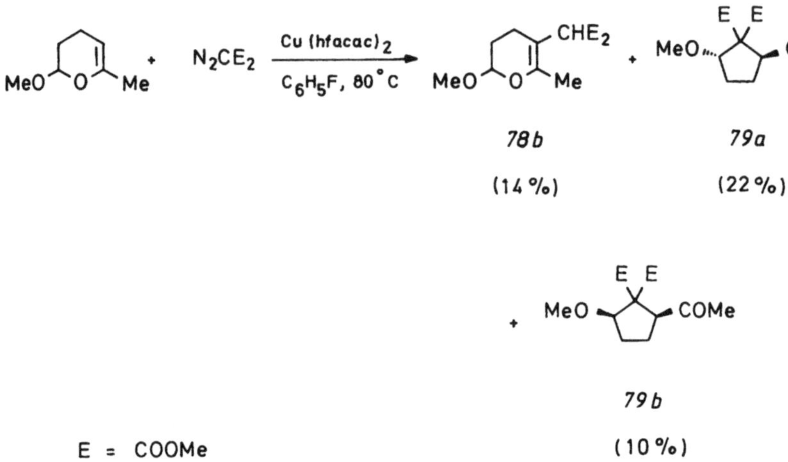

78b (14%) 79a (22%) 79b (10%)

silyl enol ethers [79,135] (see Scheme 5), 1-diazo-2-octanone and isopropenyl acetate [102], methyl 6-diazo-5-oxohexanoate and isopropenyl acetate as well as *n*-butyl vinyl ether [136], methyl 10-diazo-9-oxodecanoate and *n*-butyl vinyl ether [102] and finally between benzyl 6-diazopenicillanate as well as *t*-butyl-1-aza-7-diazo-3-methyl-8-oxo-5-thiabicyclo[4.2.0]oct-2-ene-1-carboxylate and ethyl vinyl ether [137]. The $Cu(acac)_2$-catalyzed reaction between 1-trimethylsilyloxycyclohexene and diazoacetone furnished the silyl enol ether **80** rather than the desired cyclopropane **81** [138]. A 1,5-silicon shift in a primarily formed 7-*exo*-acetylnorcarane could account for this result.

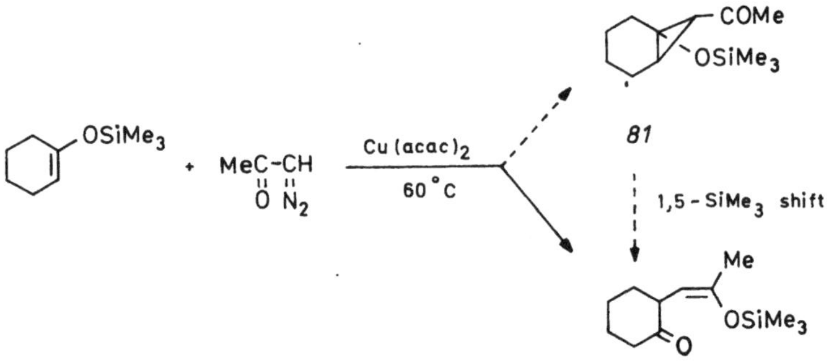

81

80 (40%)

Considering the above-mentioned facts, according to which simple diazoketones yield dihydrofurans with ketene acetals but cyclopropanes with enol ethers, one expects an interlink between these clear-cut alternatives to exist, i.e. substrates from which both cyclopropanes and dihydrofurans result. In fact, providing an enol ether with a cation-stabilizing substituent in the α-position creates such a situation: The $Rh_2(OAc)_4$-catalyzed decomposition of ω-diazoacetophenone in the presence of ethyl vinyl ether produces mainly cyclopropane **82** (R = H), but a small amount of dihydrofuran **83** (R = H) is formed. The products **82** and **83** are not interconvertible under the reaction conditions, and the product ratio is not affected by catalyst concentration (0.5–2.0 mol %) and reaction time (1–10 h). The heterocycle is obtained in significantly improved yield in reactions with 2-methoxypropene and α-methoxystyrene [45], in agreement with the more efficient stabilization of the positive charge in the presumed dipolar intermediate **84**. The similarity to the mechanistic picture given in Scheme 15 is completed by the proton transfer step **84** → **85**, on this occasion involving the α-methyl group.

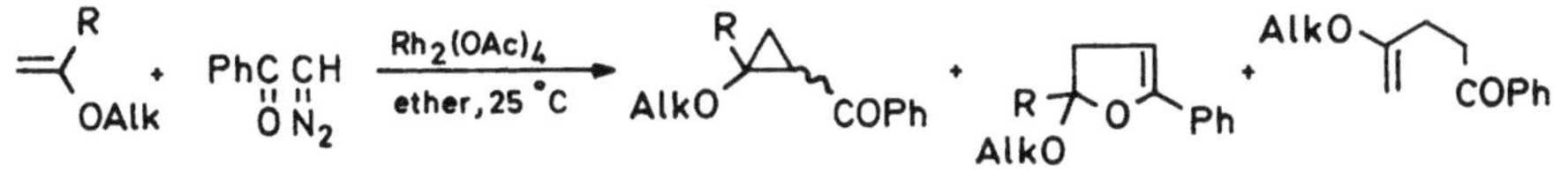

	82	*83*	*85*
R = H, Alk = Me	76% (*E*/*Z* = 2.0)	6.6%	
R = Ph, Alk = Et	67% (*Z*/*E* = 2.1)	29%	
R = Me, Alk = Et	—	32%	28%

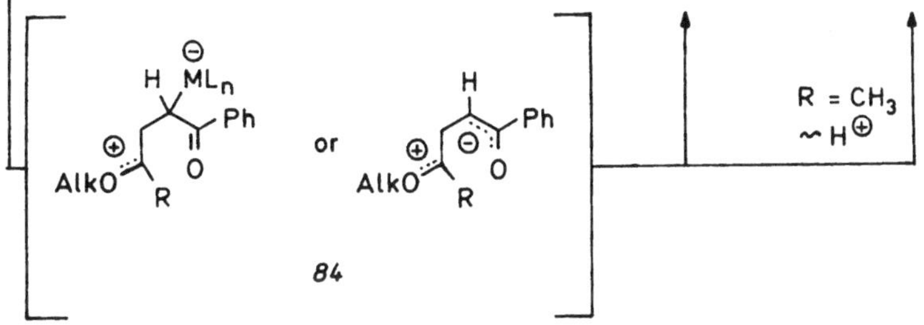

84

The comparison between the cycloaddition behavior of simple diazoketones and of ethyl diazopyruvate **56** towards the same olefin underlines the crucial influence of the ethoxycarbonyl group attached to the carbonyl function. This becomes once again evident when COOEt is replaced by an acetal function, such as in 1-diazo-3,3-dimethoxy-2-butanone **86**; with enol ethers and acetates, cyclopropanes rather than dihydrofurans are now obtained [113].

Cyclopropane **87**, obtained from *n*-butyl vinyl ether, rearranges to dihydrofuran **88** only at elevated temperature, and also partly during work-up on silica gel [113]. The complete conversion of **87** into veratrole by the action of HCl/CH_3OH gave rise to the analogous two-step synthesis of hydrourushiol monomethyl ether from 1-diazo-3,3-dimethoxy-2-nonadecanone **89** [113]. Ether cleavage of the product yields hydrourushiol, one of the vesicant components of, inter alia, poison ivy.

With a less reactive olefin such as isopropenyl acetate, diazoketone *86* gives only a low yield of cyclopropane **90**; α-acyl enol ether **92**, resulting from an intramolecular rearrangement of the ketocarbenoid, becomes the favored reaction product. If **91** is not wanted, but a higher cyclopropane yield instead, the diazo compound must be decomposed by sensitized irradiation [113].

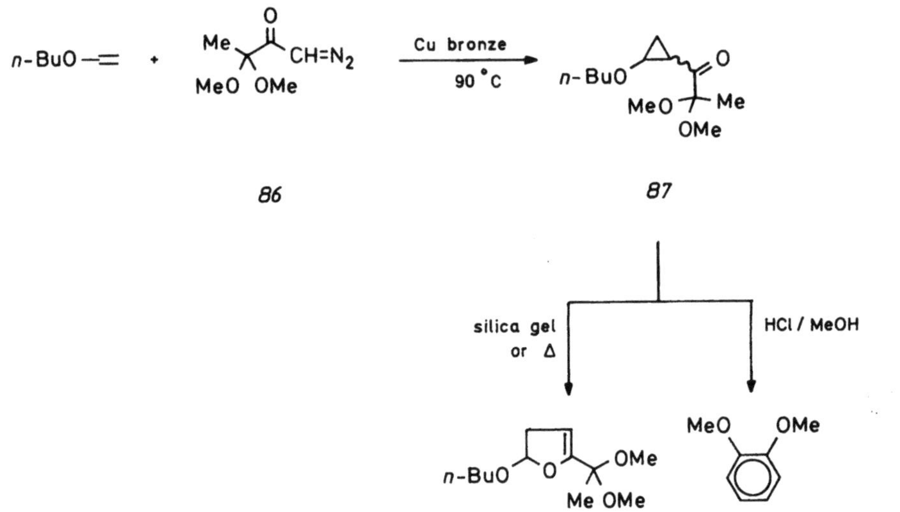

88

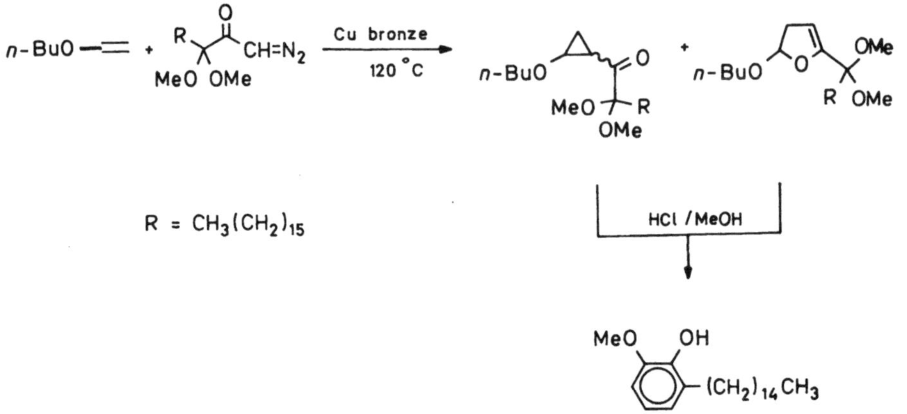

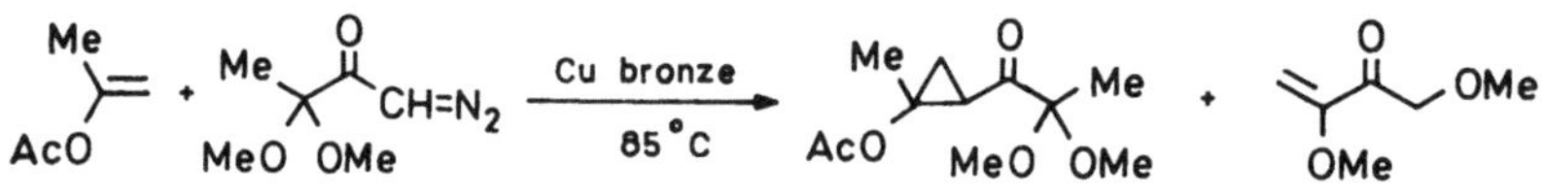

86 90 (19%) 91 (57%)

2.3.2 α,β-Unsaturated Carbonyl Compounds and Nitriles

Diazocarbonyl compounds readily undergo [3 + 2] cycloaddition to electron-poor alkenes [139]. The 1-pyrazolines thus formed usually tautomerize to 2-pyrazolines if there is a hydrogen in an α-position to one of the nitrogen atoms; otherwise, thermally induced ring contraction with evolution of nitrogen to give cyclopropanes can occur (Scheme 18).

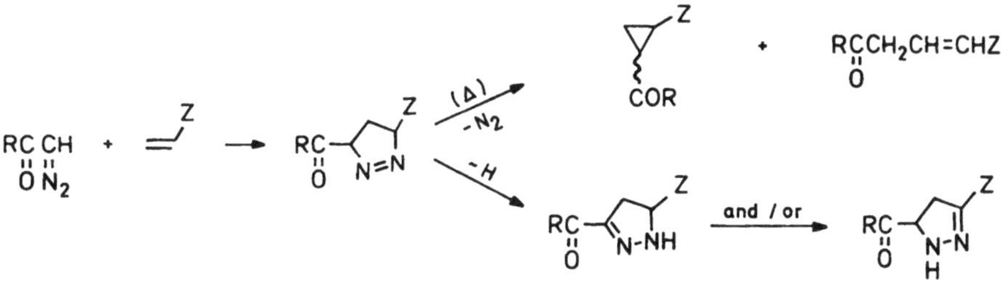

Z = CN, COR

Scheme 18

As it is known from experience that the metal carbenes operating in most catalyzed reactions of diazo compounds are electrophilic species, it comes as no surprise that only a few examples of efficient catalyzed cyclopropanation of electron-poor alkenes exist. One of those examples is the copper-catalyzed cyclopropanation of methyl vinyl ketone with ethyl diazoacetate [140], contrasting with the 2-pyrazoline formation in the purely thermal reaction (for failures to obtain cyclopropanes by copper-catalyzed decomposition of diazoesters, see Table VIII in Ref. 6).

Simultaneous occurence of the pyrazoline and carbenoid route is observed in the presence of bis(campherquinone-α-dioximato)cobalt(II) [95], but the cyclopropanes derived from ethyl diazoacetate and $H_2C{=}CHX$ (X = COOMe, CN) were obtained only in low yield.

Palladium(II) acetate was found to be a good catalyst for such cyclopropanations with ethyl diazoacetate (Scheme 19) by analogy with the same transformation using diazomethane (see Sect. 2.1). The best yields were obtained with monosubstituted alkenes such as acrylic esters and methyl vinyl ketone (64–85%), whereas they dropped to 10–30% for α,β-unsaturated carbonyl compounds bearing alkyl groups in α- or β-position such as ethyl crotonate, isophorone and methyl methacrylate [141]. In none of these reactions was formation of carbene dimers observed. *Trans*-benzalacetophenone was cyclopropanated stereospecifically in about 50% yield; $PdCl_2$ and palladium(II) acetylacetonate were less efficient catalysts [34]. Diazoketones may be used instead of diazoesters, as the cyclopropanation of acrylonitrile by diazoacenaphthenone/$Pd(OAc)_2$ (75% yield) shows [142].

Even $Pd(OAc)_2$ is not effective in catalyzing the cyclopropanation of α,β-unsaturated nitriles by ethyl diazoacetate. Instead, vinyloxazoles **92** are formed from acrylonitrile or methacrylonitrile by carbenoid addition to the C≡N bond [143]. Diethyl maleate and diethyl fumarate as well as "polyketocarbenes" are by-products in these reactions; the 2-pyrazoline which would result from initial [3 + 2] cycloaddition at the C=C bond and which is the sole product of the uncatalyzed reaction at room temperature, can be avoided completely by very slow addition of the diazoester

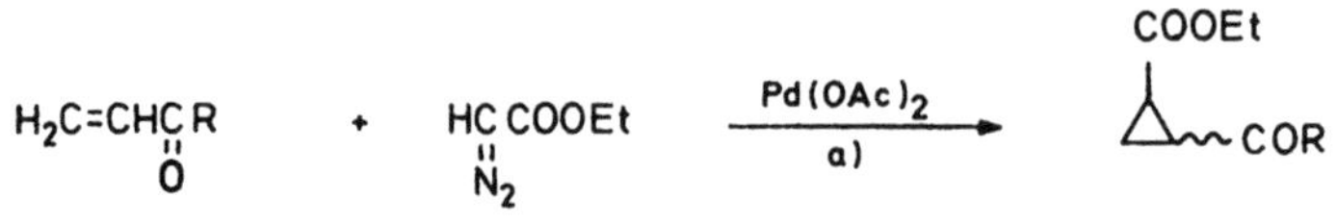

R	yield [%]	*cis*:*trans*
OMe	85	1:2.2
OEt	64	1:2.2
Me	76	1:4.1

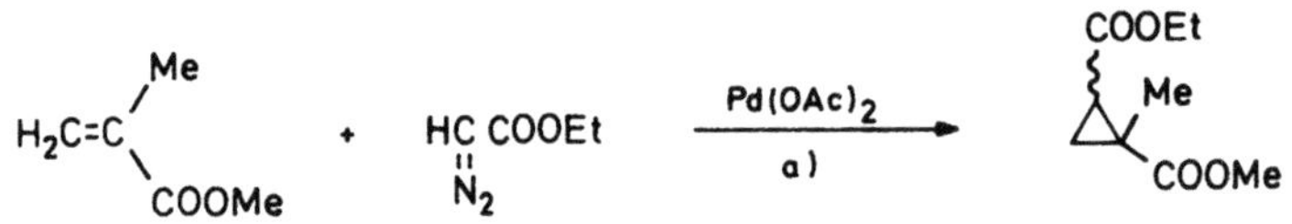

yield: 30%
Z : *E* = 1:1.2

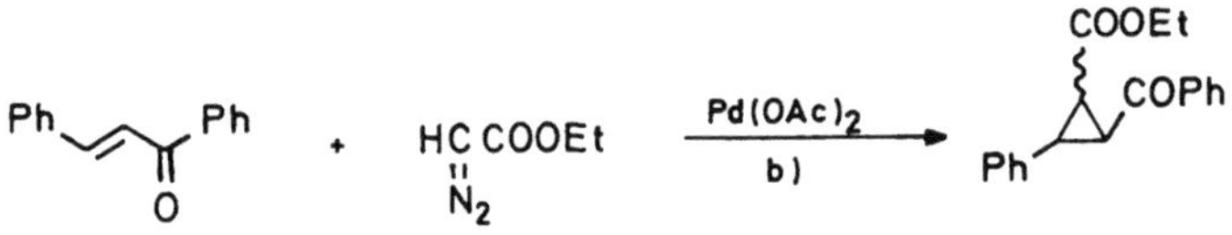

(50%)

a) 40 °C; benzene; EDA added slowly; molar ratio 75 (alkene)/100 (EDA)/1 (catalyst).
b) 0 °C, ether; EDA added slowly; molar ratio not given.

Scheme 19

to the nitrile/catalyst mixture. Vinyloxazoles are also obtained in the $Cu(OTf)_2$-catalyzed reactions; once again, no cyclopropanation of the C=C bond occurs [144].

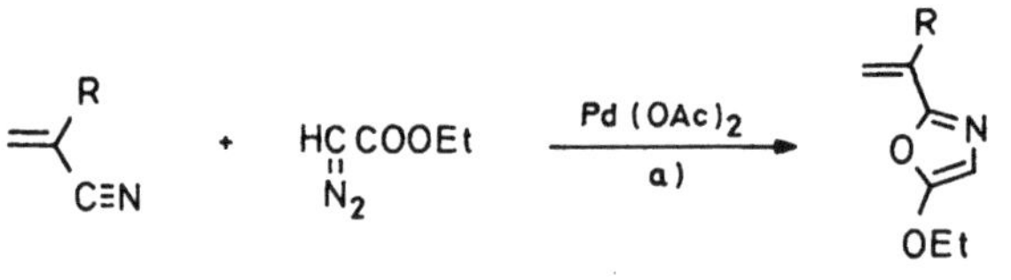

92

R = H, Me (30%)

a) Room temp., nitrile as solvent; EDA added slowly; molar ratio 500 (nitrile)/50 (EDA)/1 (catalyst).

Cyclopropanes *93* are obtained in satisfactory yields from both α,β-unsaturated carbonyl compounds and nitriles, when these substrates are allowed to react with ethyl diazoacetate or ω-diazoacetophenone in the presence of molybdenum hexacarbonyl or molybdenum(II) acetate [145,146]. Vinylic C—H insertion products **94**, 2-pyrazolines **95** and carbene dimers **96** are formed competitively. Only minor amounts of cyclopropanes were found in the complex product mixtures obtained from $Mo(CO)_6$-promoted reactions with ethyl crotonate or acrolein. Similar results as with $Mo(CO)_6$ and $Mn_2(OAc)_4$ were obtained with $V(CO)_6$ and $(\eta^5\text{-}C_5H_5)_2ZrCl_2$, whereas $Fe(CO)_5$ and $Mn_2(CO)_{10}$ were somewhat less effective. Rhenium, iridium and tungsten carbonyls proved to be totally unsuited for cyclopropane synthesis, favoring high-yield formation of 2-pyrazolines **95**.

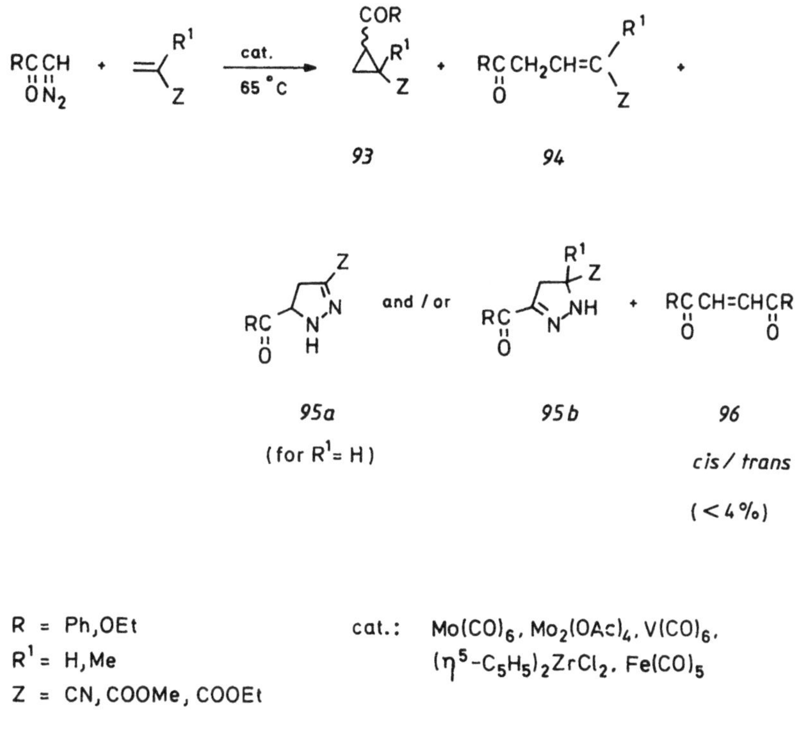

Furthermore, $Rh_6(CO)_{16}$, which can be used advantageously for cyclopropanation of more electron-rich alkenes, furnished only insignificant amounts of cyclopropane from acrylonitrile or ethyl acrylate and ethyl diazoacetate; from methacrylonitrile and ethyl diazoacetate, equally low yields of vinyloxazole, cyclopropane and carbene dimers resulted (Scheme 20) [145]. The use of $Rh_2(OAc)_4$ or $[Rh(CO)_2Cl]_2$ as catalysts did not change this situation.

The relative yields of **93**, **94** and **95** in the molybdenum-catalyzed reactions turned out to be exceptionally sensitive towards catalyst concentration, with different characteristics for different reaction partners. For example, the following yields of **93, 94** and **95b** were obtained when ω-diazoacetophenone reacted with acrylonitrile in the presence of different amounts of $Mo(CO)_6$: 46, 2, 50% (0.2 mol-% catalyst); 68, 3, 28% (1 mol-%); 83, 4, 0% (15 mol-%). In contrast, the yield of cyclopropane

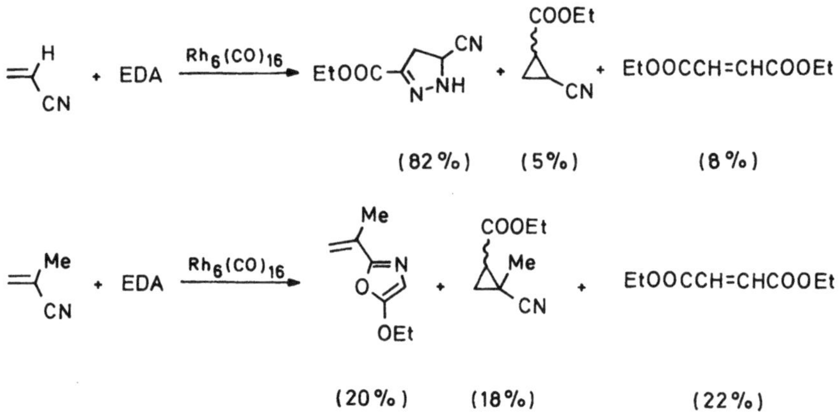

Scheme 20

from ω-diazoacetophenone and methacrylonitrile decreases sharply when increasing the catalyst concentration from ca. 0.1 to 5 mol-%.

Based on a detailed investigation, it was concluded that the exceptional ability of the molybdenum compounds to promote cyclopropanation of electron-poor alkenes is not caused by intermediate nucleophilic metal carbenes, as one might assume at first glance. Rather, they seem to interfere with the reaction sequence of the uncatalyzed formation of 2-pyrazolines (Scheme 18) by preventing the 1-pyrazoline → 2-pyrazoline tautomerization from occurring. Thereby, the 1-pyrazoline has the opportunity to decompose purely thermally to cyclopropanes and formal vinylic C—H insertion products. This assumption is supported by the following facts: a) Neither $Mo(CO)_6$ nor $Mo_2(OAc)_4$ influence the rate of [3 + 2] cycloaddition of the diazocarbonyl compound to the alkene. b) Decomposition of ethyl diazoacetate is only weakly accelerated by the molybdenum compounds. c) The latter do not affect the decomposition rate of and product distribution from independently synthesized, representative 1-pyrazolines, and 2-pyrazolines are not at all decomposed in their presence at the given reaction temperature.

When the electron demand of α,β-unsaturated carbonyl compounds is weakened by additional substituents in β-position, the double bond returns to normal behavior as far as cyclopropanation is concerned. Some recent examples, displayed in Scheme 21, may illustrate this point.

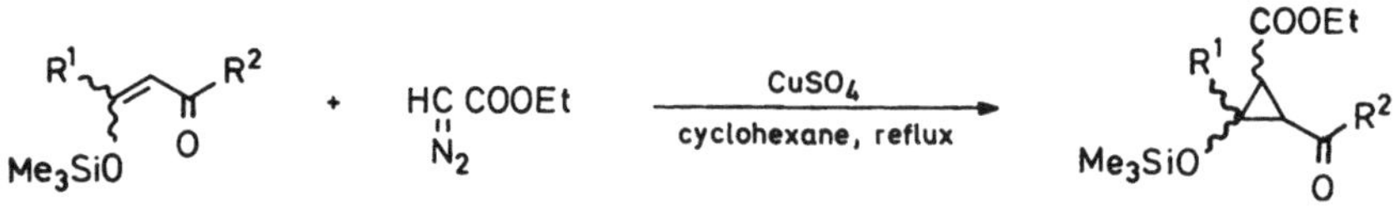

R^1 = Ph, Me, *n*-Bu, *i*-Bu, CH_2CH_2Ph, C_5H_{11}, C_6H_{13} Ref. 117
R^2 = OEt, *i*-Bu, *t*-Bu, C_5H_{11}

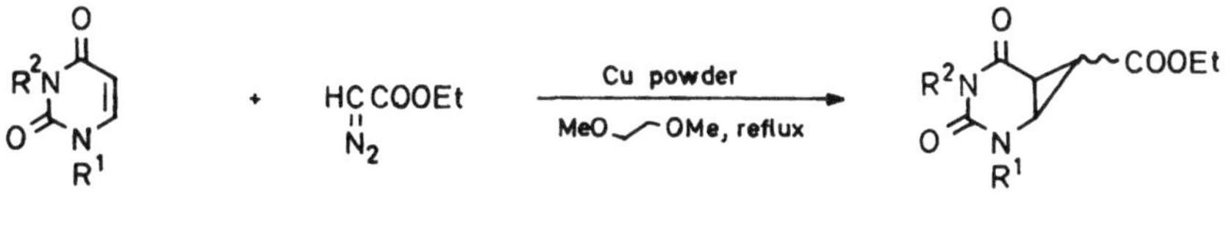

97 Ref. 147

R^1	R^2	*anti-97*	*syn-97*
Me	Me	23	33
$PhCH_2$	$PhCH_2$	14	27
$PhCH_2OCH_2$	Me	7	7
$Ph(CH_2)_3$	Me	40	23
$PhCH_2OCH_2$	$PhCH_2$	2.5	4

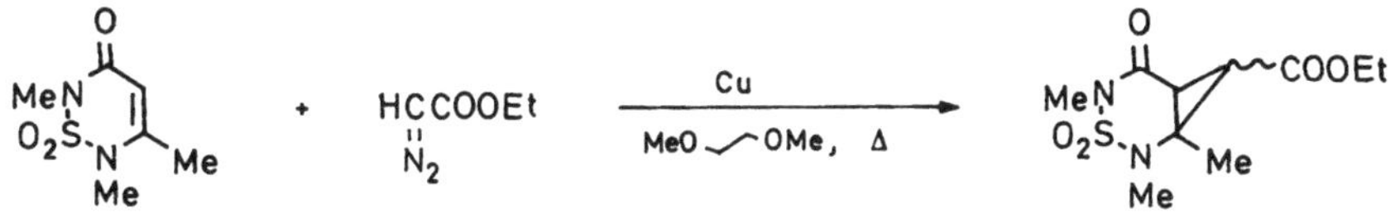

(30 %) Ref. 148

Scheme 21

2.3.3 The Question of Allylic C/H Insertion

Alkyl diazoacetates undergo little or no allylic C/H insertion when decomposed catalytically in the presence of appropriate olefins [6,13,14]. In contrast, such insertions occur with diazomalonates or α-diazoketones. From the available facts, the conclusion can be drawn that different pathways may lead to what finally looks like the "direct" or "rearranged" allylic insertion product, but convincing evidence for one or the other mechanism is available only in a few cases. As Scheme 22 shows, the C/H insertion products **98–100** may arise from one of three major sources:

a) A primarily formed acylcyclopropane may suffer thermocatalytic ring-opening [97,120] (**101** → **98**) or a thermal 1,5-homo-hydrogen shift [149] (**101** → **99**).
b) An abstraction/recombination mechanism, giving rise to an intermediate allyl cation/hydrocarbenoid pair or the corresponding η^3-allyl complex **103**, may lead to **99** and **100** [55].
c) Formation of only one new C—C bond between olefin and ketocarbenoid would create a dipolar intermediate **102** which then could yield both **98** and **99** by proton transfer. The chance for this pathway to occur increases when X is a cation-stabilizing substituent, such as the alkoxy group of an enol ether. A diradical intermediate instead of **102** would not be unlikely *a priori*, but at least for the copper-catalyzed diazomalonate reactions, it is excluded by an experiment with 1,1-dicyclopropylethylene, from which only the cyclopropanation product was obtained. A diradical intermediate $(C_3H_7)_2C^{\bullet}$—CH_2—$^{\bullet}C(OOR)_2$ would have

revealed itself by spontaneous fragmentation of one of the three-membered rings [128,150].

Further precursors to **98–100** can be envisaged [55,133] without having been proven so far. The simple picture of Scheme 22 is additionally complicated when an olefin contains two allylic centers instead of one.

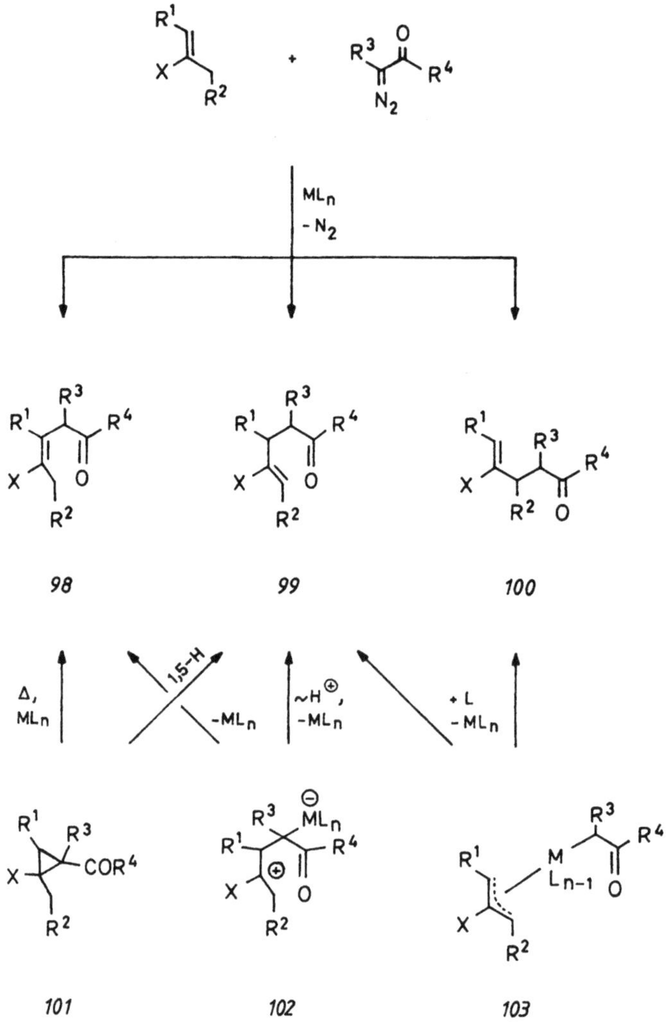

Scheme 22

C/H-insertions have been reported to occur in copper-catalyzed reactions between diazomalonates and cyclohexene as well as some alkylated derivatives [9,57]. Some acyclic alkenes behave similarly [9], but not so 1,1-dicyclopropylethylene [150]. An abstraction/recombination mechanism *via* intermediates of type **103** has been proposed [55] which would account not only for the three insertion products **104–106** obtained from 1-methylcyclohexene but also for some of the cyclopropane **107**, according to kinetic data.

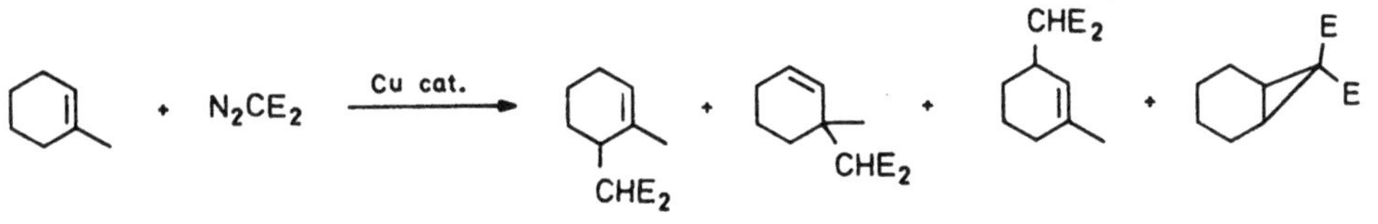

E = COOMe *104* *105* *106* *107*

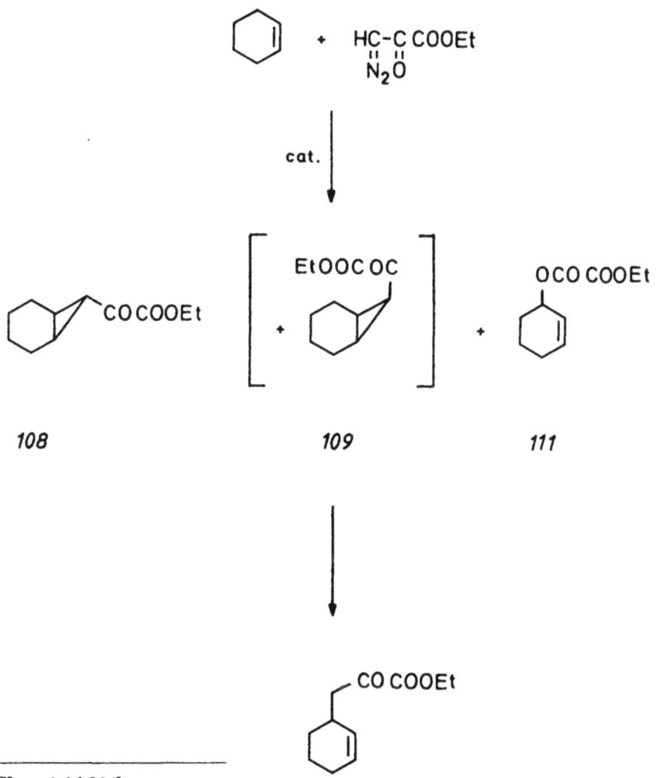

cat.	temp. [°C]	yield [%]		
		108	*110*	*111*
$Rh_2(OAc)_4$	125	70	10	—
$CuCl \cdot P(OMe)_3$	85	42	15	18[2]
Cu bronze	100	58	3	—

[2] Oxalate **111** is formed when the reaction is carried out in the presence of air. In that case, catalytic oxidation of cyclohexene to cyclohexen-3-ol takes place. The alcohol reacts with ethoxycarbonyl ketene, formed by Wolff rearrangement of the ketocarbenoid. $Rh_2(OAc)_4$ seems to suppress this rearrangement, whereas it becomes the sole reaction mode in the presence of $[(\eta^3-C_3H_5)PdCl]_2$ as well as Ag_2O [126].

Whereas this mechanistic proposal seems reasonable and no reason can be seen why it should not be cited to explain the allylic C/H insertion product from cyclohexene, other cases exist where cyclohexene is not the best substrate to distinguish between this and one of the other alternatives of Scheme 22.

For example, reaction of ethyl diazopyruvate with cyclohexene in the presence of rhodium [126] or copper [113,126] catalysts furnishes, besides the 7-*exo*-substituted norcarane **108**, a small amount of **110**, which may arise either from allylic insertion or from the 7-*endo*-substituted norcarane **109** by a thermal 1,5-homo-hydrogen shift.

In order to distinguish between these alternatives the reaction was repeated with 4-*t*-butyl-1-methylenecyclohexane [125]. In this case, cyclopropane **112** and a small amount of **113** were obtained from the $Cu(acac)_2$-catalyzed reaction at 80 °C, but not product **114** which would result from direct allylic C/H insertion. **112** rearranges thermally to give **113** by a 1,5-homo-hydrogen shift. It seems, therefore, safe to say that the small amount of olefin **110** formed in the reaction with cyclohexene also arises from a 1,5-homo-hydrogen shift of the non-isolated cyclopropane derivative **109**, which has the right geometry to realize the six-membered transition state of such a rearrangement.

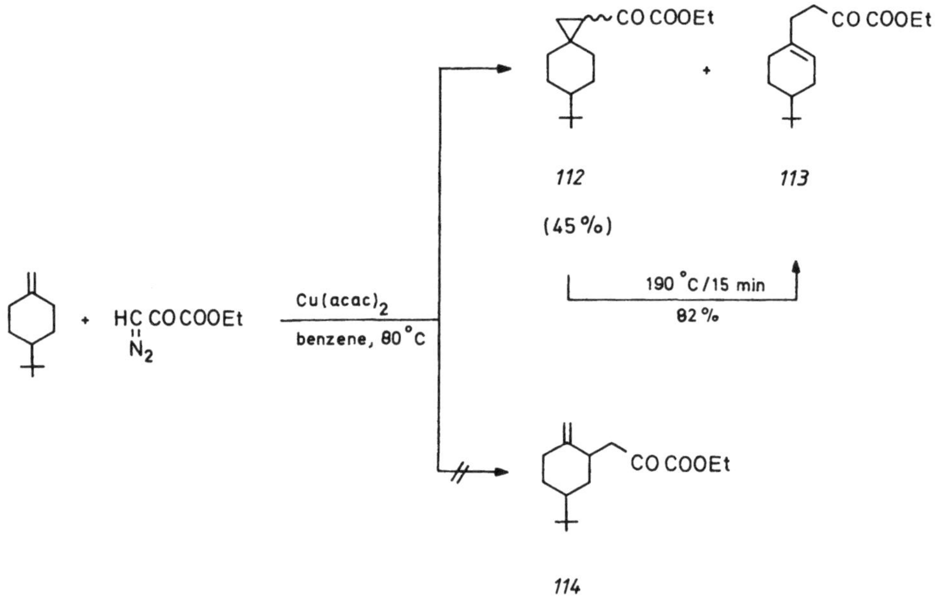

The question as to whether enol ether **72**, the "insertion product" derived from diethyl diazomalonate and 1-methoxycyclohexene, has a similar origin or arises from a dipolar intermediate of type **102**, has already been discussed (Sect. 2.3.1). Interestingly enough, only one formal C/H insertion product was reported in that case, rather than three as in the reaction with 1-methylcyclohexene.

A 1,5-homo-hydrogen shift also accounts for the formation of olefin **116** from the $CuSO_4$-catalyzed reaction of methyl diazopyruvate with *cis*-2-butene [127]. The all-*cis*-cyclopropane **115** rearranges quantitatively to **116** under the work-up conditions.

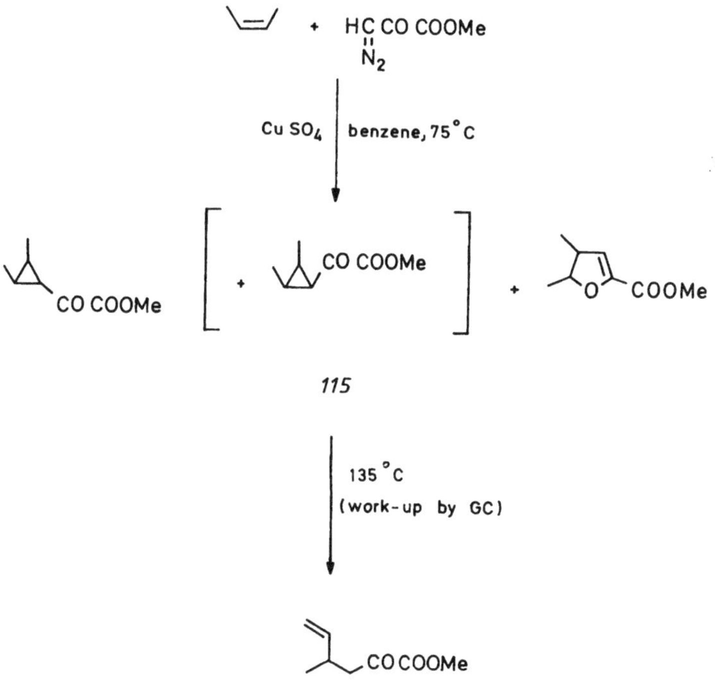

115

116

Allylic C/H insertion accompanied by an allylic rearrangement has been observed for carbenoid reactions of ethyl diazoacetate with allylamines (Scheme 23) [151]. Apparently, metal-catalyzed isomerization **117** ⇌ **118** preceeds the C/H insertion process. Although mechanistic details have not yet been unraveled, η^3-allyl complexes

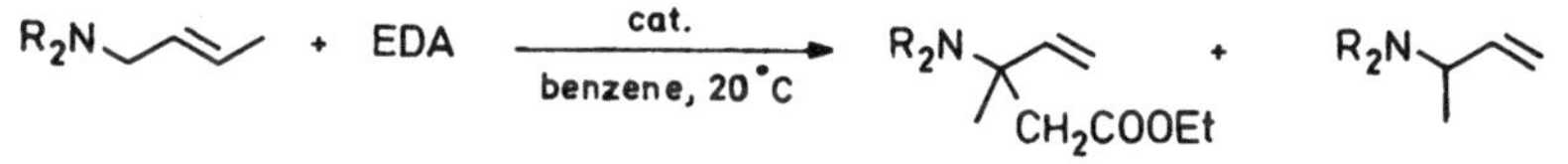

117

catalyst	yield [%]	
$Cu(acac)_2$	55	26
$P(OAn)_3/AlEt_3/(1:2:4)$		
$Rh_2(OAc)_4$	38	12

An = C_6H_4—4—OMe.

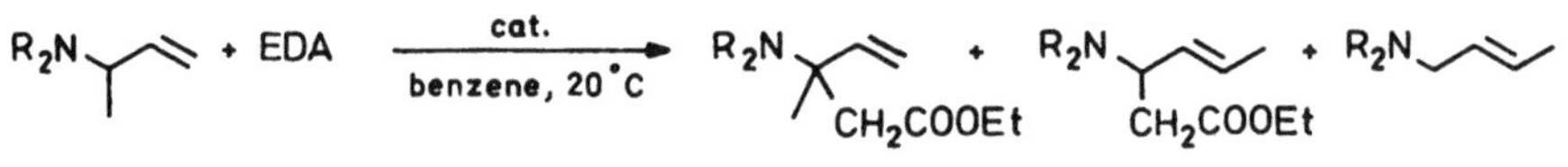

118 121

catalyst	yield [%]		
Cu catalyst as above	3	15	2
$Rh_2(OAc)_4$	19.5	1.3	1.2

119 120

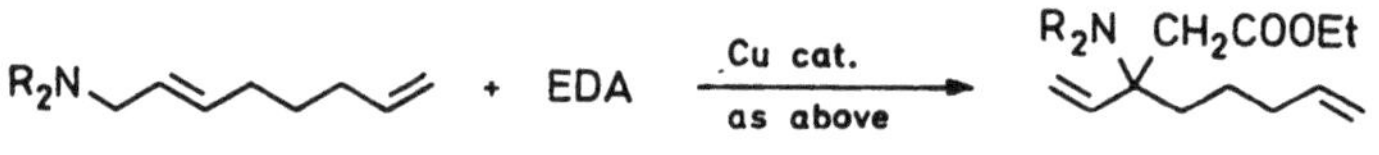

122 (20%)

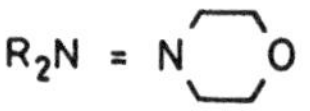

Scheme 23

119 and **120** were assumed to participate in the insertion step. However, a look at the product pattern obtained from **117** and **118** (compare the yields of **121** which is the "unrearranged" insertion product from **117**, but is only formed as the rearranged insertion product from **118**) leads to the conclusion that the transformation **118** → **121** follows a pathway which is different from the sequence **118** → **117** → **119** → **121**.

When a second, non-activated double bond [15] is present in the allylamine, cyclopropanation can still not compete with allylic insertion, as the example of **122** shows.

2.3.4 Allyl Halides, Sulfides, Amines, Acetals and Dithioketals

Interaction of carbenes or carbenoids with heteroatoms may lead to reactive ylide intermediates which then undergo rearrangement to stable products. If the heteroatom occupies an allylic position, a [2,3] rearrangement may be one of those processes, siphoning off the carbenoid and creating a serious competition to the cyclopropanation reaction. Carbenes generated photochemically or strictly thermally do not discriminate efficiently between these two reaction paths, and practical application of ylide generation, until very recently, has been restricted to copper-catalyzed reaction of diazoesters with thioethers [152,14]. The scope of ylide formation, the influence of catalysts on the ylide/cyclopropane ratio and the mechanistic pathways have been investigated in greater detail by Doyle's group [153,154]; some results are presented in the following.

Allyl halides [153]. The competition between insertion product **123** and cyclopropane **124** depends on the halogen atom and on the catalyst. In the presence of $Rh_2(OAc)_4$, no cyclopropane **124** at all is obtained from allyl iodide, but mainly cyclopropane

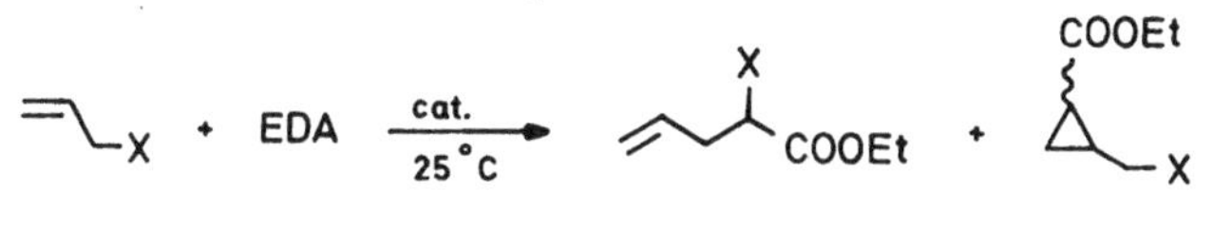

123 *124*

X	catalyst	*123* + *124* [%][a]	*123*:*124*
Cl	$Rh_2(OAc)_4$	95	0.05
	$CuCl \cdot P(O\text{-}i\text{-}Pr)_3$	86	0.73
	Cu bronze	23	0.22
Br	$Rh_2(OAc)_4$	76	0.40
	$CuCl \cdot P(O\text{-}i\text{-}Pr)_3$	49	5.8
	Cu bronze	56	7.8
I	$Rh_2(OAc)_4$	98	(only *123*)

[a] Conditions: 0.5 mol% of catalyst; five- to tenfold molar excess of allyl halide.

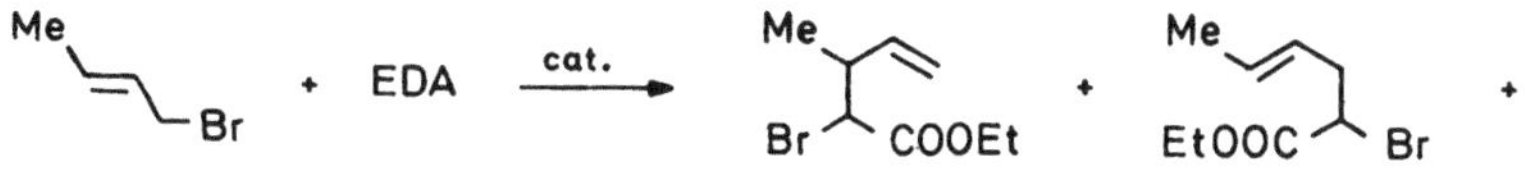

125 *126*

127 *128*

catalyst	temp. [°C]	total yield [%][b] (*125*–*128*)	relative yield			
			125	*126*	*127*	*128*
$Rh_2(OAc)_4$	25	77	67	0	33	<1
	50	26	17	13	16	54
$Cu(acac)_2$	25	61	43	45	7	5
	50	31	40	47	5	8
$CuCl \cdot P(O\text{-}i\text{-}Pr)_3$	25	68	32	53	6	9
	50	13	39	39	6	16
Cu bronze	25	67	34	52	7	7
	50	50	28	29	4	39

[b] Carbene dimers are obtained additionally (<8%).

results from allyl chloride. The use of copper catalysts greatly promotes product **123**, which is derived from a [2,3] sigmatropic rearrangement of an intermediary halonium ylide. As the nucleophilicity of the halide decreases in the order allyl iodide, bromide, chloride, the different product ratios can be taken as reflecting the electrophilicity of presumed metal carbene intermediates. According to this, copper carbenes should be more electrophilic than rhodium carbenes. Furthermore, as the product ratio **123**:**124** is rather similar for the homogeneous copper catalysts and copper bronze, the copper carbenes generated in both cases should be very similar as well.

In contrast to ethyl diazoacetate, diethyl diazomalonate reacts with allyl bromide in the presence of $Rh_2(OAc)_4$ to give the ylide-derived diester favored by far over the cyclopropane (at 60 °C: 93:7 ratio). This finding bespeaks the greater electrophilic selectivity of the carbenoid derived from ethyl diazomalonate. For reasons unknown, this property is not expressed, however, in the reaction with allyl chloride, as the carbenoids from both ethyl diazoacetate and diethyl diazomalonate exhibit a similarly high preference for cyclopropanation.

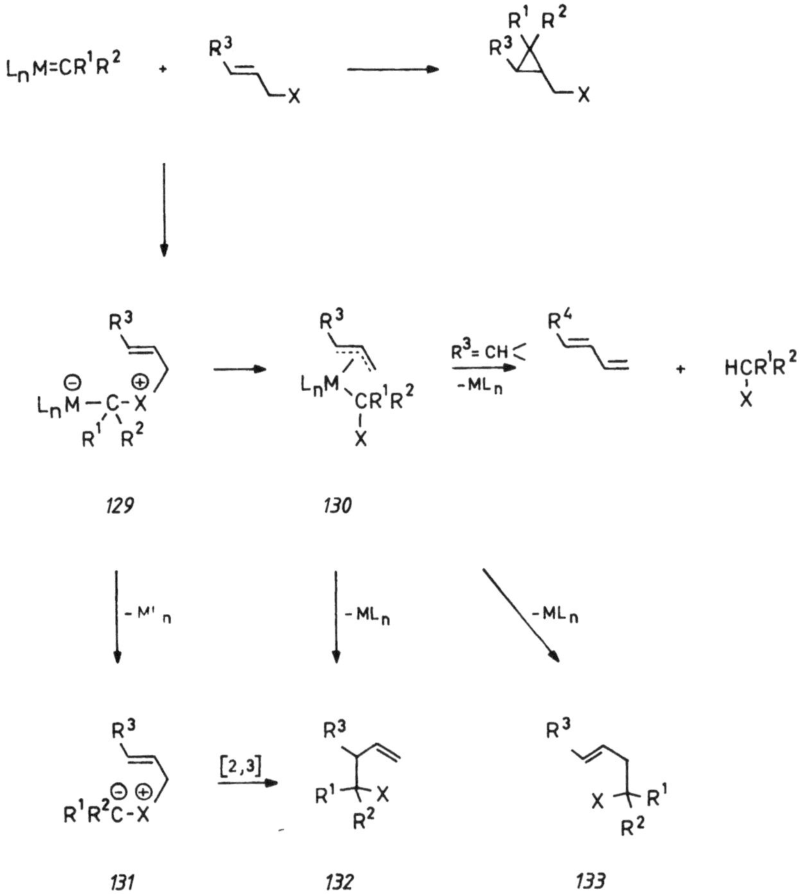

Scheme 24

The reaction of crotyl bromide with ethyl diazoacetate once again reveals distinct differences between rhodium and copper catalysis. Whereas with copper catalysts, the products **125** and **126**, expected from a [2,3] and a [1,2] rearrangement of an intermediary halonium ylide, are obtained by analogy with the crotyl chloride reaction [152a)], the latter product is absent in the rhodium-catalyzed reaction at or below room temperature. Only when the temperature is raised to ca. 40 °C, **126** is found as well, together with a substantial amount of bromoacetate **128**. It was assured that only a minor part of **126** arose from [2,3] rearrangement of an ylide derived from 3-bromo-1-butene which is in equilibrium with the isomeric crotyl bromide at 40 °C.

A mechanistic picture which reconciles the experimental results is given in Scheme 24. It is assumed that both the heteroatom and the double bond of the allyl halide compete for an electrophilic metal carbene. Heteroatom attack yields a metalated ylide **129**, which may go on to ylide **131** by demetalation and/or to allylmetal complex **130**. Symmetry-allowed [2,3] rearrangement of **131** accounts for product **132**, and metal elimination from **130** gives rise to products **132** and **133**, corresponding to [2,3] and [1,2] rearrangement, respectively, as well as haloacetate (if R^3 = CH$<$).

The $Rh_2(OAc)_4$-catalyzed reaction between crotyl bromide and ethyl diazoacetate at or below room temperature follows the pathway **129** → **131** → **132** exclusively. At higher temperature, when ethyl bromoacetate and increasing amounts of the [1,2] rearrangement product **126** are found additionally, the **129** → **130** → **132** + **133** route becomes a competing process. With copper catalysts, this situation may be applicable at all temperatures, but it has been suggested that the route via complex **130** operates solely, when copper bronze is the catalyst [154)].

The $Cu(acac)_2$-catalyzed decomposition of 6-diazopenicillanates **134** in the presence of allyl bromide furnishes the rather labile 6α-allyl-6β-bromopenicillanate **135** via the ylide pathway discussed above [155)].

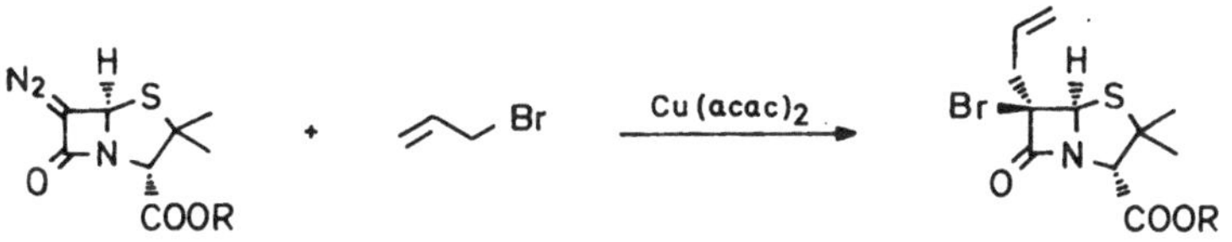

134 135

R = CH_2CCl_3

Allyl sulfides and allyl amines. Rhodium-catalyzed decomposition of ethyl diazoacetate in the presence of these allyl compounds generates products **136** and **137**, respectively, derived from [2,3] rearrangement of an S- or N-ylide intermediate, besides small amounts of carbene dimers [153)]. No cyclopropane and no product resulting from the ylide by [1,2] rearrangement were detected. Besides $Rh_2(OAc)_4$ and $Rh_6(CO)_{16}$, the rhodium(I) catalysts $[(cod)RhCl]_2$ and $[(CO)_2RhCl]_2$ were found to behave similarly, but yields with the only allyl amine tested, $CH_2{=}CH{-}CH_2NMe_2$, were distinctly lower with the latter two catalysts. Reaction temperatures are higher than usually needed in rhodium-promoted diazoalkane decomposition, which is certainly due to competition between the diazo compound and the allylic heteroatom substituent for coordination to the metal. The high-yield formation of homoallyl methyl sulfide **136** parallels the results obtained in copper-catalyzed reactions [152a)], whereas small amounts of the cyclopropanecarboxylate were formed additionally upon photochemical decomposition of the diazo compound [152b)].

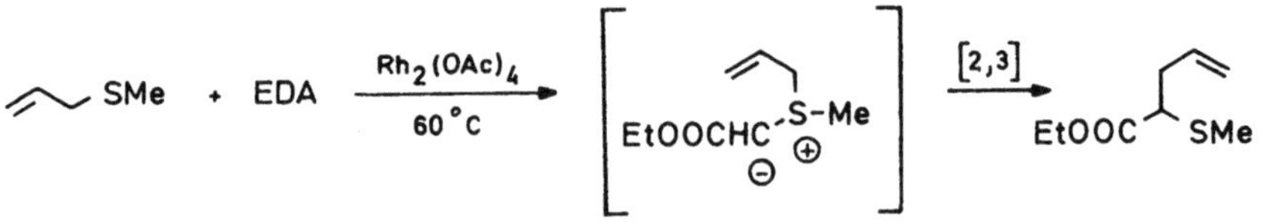

136

(96%)

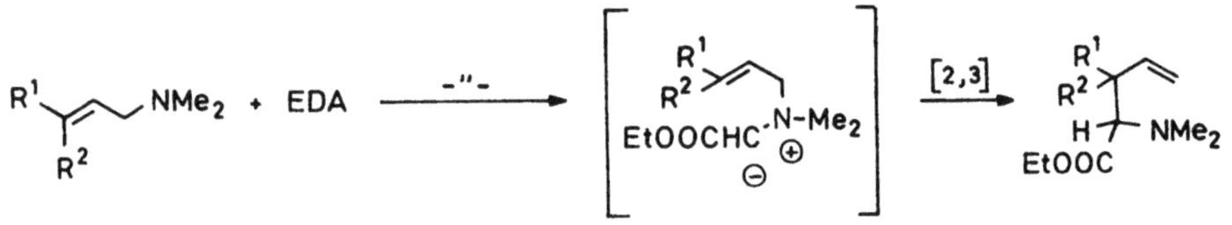

137

(2 diastereomers)

R^1	R^2	yield [%] of *137*
H	H	60
Me	Me	37
Me	H	79
Ph	H	59

Optimized reaction conditions: 0.5 to 1.0 mol% of catalyst; olefin:EDA molar ratio 5–10.

Contrary to the allyldimethylamines, the less nucleophilic 1-morpholino-2-butene **117** and 3-morpholino-1-butene **118** do not yield products derived from an intermediary N-ylide; rather, allylic C/H insertion products were isolated (see Sect. 2.3.3) [151)].

Vedejs has used the [2,3] rearrangement of an intermediary S-ylide for the

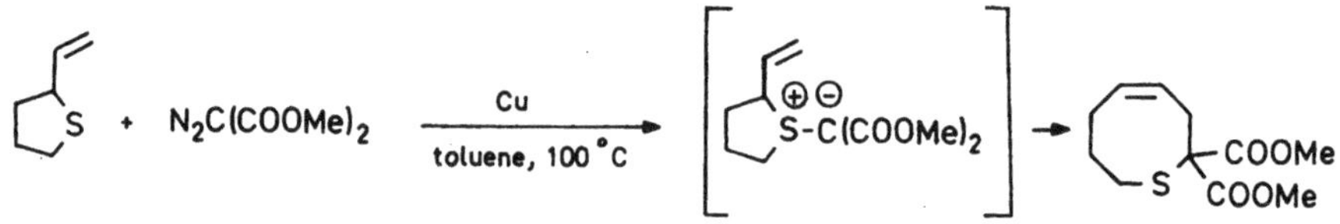

138 *139*

(53%)

three-carbon ring expansion **138** → **139** [156,157]. Initiating the reaction by copper-catalyzed decomposition of ethyl diazoacetate or diazoketones obviously was less efficient than in the case of dimethyl diazomalonate, since the yield of the thiacyclooctene dropped to 15 and 0%, respectively. In these cases, non-carbenoid routes to the intermediary S-ylide had to be taken.

[2,3] Rearrangement of an intermediary S-ylide is the key step of a synthesis of **142** from the 6-phenylthiomethyl-2-pyrone **140** and ethyl diazoacetate in the presence of a catalytic amount of $Cu(acac)_2$ [158]. The primary rearrangement product **141** is smoothly isomerized to **142** by treatment with silica gel.

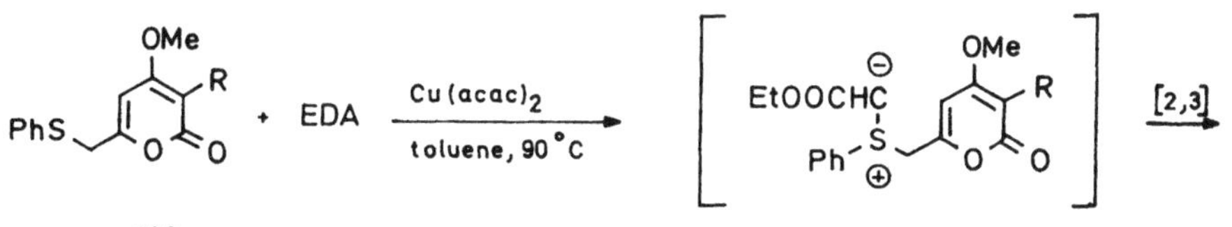

140

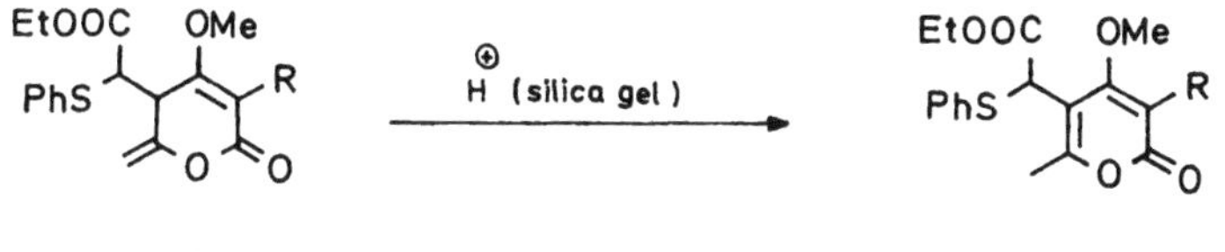

141

142

R = H : 41%
R = Br : 40%

As has been described for allyl bromide (see preceding paragraph), allyl sulfides and allyl phenyl selenide react with 6-diazopenicillanates **134** under $Cu(acac)_2$ catalysis to give the products of ylide formation and subsequent [2,3] rearrangement [155,159]. Both C-6 epimers are formed. The yields are better than with $BF_3 \cdot Et_2O$ catalysis, and, in contrast to the Lewis acid case, no 6α-monosubstituted (RS or RSe) penicillanate is observed.

Allyl acetals [154]. Allyl ethers give no or only trace amounts of ylide-derived products in the $Rh_2(OAc)_4$-catalyzed reaction with ethyl diazoacetate, thus paralleling the reactivity of allyl chloride. In contrast, cyclopropanation must give way to the ylide route when allyl acetals are the substrates and ethyl diazoacetate or dimethyl diazomalonate the carbenoid precursors.

In addition to cyclopropane **145** and the expected [2,3] rearrangement product **143** of an intermediary oxonium ylide, a formal [1,2] rearrangement product **144** and small amounts of ethyl alkoxyacetate **146** are obtained in certain cases. Comparable results were obtained when starting with dimethyl diazomalonate. $Rh_2(CF_3COO)_4$ displayed an efficiency similar to $Rh_2(OAc)_4$, whereas reduced yields did not recommend the use of $Rh_6(CO)_{16}$ and several copper catalysts. Raising the reaction temperature had a deleterious effect on total product yield, as had already been the case with allyl halides [153], but the (**143**+**144**)/**145** ratio remained rather constant.

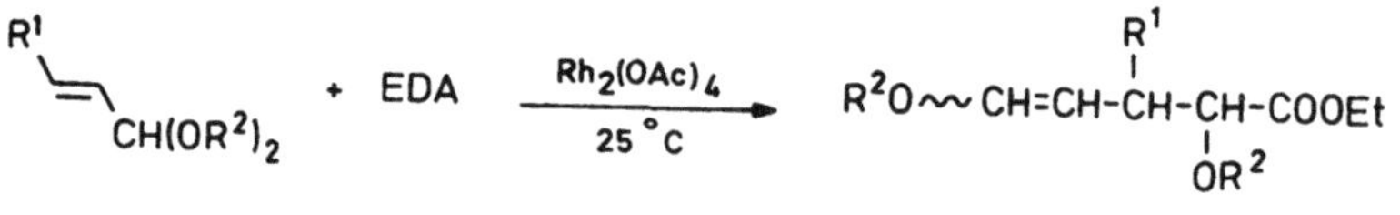

143

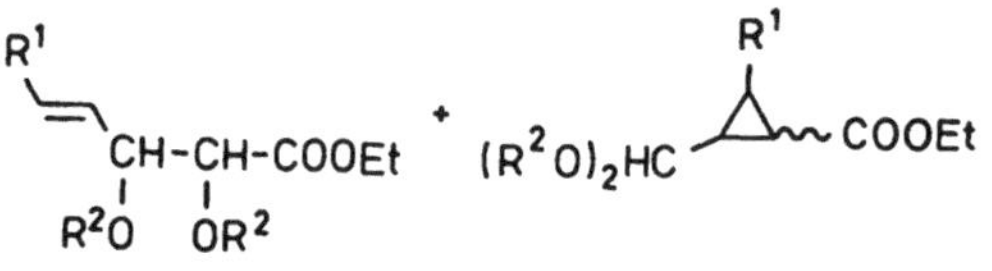

144 *145*

+ R^2OCH_2COOEt

146

R^1	R^2	total yield [%]	relative yield			
		(*143–146*)	*143*	*144*	*145*	*146*
H	Me	60	75	0	23	2
H	Et	78	42	0	49	9
H	$-(CH_2)_2-$	44	61	24	15	<1
Ph	Me	62	72	18	10	0

Assuming a reactive oxonium ylide **147** (or its metalated form) as the central intermediate in the above transformations, the symmetry-allowed [2,3] rearrangement would account for all or part of **148**. The symmetry-forbidden [1,2] rearrangement product **150** could result from a dissociative process such as **147** → **149**. Both as a radical pair and an ion pair, **149** would be stabilized by the respective substituents; recombination would produce both [1,2] and additional [2,3] rearrangement product. Furthermore, the ROH-insertion product **146** could arise from **149**. For the allyl halide reactions, the [1,2] pathway was envisaged as occurring via allyl metal complexes (Scheme 24) rather than an ion or radical pair such as **149**. The remarkable dependence of the yield of [1,2] product **150** on the allyl acetal substituents seems, however, to justify a metal-free precursor with an allyl cation or allyl radical moiety.

For the $Cu(OTf)_2$-promoted reaction between ethyl diazoacetate and cinnamaldehyde dimethyl acetal, products **143–145** account for only 35% the total yield. C/C and C/H insertion products **151** and **152** are obtained additionally in 49 and 14% yield, respectively [154]. It was assumed that the copper compound acts through Lewis-acid catalysis here, just as it is believed to do when orthoesters are used as substrates [160]. According to this, catalyst-induced formation of a methoxycarbenium ion from the acetal would be followed by addition of the diazoester to give **153**, which then could go on to the products by N_2 loss, rearrangement and methoxide addition.

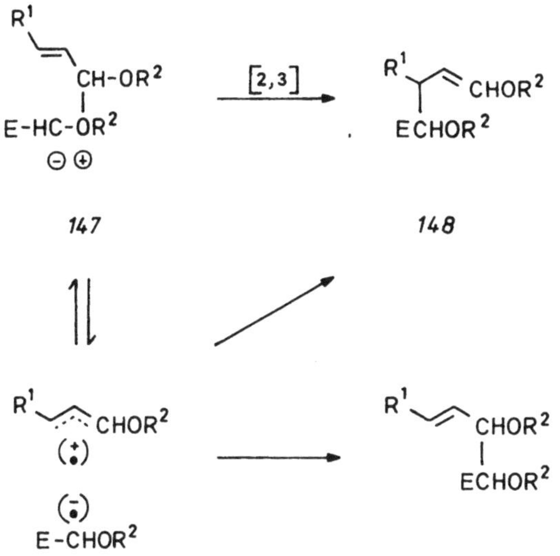

147 *148* *149* *150*

E = COOEt

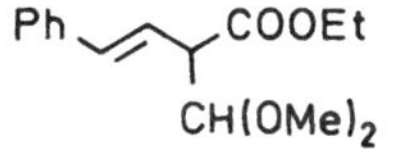

151

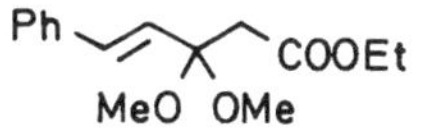

152

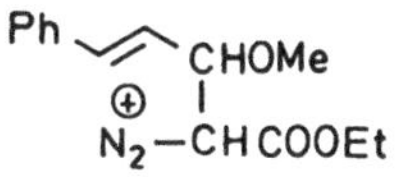

153

Allyl dithioketals [154]. By analogy with simple allyl sulfides, rhodium-catalyzed decomposition of ethyl diazoacetate in the presence of allyl thioketals **154** and **157** gives rise to intermediary sulfur ylides which are "trapped" by a [2,3]-sigmatropic rearrangement. As cyclic dithioketals are used, this reaction constitutes a three-carbon ring expansion (**154** → **155** and **157** → **158/159**). Intramolecular proton abstraction at the ylide stage is responsible for butadienes **156** and **160**. No products stemming from a [1,2] rearrangement were detected and cyclopropanation products derived from the allyl dithioketals were also absent.

Olefins analogous to **158** and **159** were also isolated from the $CuSO_4$-catalyzed decomposition of ethyl diazoacetate in the presence of 2-isopropenyl-2-methyl-1,3-dithiane (total yield 56%, $E:Z = 4:1$); a butadiene was absent from the reaction mixture [161]. With dimethyl diazomalonate instead of ethyl diazoacetate, only the Z-olefin resulting from a [2,3]-sigmatropic rearrangement of the corresponding sulfur ylide was obtained in 36% yield [161]. When the same procedure was applied to

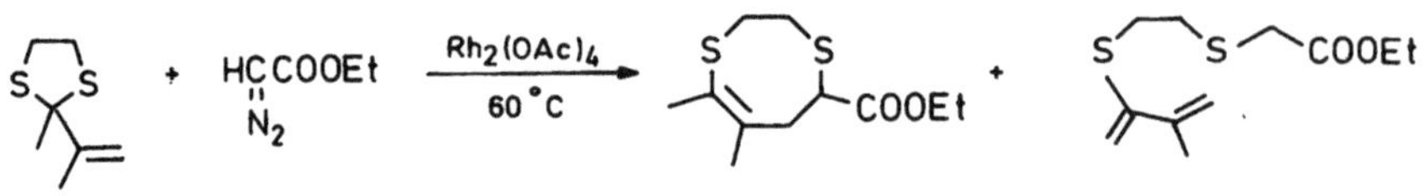

154 155 156

[154] : [EDA] = 4 : 1 (34%) (58%)

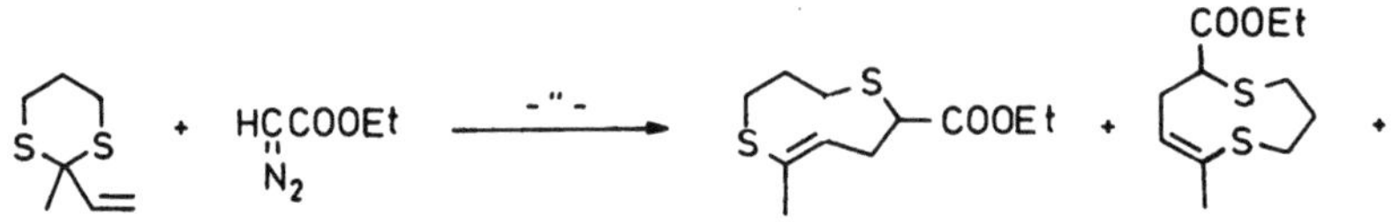

157 158 159

(54%) (21%)

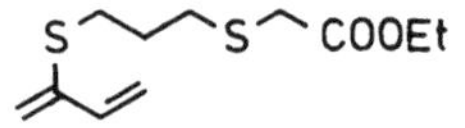

160

[157] : [EDA] = 5:1 (22%)

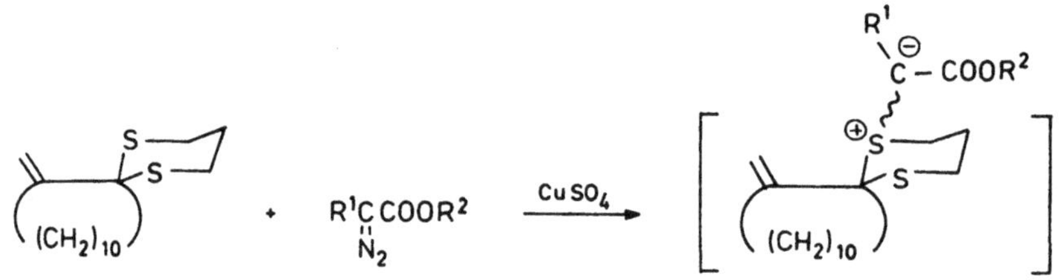

161

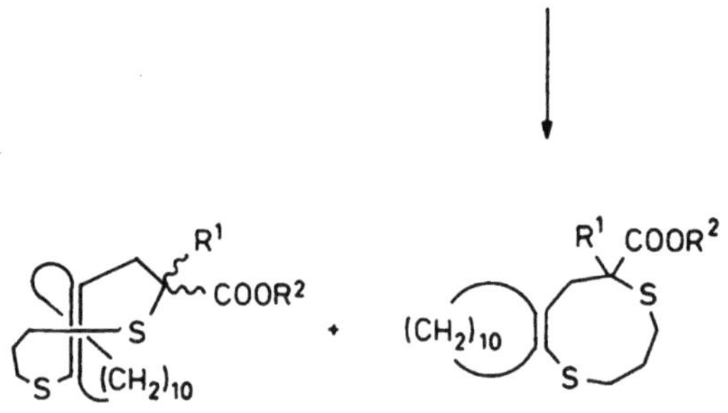

162 163

R^1	R^2	conditions	*162* + *163* [%]	*162*:*163*
H	Et	benzene, 55–60 °C	58	4:1
COOMe	Me	toluene, 100–110 °C	71	5:1

spirocyclic allyl dithioketal **161**, "betweenanenes" **162** and their *Z*-isomers **163** were produced [161].

2.3.5 Allyl and Other Unsaturated Alcohols

Competition of the double bond and the O—H function in an unsaturated alcohol for a ketocarbenoid invariably results in favor of the O/H insertion product, but cyclopropanation can be promoted to a certain extent by correct choice of the catalyst [162]. Thus, the following yields of ether **164** vs. cyclopropane **165** and cyclopropane-derived γ-lactone **166** were reported for $Rh_2(OOCR)_4$-catalyzed reaction of allyl alcohol with ethyl diazoacetate: R = Me: **164**(**165** + **166**) = 66/15%; *t*-Bu: 46/19; *n*-C_6H_{13}: 63/14; CH_2OMe: 46/11; CF_3: 64/6; C_7H_{15}: 63/16; C_6F_5: 62/19; ferrocenyl: 66/34; 1-adamantyl: 53/17; OOCR = L(+)-2-(tetrachlorophthalimido)propionato = 62/32. No general trend concerning the selectivities of different diazoesters (methyl, ethyl or *t*-butyl diazoacetate) in the presence of a given rhodium catalyst could be detected. It may be true, however, that the combination of *t*-butyl diazoacetate with a more lipophilic rhodium carboxylate gives the best yields of cyclopropanes [162].

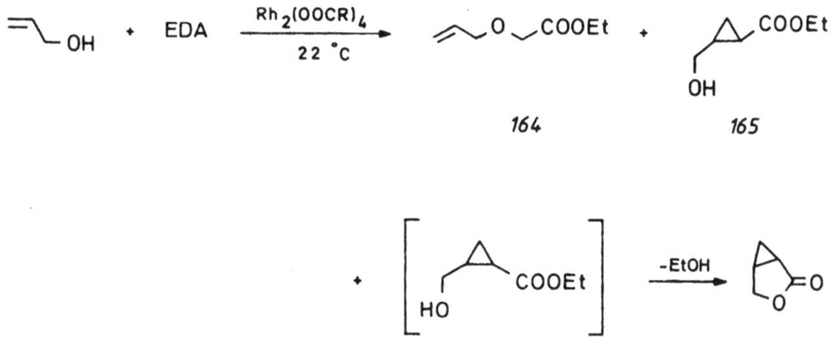

164 165

166

molar ratio: 5000 (alcohol) / 300 (EDA) / 1 (catalyst)

Copper(II) triflate is quite inefficient in promoting cyclopropanation of allyl alcohol, and the use of *t*-butyl diazoacetate [**164**/(**165**+**166**) = 97/3%] brought no improvement over ethyl diazoacetate (67/6%) [162]. If, however, copper(I) triflate was the catalyst, cyclopropanation with ethyl diazoacetate increased to 30% at the expense of O/H insertion (55%). As has already been discussed in Sect. 2.2.1, competitive coordination-type and carbenoid mechanisms may be involved in cyclopropanation with copper catalysts, and the ability of Cu(I) to coordinate efficiently with olefins may enhance this reaction in the intramolecular competition with O/H insertion.

Olefinic alcohols other than allyl alcohol display a preference for O/H insertion which is quite similar to that of the latter and rather independent of the particular compound [162]. Relative reactivity studies show, however, that an allylic O—H bond reacts faster than a non-allylic one, and that steric hindrance slows down the O/H insertion [162]. With $Cu(OTf)_2$ as catalyst, the yield of O/H insertion is generally lower than with $Rh_2(OAc)_4$, but the cyclopropane yield remains at a consistently low level, too.

Exclusive O/H insertion takes place in the $Rh_2(OAc)_4$-catalyzed reaction of diethyl diazomalonate with α,β-unsaturated γ-hydroxyesters **167a–c** [163]. This is not surprising in view of the reluctance of electrophilic metal carbenes to add to electron-poor double bonds (see Sect. 2.3.2). However, the more electron-rich double bond of p-methoxybenzyl clavulanate **168** also cannot compete with the O—H function for the same carbenoid [164]. The steric situation at the trisubstituted double bonds of **167** and **168** may be reason enough to render an attack there highly unfavorable as compared to the easily accessible O—H function, no matter how nucleophilic the double bond is.

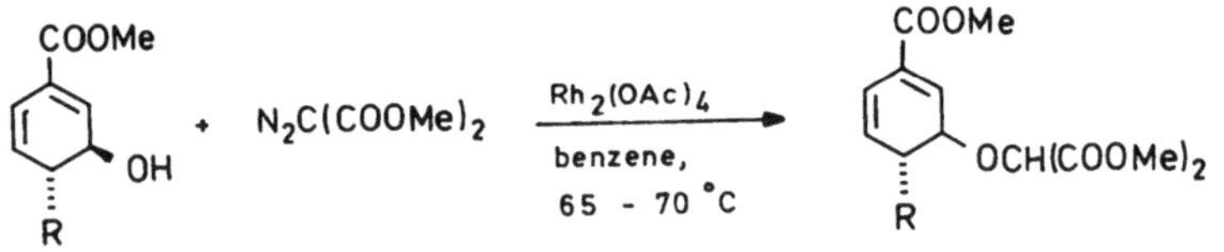

167a: R = OMEM (75%)

167b: R = NHCOO*t*-Bu (≥ 94%)

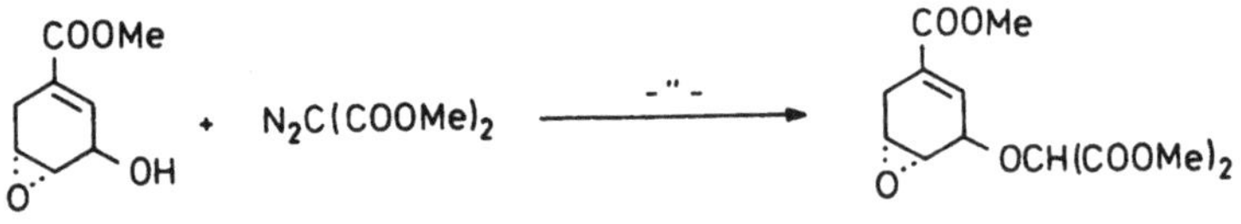

167c (82%)

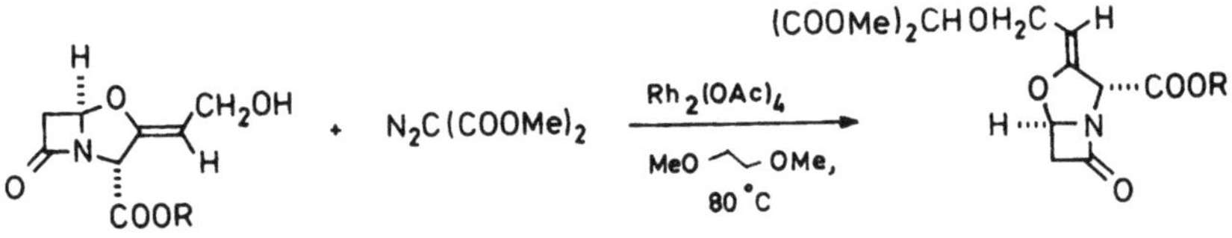

168

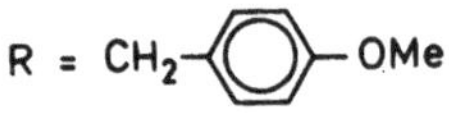

2.4 Intramolecular Cyclopropanation

This topic has been treated in an extensive review, covering the literature up to 1977 [12]. In several hundred cases, copper-catalyzed decomposition of unsaturated α-diazoketones or diazoesters has given access to bicyclic and polycyclic molecules [12,14]. The cyclopropane moiety formed by this intramolecular carbenoid-to-olefin addition was often used for further transformations, such as thermal rearrangement, acid-catalyzed or nucleophilic ring-opening and hydrogenolysis. The methodology being well developed, olefinic α-diazocarbonyl and α-diazophosphoryl compounds continue to be used, inter alia, for the construction of strained small-ring compounds as well as non-trivial cyclic natural products; intramolecular cyclopropanation starting with a vinyldiazo unit has also been reported (Table 10).

Danishefsky has shown that for the intramolecular transformation **169** → **170** stereocontrol at C-4 of the bicyclo[3.1.0] system is possible because of the bulky nature of the phthalimido group (NPht) [179]. Steric hindrance between NPht and the OCH_2THP group, which is directed into the *endo*-position in the bicyclic product because of stereospecific cyclopropanation, favors the transition state leading to **170** (NPht in the *exo*-position) rather than the one leading to **171**. Although the buttressing effect of the OCH_2THP group is absent in the *trans*-1,2-disubstituted olefin, there is still a high preference for the *exo*-NPht C-4 epimer. Further reduction of stereocontrol is observed if a methylene group is placed between the chiral allylic carbon and the ester oxygen atom, thus allowing for a more flexible transition state leading to a bicyclo[4.1.0] system (see Table 10 for examples).

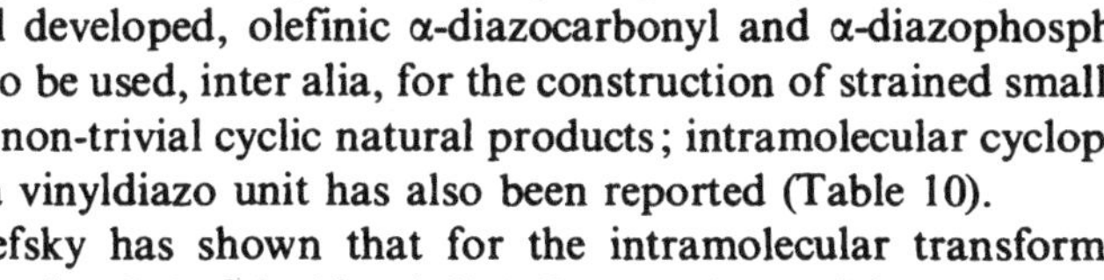

169 *170*

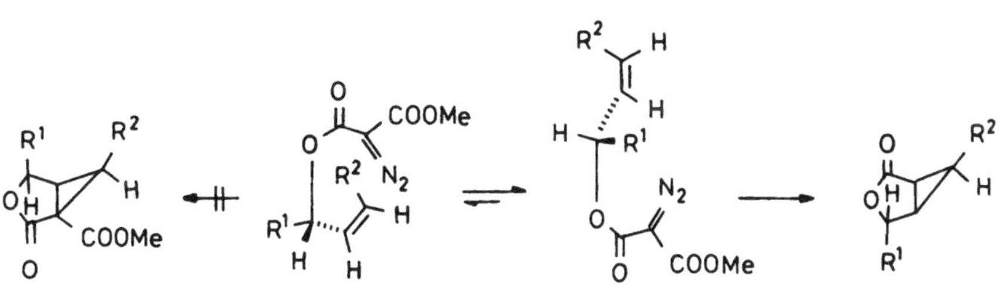

171 *170*

R^1 = NPht, phthalimido;

R^2 = OCH_2THP (THP = tetrahydropyranyl)

Table 10. Intramolecular cyclopropanation of unsaturated α-diazocarbonyl, α-diazophosphoryl and vinyldiazo compounds

Diazo compound	Conditions	Product	Yield [%]	Further transformation of cyclopropane to α	Ref.
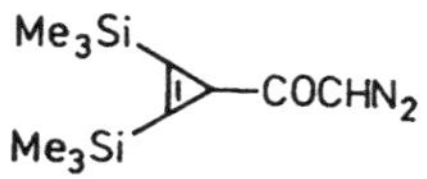	CuBr $CHCl_3$, 60°C	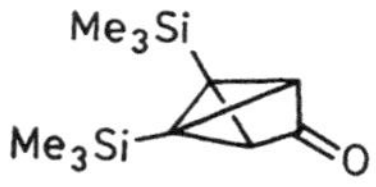	24	 (KF, crown ether)	165)
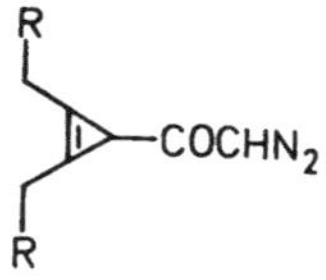	$Rh_2(OAc)_4$ $CHCl_3$, 60°C	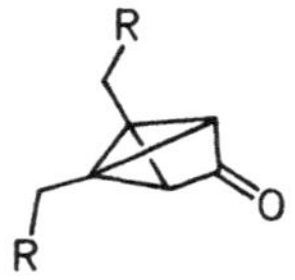	R = OAc : 37 R = Cl : 37		166) 167)
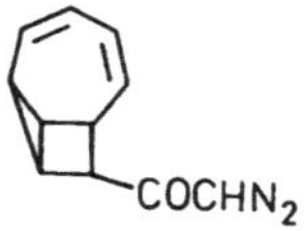	Cu THF, reflux	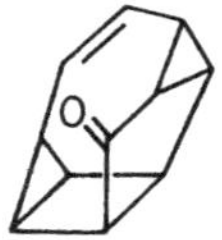	35	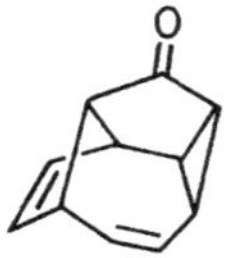 (200°C)	168)
		+ 	47	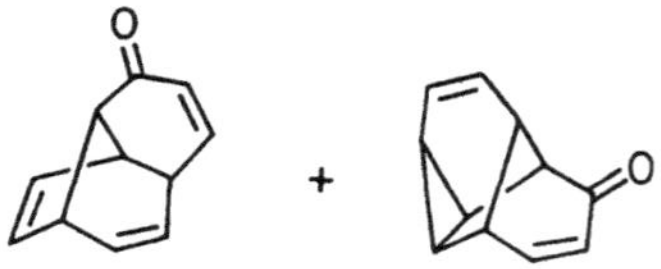 (160°C)	

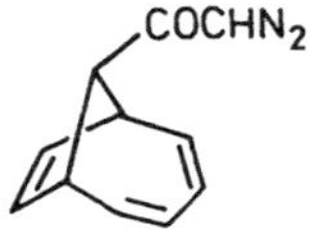	Cu, Δ (no details)	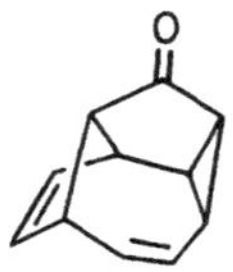	–		168)
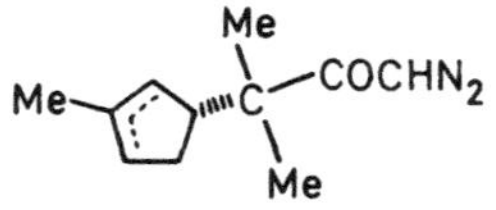	$CuSO_4$ Benzene/C_6H_{12}, reflux 12 mmol of diazoketone	 50 +  35		 (+) - homofenchone (two steps)	169)
	as above, but > 20 mmol of diazoketone	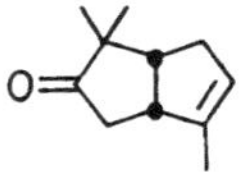	~50		
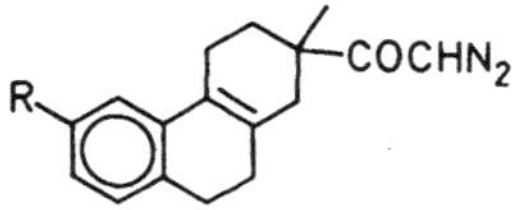	Activated CuO, *hν*, cyclohexane, reflux	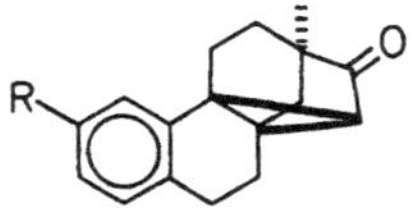	R = H : 62[b] R = Me : 80[b]	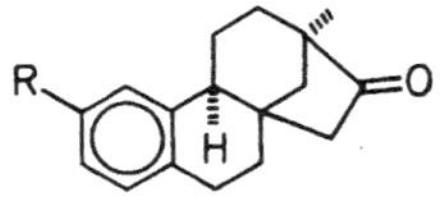 (Pd – C, EtOH, H_2)	170)
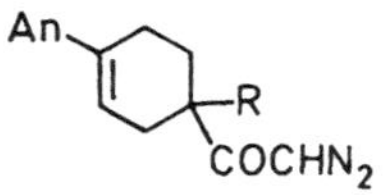 An = 4 - MeO - C_6H_4	$Ni(acac)_2$, *hν*, C_6H_{12}/THF, reflux	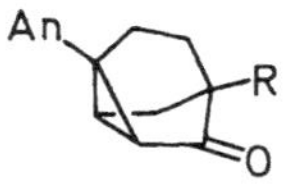	R = H : 85 R = Me : >72		171)

Diazo compound	Conditions	Product	Yield [%]	Further transformation of cyclopropane to a	Ref.
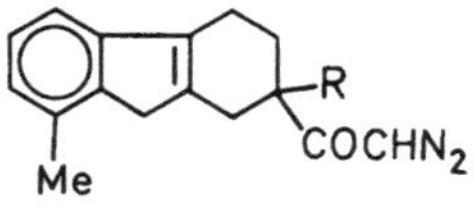	as above	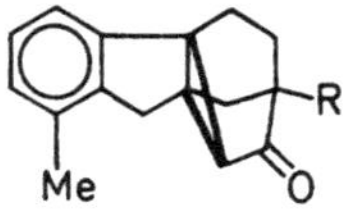	R = H : 59[b] R = Me : 63[b]	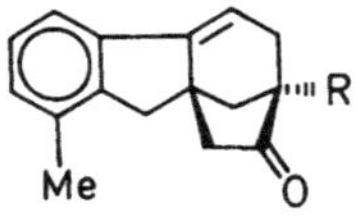 R = Me (±)-gibberone (HCl(g), $CHCl_3$)	172)
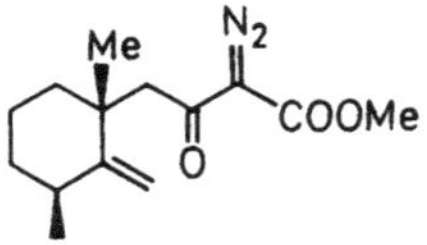	Anhydrous $CuSO_4$, cyclohexane, reflux	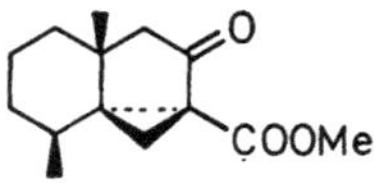	54	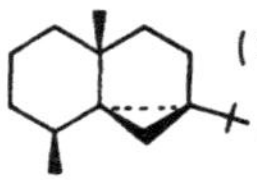 (±)-cycloeudesmol (several steps)	173)
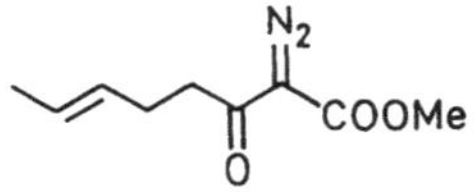	Copper bronze toluene, reflux	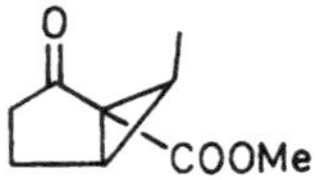	57		174)
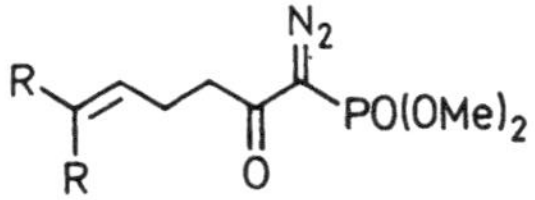	Copper cyclohexane, reflux[c]	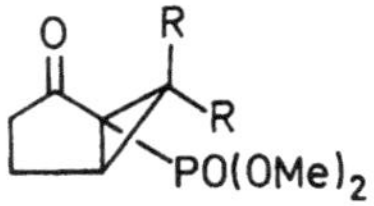	R = H : 76 R = Me : 49	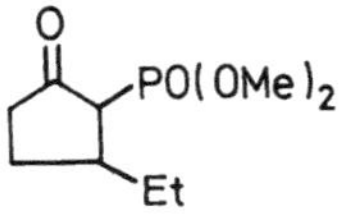 (Me_2CuLi; from R = H)	175)
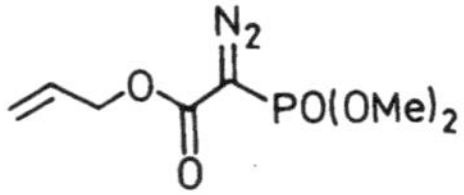	as above		67		175)

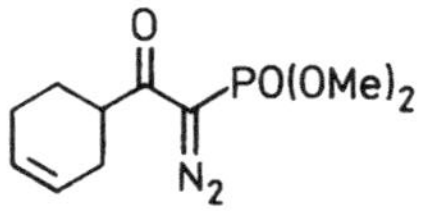	as above	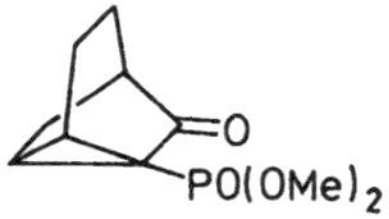	63		175)
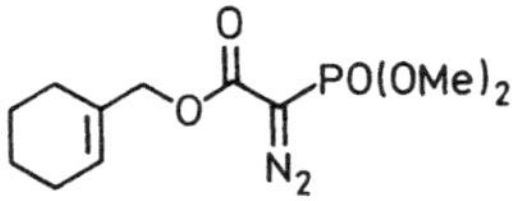	as above	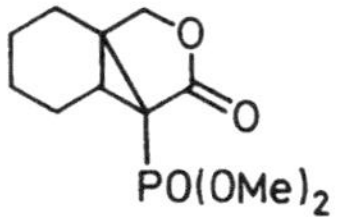	51		175)
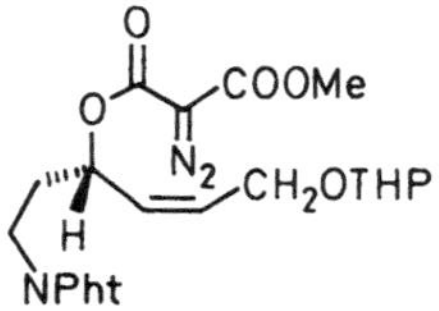 (THP = tetrahydropyranyl, Pht = Phthalimido)	Copper bronze, 110°C	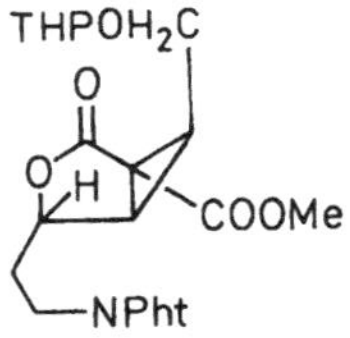	48	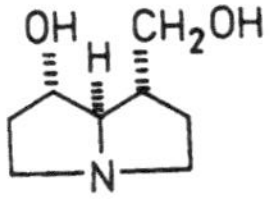 (several steps) (±)-hastanecine	156)
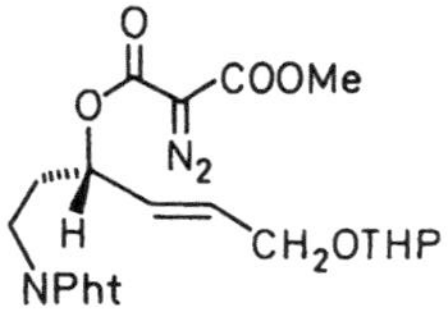	Copper bronze, 110°C	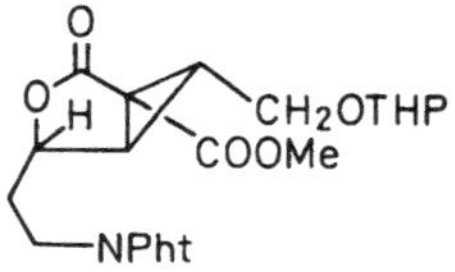	44 (+ traces of C-4 epimer)	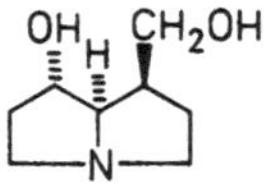 (several steps) (±)-dihydroxy-heliotridane	176)

Diazo compound	Conditions	Product	Yield [%]	Further transformation of cyclopropane to [a]	Ref.
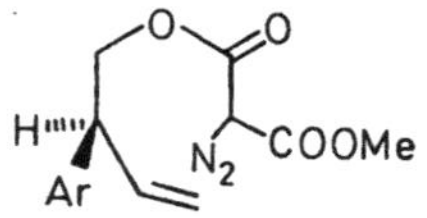	Cu(acac)$_2$, chlorobenzene, 125°C	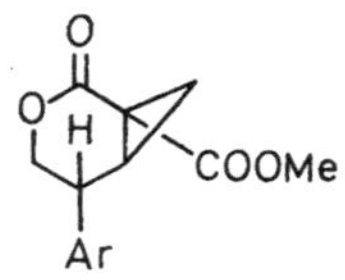	35	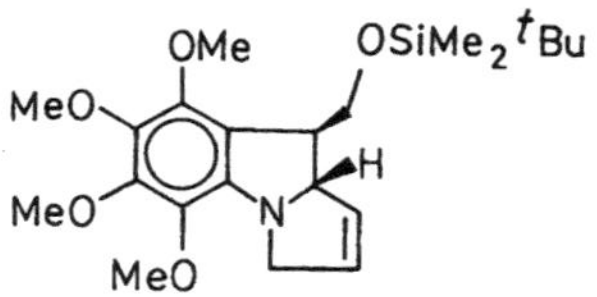 (several steps)	[177]
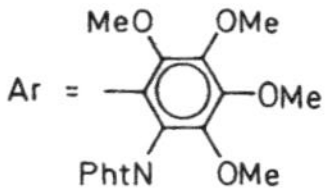		+ 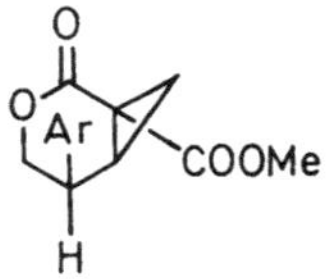	7		
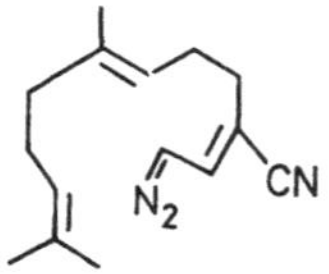	CuI, THF, r.t.	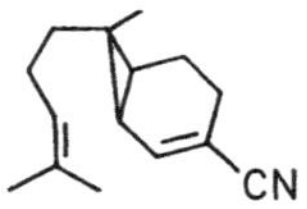		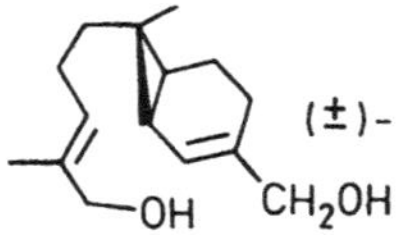 (±)- sirenin	[178]

[a] Conditions for transformations are given in parentheses.
[b] With Ni(acac)$_2$ instead of CuO as catalyst, yields of 72–90 % were reported, but not specified for individual compounds.

Intramolecular cyclopropanation of [(7 H-benzocycloheptene-7-yl)diazomethyl]-diphenylphosphinoxide **172** represents the first synthesis of a benzo-annulated octavalene (**173**)[180]. Not unexpectedly, carbenic rearrangement products **174** and **175** are also formed.

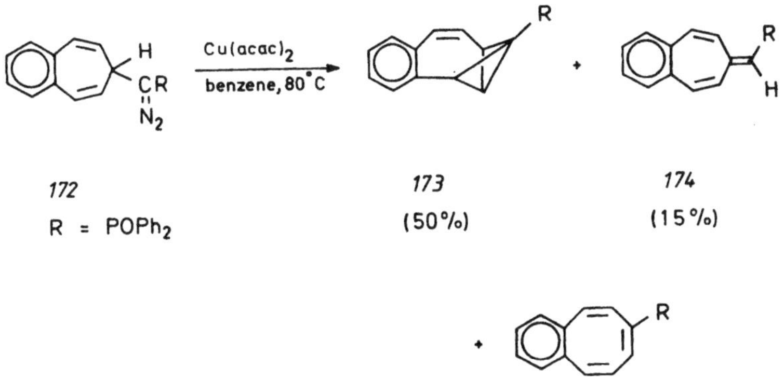

172 R = POPh₂

173 (50%)

174 (15%)

175 (not isolated purely)

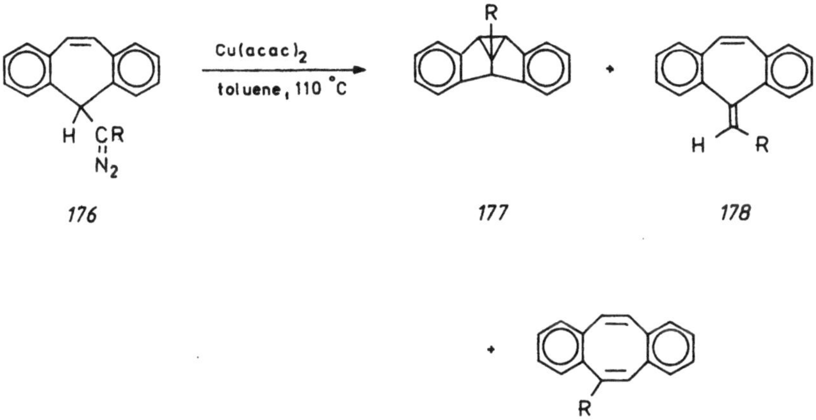

176

177

178

179

176–179	R	% 177	% 178	% 179
a	$POPh_2$	5	—	76
b	$PO(OMe)_2$	71	—	9
c	PO(Ph)Me	15	—	52
d	COOMe	~30	—	~46
e	COPh	~30	—	44
f	CO*t*-Bu	—	78	—

5-(Diazomethyl)-5*H*-dibenzo[a,d]cycloheptenes **176a–f** are somewhat more stable towards copper-catalyzed decomposition than **172**. In refluxing toluene, competitive cyclopropanation, yielding dibenzosemibullvalenes **177**, and ring enlargement to dibenzo[a,e]cyclooctenes **178** occur for **176a–e**, but exclusively dibenzoheptafulvene **178** is obtained from **179f** [180]. For the phosphoryl-substituted diazo compounds **176a–c**, the yields of cyclopropanation very inversely with the steric demand of the phosphoryl substituent, and an analogous correlation may be inferred from the results in the carbonyl series **176d–f**.

Intramolecular cyclopropanation of 1,3-dienic diazoketones and diazoesters (**180** → **181**) has been known for some years, but this transformation was revalorized only recently by demonstrating that the resulting bicyclo[n.1.0]-alkanones can be put to good use for further synthetic steps (Scheme 25): Regioselective nucleophilic ring-opening by thiophenolate paves the way into the prostaglandin area, with the possibility of stereoselective introduction of a hydroxyl group at C-15 by a [2,3]-sigmatropic allyl sulfoxide rearrangement. The keto function in **181** gives access to a 6-vinyl bicyclo[3.1.0]hex-2-ene → bicyclo[3.2.1]octa-2,6-diene rearrangement via the silyl enol ether. Finally, the vinylcyclopropane unit for **181** can undergo thermal ring enlargement to the cyclopentene, so that a two-step intramolecular cyclopentene annulation sequence starting with the diazo compounds is available. Examples of all these procedures are given in Table 11. All cyclopropanations occur stereo- and regiospecifically. By analogy with open-chain olefinic diazocarbonyl compounds, the reaction works best when a bicyclo[n.1.0]alkane with n = 3,4 can be constructed: 2-Oxo-5-vinyl-bicyclo[2.1.0]pentane was obtained only in trace amounts from the diazo precursor, while, on the other hand, 2-oxo-8-vinyl-bicyclo[5.1.0]octane resulted from 10-diazo-1,3-decadiene-9-one in unreproducible yields (20–91 %) [181].

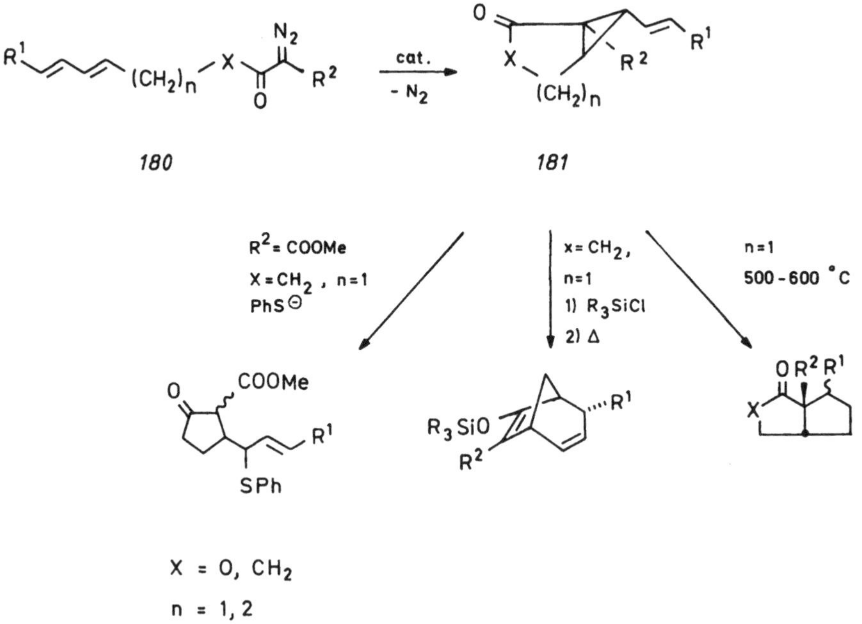

Scheme 25

It is notable that Table 11 contains examples of intramolecular cyclopropanation of an acrylate. It was found that $Cu(acac)_2$ was not an efficient catalyst for this transformation; cosiderable improvement was achieved by using catalytic amounts of $Cu(acac)_2$ and excess $CuSO_4$ [186]. A similar observation was made with (2,4-pentadien-1-yl)diazoacetates or diazomalonates [191].

The dominant role of the traditional copper catalysts, generally used under heterogeneous conditions, has not been challenged as yet. Only a few reports shed light on the efficiency of alternative catalysts. Copper(II) triflate allows high-yield intramolecular cyclopropanation of γ,δ-unsaturated diazoketone **182** [160]; it is superior to $CuSO_4$ (53% yield [192]) or $Rh_2(OAc)_4$ [160]. The solvent is crucial for an efficient conversion: If the reaction is carried out in ether, the solvent competes with the double bond for the electrophilic metal carbene to give **184**, presumably via an oxonium ylide intermediate.

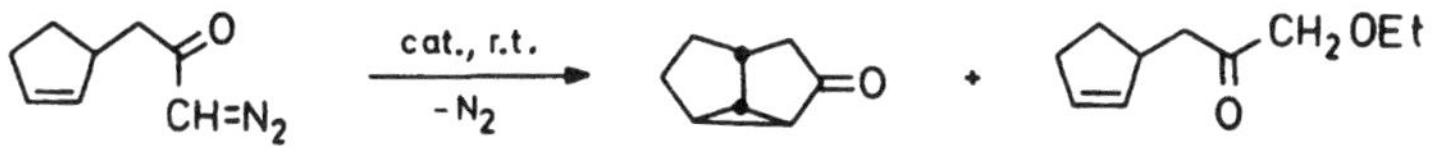

182 *183* *184*

catalyst	solvent	% *183*	% *184*
$Cu(OTf)_2$	CH_3NO_2	90	—
	Et_2O	48	10
$Rh_2(OAc)_4$	CH_3NO_2	36	—
	Et_2O	24	18

Products of a so-called vinylogous Wolff rearrangement (see Sect. 9) rather than products of intramolecular cyclopropanation are generally obtained from β,γ-unsaturated diazoketones [193], the formation of tricyclo[2.1.0.$0^{2,5}$]pentan-3-ones from 2-diazo-1-(cyclopropene-3-yl)-1-ethanones being a notable exception (see Table 10 and reference [12]). The use of $Cu(OTf)_2$ does not change this situation for diazoketone **185** in the presence of an alcohol [193]. With $Cu(OTf)_2$ in nitromethane, on the other hand, Δ^3-hydrinden-2-one **186** is formed [160]. As **186** also results from the $BF_3 \cdot Et_2O$-catalyzed reaction in similar yield, proton catalysis in the $Cu(OTf)_2$-catalyzed reaction cannot be excluded, but electrophilic attack of the metal carbene on the double bond (Scheme 26) is also possible. That $Rh_2(OAc)_4$ is less efficient for the production of **186**, would support the latter explanation, as the rhodium carbenes rank as less electrophilic than copper carbenes.
Nickel(II) acetylacetonate has been recommended as a very efficient homogeneous catalyst for intramolecular cyclopropanations for unsaturated diazoketones [171]. The yields were better than with activated CuO as catalyst (see Table 10 for examples). The authors of this study seem to combine routinely thermocatalytic with photochemical (tungsten lamp) decomposition of the diazoketones. The benefit of this procedure (higher yields, shorter reaction times) has been communicated in the CuO case, but not for the $Ni(acac)_2$-catalyzed reaction.

The choice of the catalyst is crucial when it comes to competition between intramolecular cyclopropanation and intramolecular carbonyl ylide formation by a

Table 11. Intramolecular ketocarbenoid addition to 1,3-dienes

Diazo compound	Conditions	Product	Yield [%]	Further transformation of cyclopropane to a	Ref.
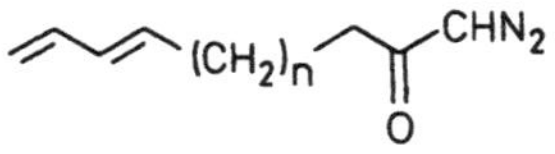	$Cu(acac)_2$, benzene, r.t. or reflux		n = 1 : 95 n = 2 : 95 n = 3 : 20 - 91	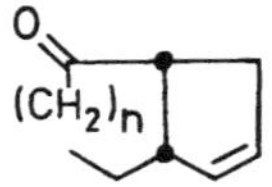 (500 - 600°C)	181, 182)
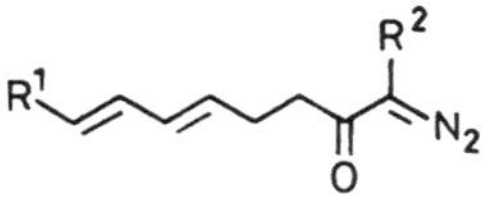	Cu(acac)$_2$, benzene, reflux	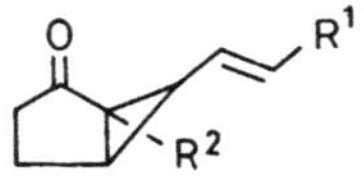	R^1 = Me, R^2 = H : 82 R^1 = Me, R^2 = Me : 75 R^1 = H, R^2 = Me : 94	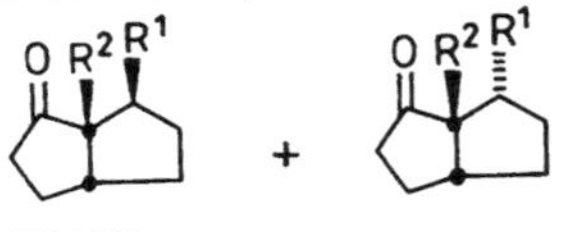 (600°C)	182)
				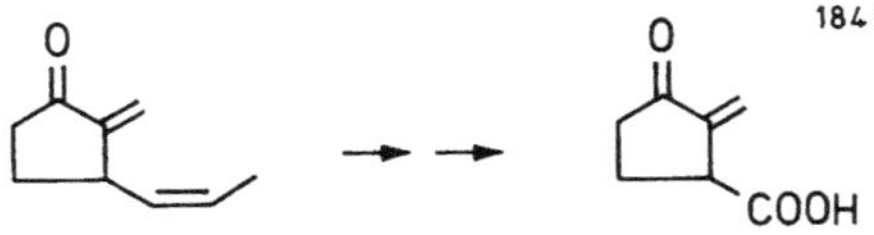 (from cyclopropane with R^1 = H, R^2 = Me at 450°C) sarkomycin	184)
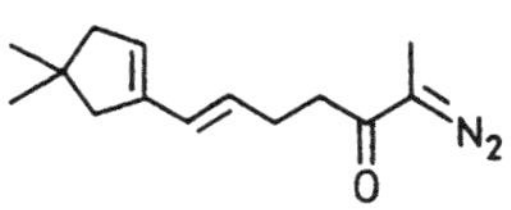	as above	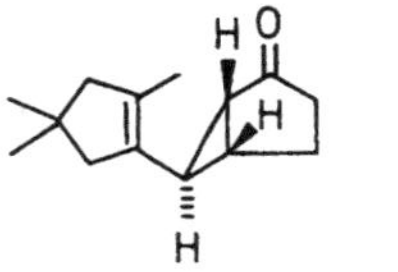	94	 (580°C) Hirsutene	182, 183)
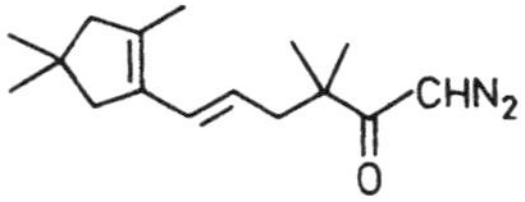	$Cu(acac)_2$ / $CuSO_4$ benzene, reflux	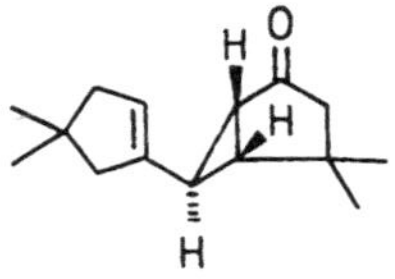	96	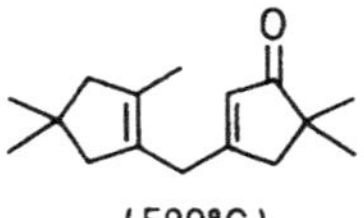 (+ others) (580°C)	185)

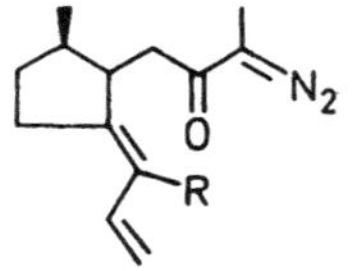	$Cu(acac)_2$ (10 mol%) / $CuSO_4$ (200 mol%), benzene, reflux	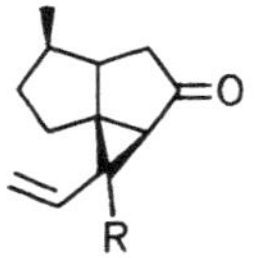 (2 diastereomers)	75	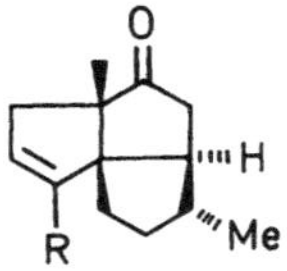 (550°C)	186)
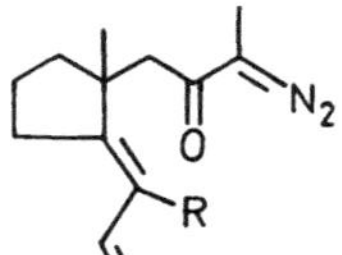 R=COOEt	as above	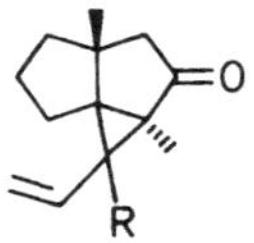 (2 diastereomers)	61	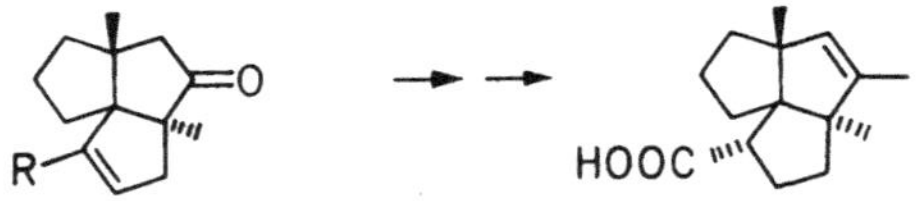 (580°C) Isocomenic acid	187)
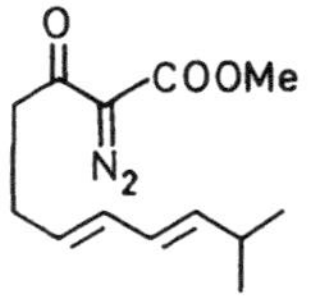	$Cu(acac)_2$, benzene, reflux	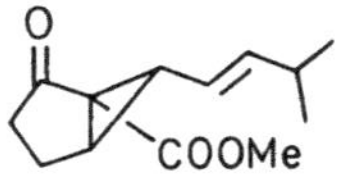	78	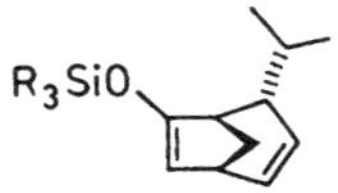	188)
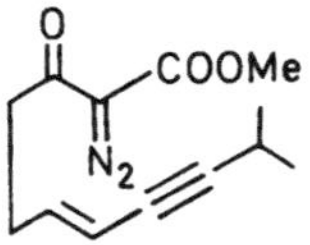	as above	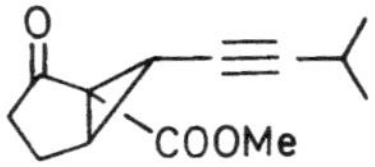	77	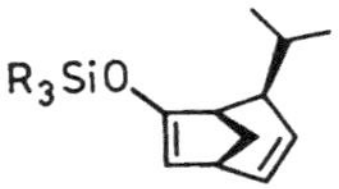	188)
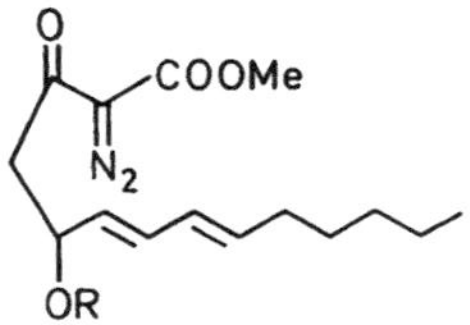 R = CH(Me)COOEt	as above	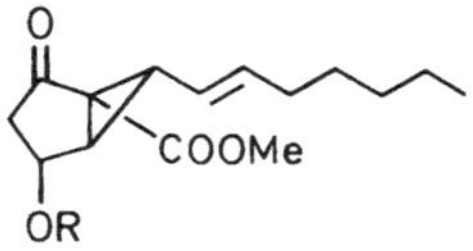 (both C4 - epimers)	61	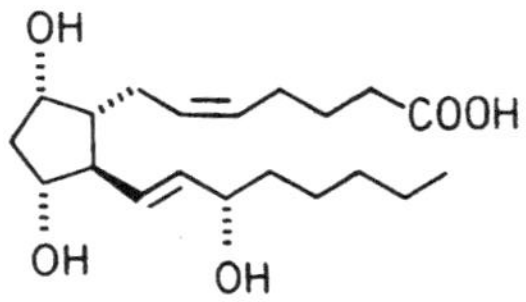 Prostaglandin F_2	189)

Diazo compound	Conditions	Product	Yield [%]	Further transformation of cyclopropane to [a]	Ref.
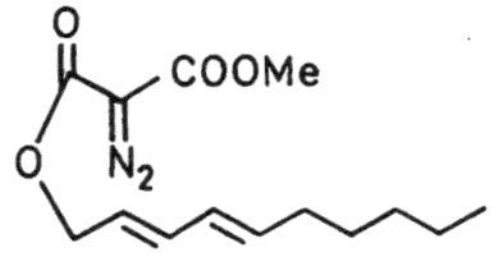	Cu powder, toluene, reflux	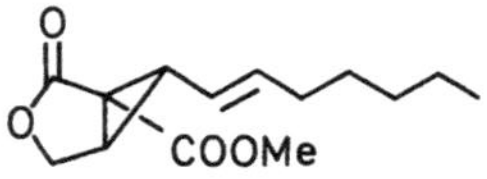	70	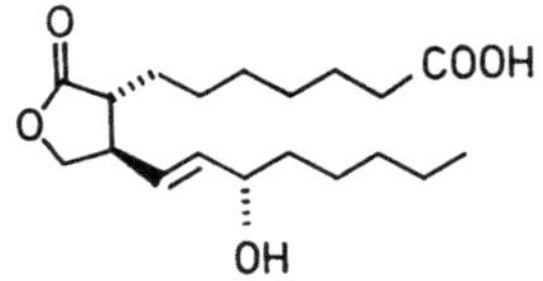 10-oxa-11-deoxy-PGE_1	190)
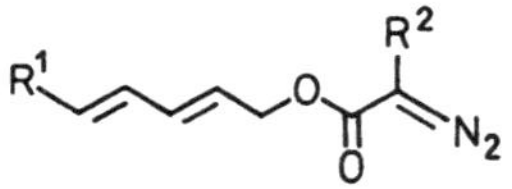	$Cu(acac)_2$ / $CuSO_4$, benzene, reflux	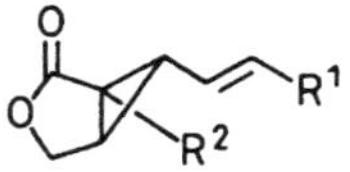	R^1 = H, R^2 = Me: 63 R^1 = Me, R^2 = H: 68	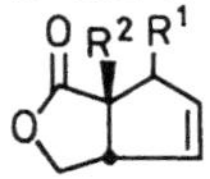 (620 °C)	191)
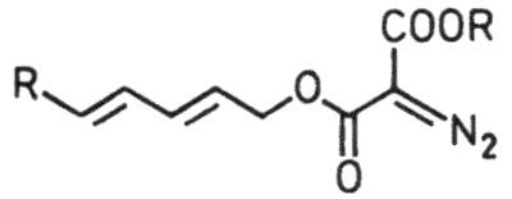	as above	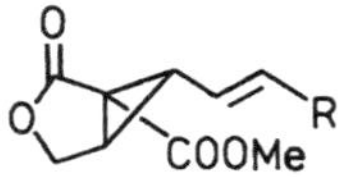	R^1 = H: 46 R^1 = Me: 84	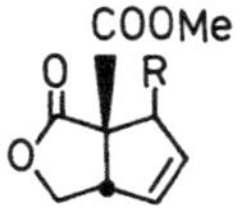 (580 °C)	191)

[a] Temperatures in parentheses refer to thermal vinylcyclopropane → cyclopentene rearrangement.

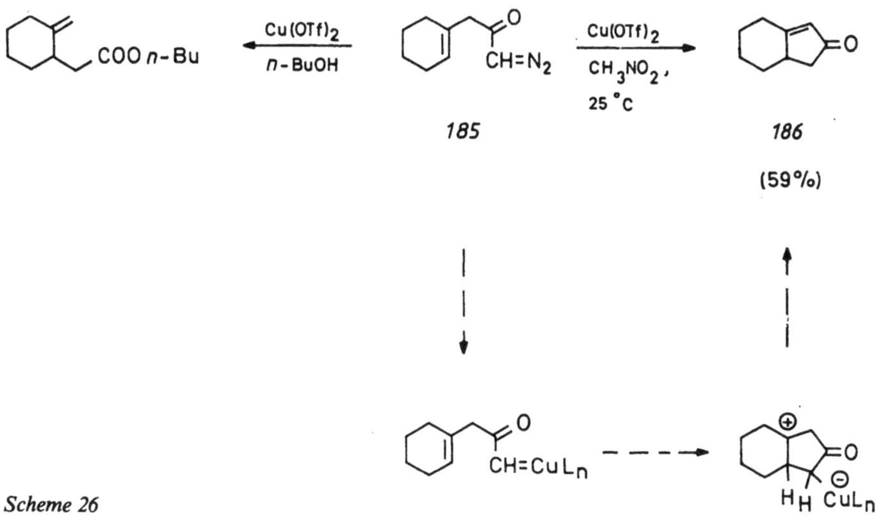

Scheme 26

ketocarbenoid. From diazoketone **187**, the cyclopropane **188** is formed with high selectivity in the presence of palladium-based catalysts, whereas $CuCl \cdot P(OMe)_3$ and $Rh_2(OAc)_4$ promote the carbonyl-ylide pathway leading to **189** with opposite selectivity [194]. The result of thermal decomposition of **187** resembles that obtained with the latter two catalysts. The different activity of the given catalysts is in line with the ideas developed to explain their different efficiency in olefin cyclopropanation (see Sect. 2.2), according to which metal-to-olefin coordination is the decisive factor in the palladium-promoted cyclopropanation. Doyle's alternative explanation of catalyst-dependent regioselectivity in cyclopropanation reactions, considering charge separation in the transition states as the differentiating factor (see Sect. 2.2.2), is

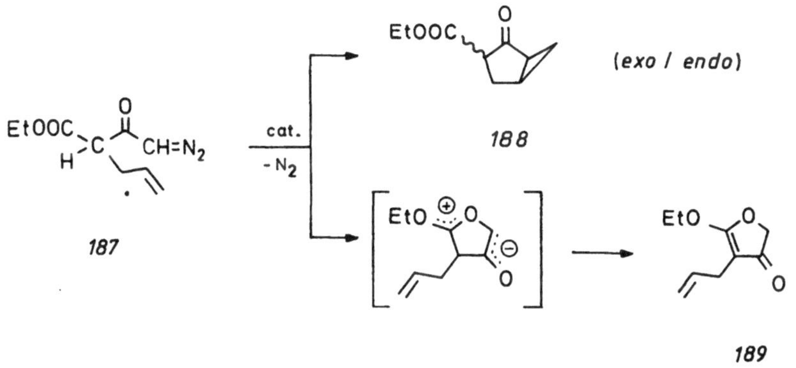

catalyst	temp. [°C]	% *188*	% *189*
$[(\eta^3-C_3H_5)PdCl]_2$	25	53	3
$Pd(OAc)_2$	25	53	<1
$CuI \cdot P(OMe)_3$	25	3	35
$Rh_2(OAc)_4$	25	1	58
none	80	15	54

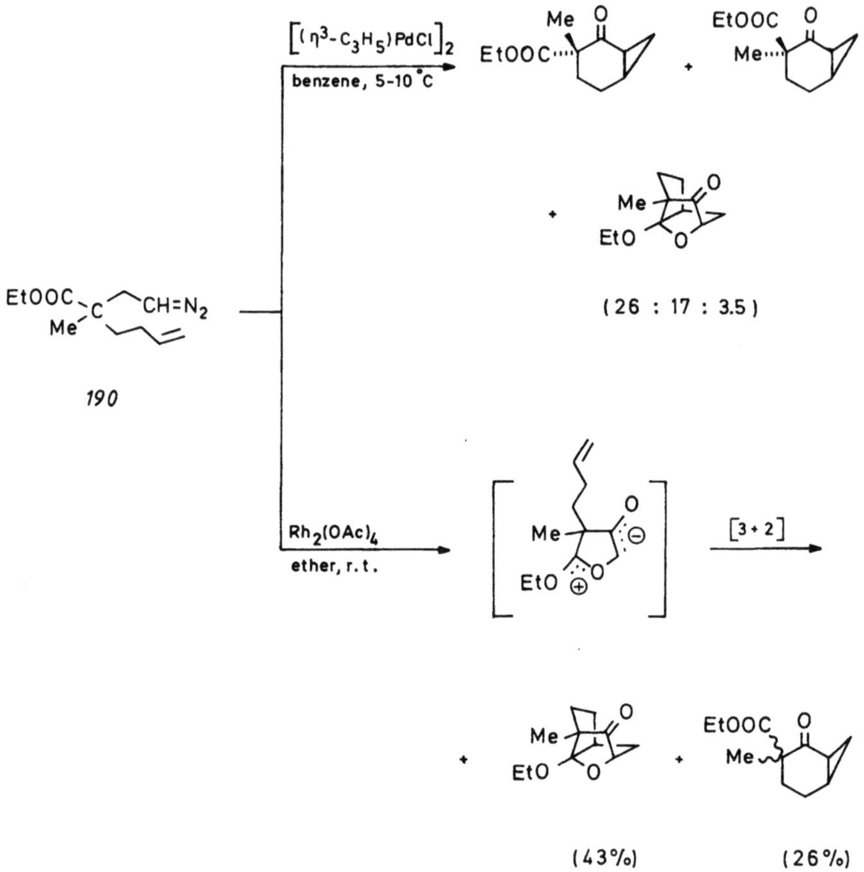
[(η3-C3H5)PdCl]2
benzene, 5-10 °C
Me O
EtOOC
EtOOC O
Me
Me
EtO
O
(26 : 17 : 3.5)
EtOOC
C
Me
CH=N2
Rh2(OAc)4
ether, r.t.
Me
O
EtO
[3+2]
Me
EtO
O
EtOOC O
Me
(43%)
(26%)

190

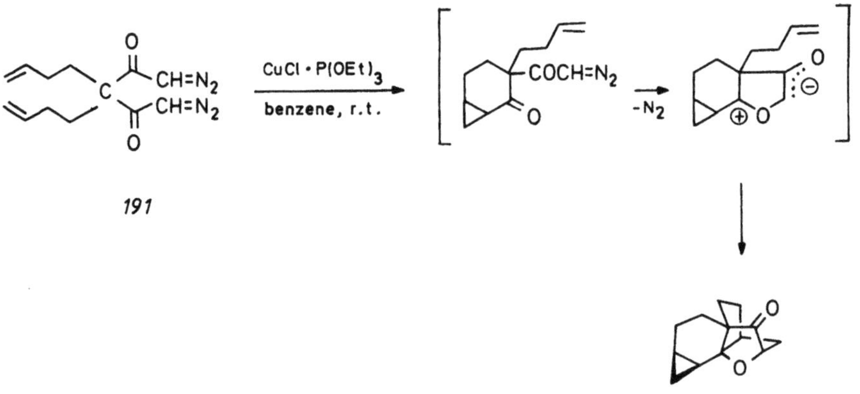
O
CH=N2
C
CH=N2
O
CuCl · P(OEt)3
benzene, r.t.
COCH=N2
O
-N2
O
O

191

192

applicable to the selective formation of **188** and **189** also: Palladium carbenes tend to be involved in transition states with little charge separation (cyclopropanation) whereas the opposite is true for rhodium carbenes which therefore favor the carbonyl ylide pathway.

The distinction between Pd and Rh catalysts was also verified for diazoketone **190**. In this case, the carbonyl ylide was trapped by intramolecular [3+2] cycloaddition to the C=C bond [195]. Decomposition of bis-diazoketone **191** in the presence of CuCl · $P(OEt)_3$ or $Rh_2(OAc)_4$ led to the pentacyclic ketone **192**; most remarkably, one diazoketone unit reacted by cyclopropanation, the second one by carbonyl ylide formation [194]. With $[(\eta^3\text{-}C_3H_5)PdCl]_2$, a non-separable mixture containing mostly polymers was obtained, although bis-diazoketones with one or two allyl side chains instead of the butenyl groups underwent successful twofold cyclopropanation [196].

2.5 Optical Induction in Cyclopropanation Reactions

Enantioselective carbenoid cyclopropanation can be expected to occur when either an olefin bearing a chiral substituent, or such a diazo compound or a chiral catalyst is present. Only the latter alternative has been widely applied in practice. All efficient chiral catalysts which are known at present are copper or cobalt(II) chelates, whereas palladium complexes [86] proved to be uneffective. The carbenoid reactions between alkyl diazoacetates and styrene or 1,1-diphenylethylene (Scheme 27) are usually chosen to test the efficiency of a chiral catalyst. As will be seen in the following, the extent to which optical induction is brought about by enantioselection either at a prochiral olefin or at a prochiral carbenoid center, varies widely with the chiral catalyst used.

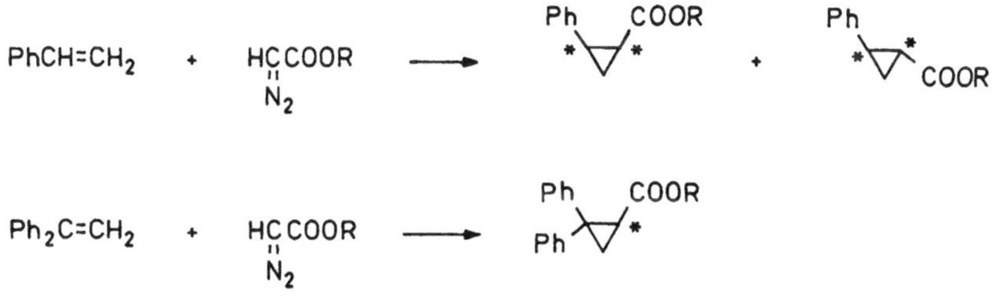

Scheme 27

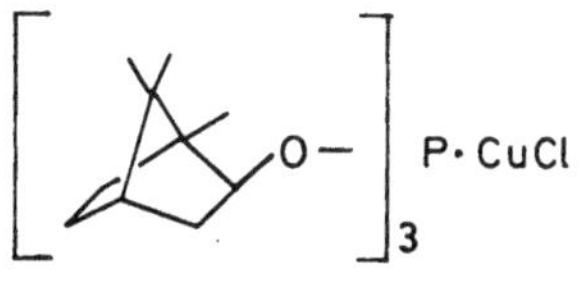

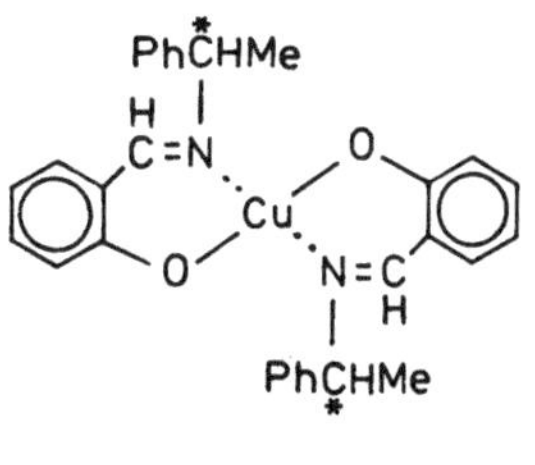

193 *194*

bornylphosphite)copper(I) chloride **193** [58] or the chiral copper chelates (*R*)-**194** and (*S*)-**194** [50], but the optical yields of 2-phenylcyclopropanecarboxylates were rather discouraging (3% with **193**, 6% with **194**).

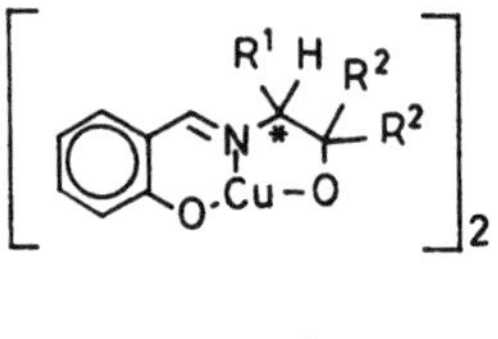

195

	R¹	R²
195a (=*38*)	Me	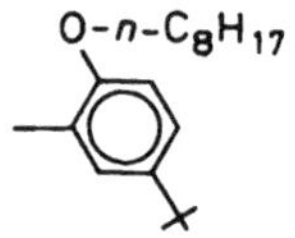
195b	Me	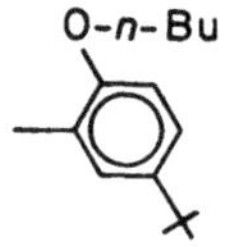
195c	PhCH₂	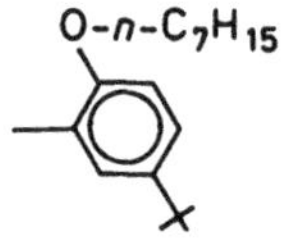
195d	Me	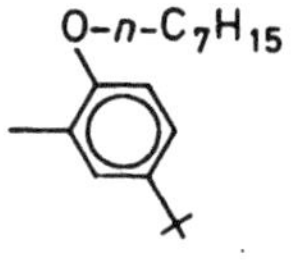

Table 12. Asymmetric cyclopropanation by **195a**–catalyzed decomposition of *l*-methyl diazoacetate in olefins[a, b]

olefin	catalyst configuration[c]	cyclopropane		
		cis/trans	% e.e.[d]	
			cis	*trans*
styrene	*R*	14/86	(+)54	(+)69
	S	18/82	(−)78	(−)81
1-octene	*R*	17/83	46	(+)76[e]
	S	22/78	64	(−)84[e]
trans-octene	*R*		(+)82	
	S		(−)84	
trans-anethole	*R*	9/91	44	(+)81
	S	12/88	60	(−)89
1,1-diphenylethylene	*R*		(+)75[e]	
	S		(−)64[e]	
α-methylstyrene	*R*	40/60	86	68
	S	36/64	68	58

[a] From Ref. [91]; [b] No experimental details given; [c] (*R*)-*195a* is dextrorotatory; [d] Determined by direct GC analysis of the *l*-menthyl cyclopropanecarboxylate if not stated otherwise; [e] Determined by GC analysis of the *d*-2-octyl cyclopropanecarboxylate.

Considerable improvement was achieved with binuclear copper(II) chelates of type **195** whose ligands are derived from salicylaldehyde and an optically active amino alcohol [91,92].

From a variety of differently substituted compounds, best results were obtained with the catalysts **195a–c**; in combination with *l*-methyl diazoacetate and monoolefins, cyclopropanes were obtained with a relatively high *trans*/*cis* ratio and enantiomeric excesses of 44–89% (Table 12). The absolute configuration at the catalyst's chiral center determines the enantioselectivity for both diastereoisomers.

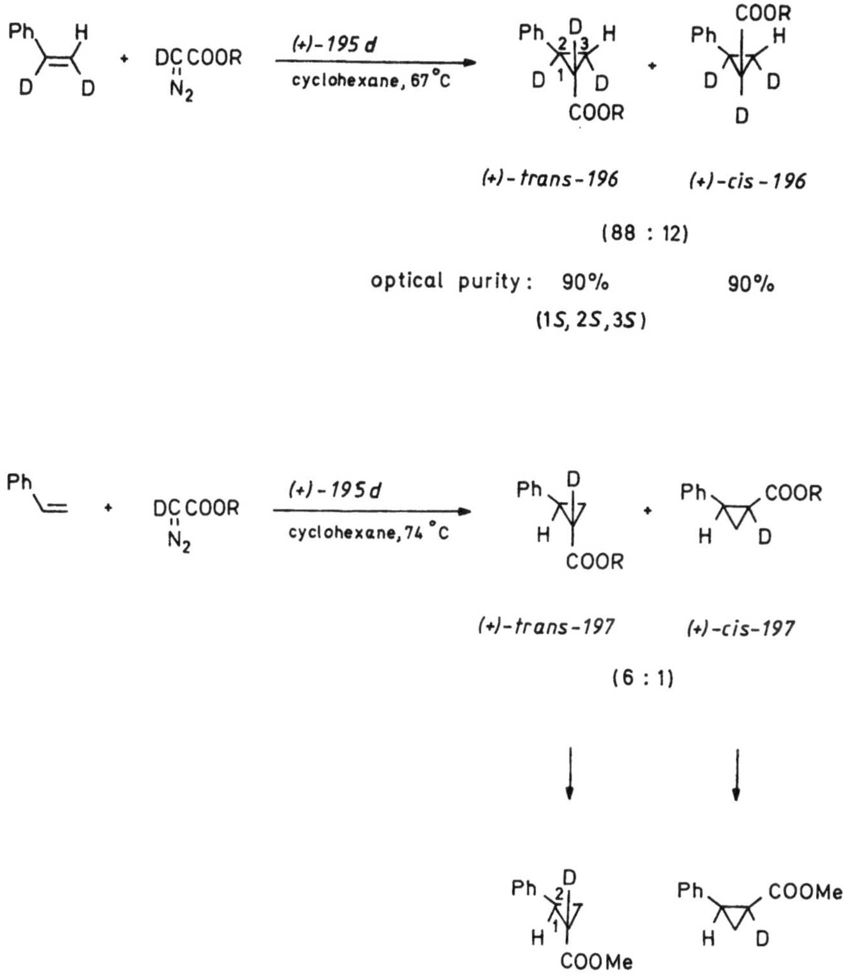

R = *dl*-menthyl

(+)-*trans*-198 (1*S*,2*S*) (+)-*cis*-198 (1*S*,2*R*)

Scheme 28 optical purity: 89% 86%

Baldwin et al. have used the same catalyst/diazo ester combination for the synthesis of optically active deuterated phenylcyclopropanes (Scheme 28) [197]. From *cis*-1,2-dideuteriostyrene, *dl*-menthyl α-deuteriodiazoacetate and (+)-**195d**, the *cis*- and *trans*-cyclopropanes **196** were obtained, both with 90% optical purity. The dominant enantiomer of *trans*-**196** had (+)-(1*S*, 2*S*, 3*S*) configuration. Analogously, the cyclopropanes *cis*-**198** and *trans*-**198**, obtained from styrene, *dl*-menthyl α-deuteriodiazoacetate and (+)-**195d** with subsequent transesterification of *cis*/*trans*-**197**, had optical purities of 86 and 89%, respectively. The major optical isomer of *cis*-**198** had (1*S*, 2*R*) configuration, that of *trans*-**198** (1*S*, 2*S*) configuration.

Another systematic search for chiral copper chelates, tested for the cyclopropanation of 1,1-diphenylethylene with ethyl diazoacetate, has met with limited success, as optical yields were generally low [198]. 37 optically active azomethine-type ligands were combined with copper(II) acetate to give in situ catalysts. As has been tested for the $Cu(OAc)_2$/**199** system, the in-situ method is equivalent to that of using the isolated catalyst, as far as product yield and optical induction are concerned. The former procedure is characterized, however, by a small but significant temperature dependency of the optical yield (65 °C: 10.8%; 35 °C: 13.4%). An excess of ligand does not influence the optical induction in the range 1.2:1 to 22:1. From the different groups of ligands (Schiff bases derived from (*S*)-(—)-1-phenylethylamine, ethylenediamines and binaphthyl-o,o'-diamines, α-amino alcohols, amino acids and esters), compounds **199**–**203** gave the highest optical yields in their respective group (% e.e. of a given configuration in ethyl 2,2-diphenylcyclopropanecarboxylate: **199**, 11% (*S*); **200**, 24% (*S*); **201**, 13% (*R*); **202a**, 52% (*S*); **202b**, 45% (*S*); **202c**, 66% (*S*); **202d**, 38% (*S*); **203**, 10% (*S*). The best results are thus achieved with ligands

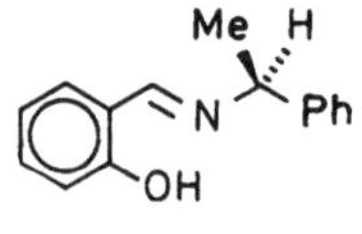

199

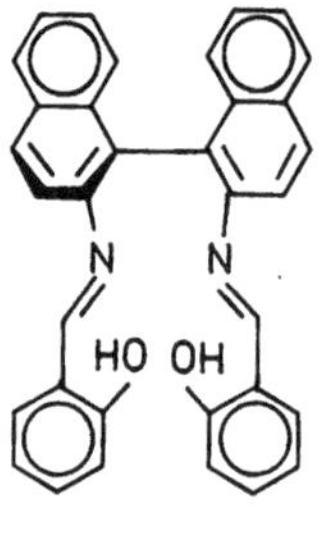

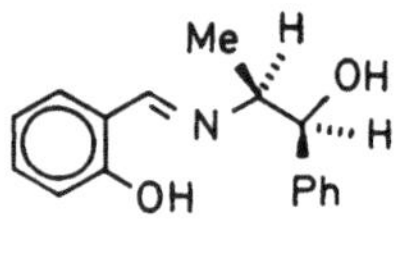

201

200

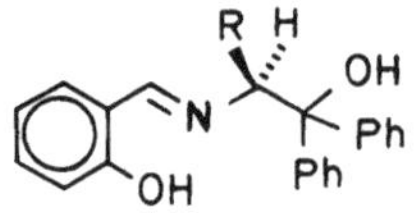

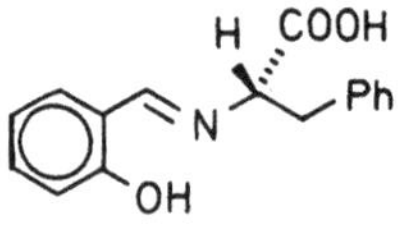

202

203

202	a	b	c	d
R	Me	$CHMe_2$	CH_2Ph	CH_2OH

202a–d, which correspond to Aratani's catalysts **195**. The efficiency of a chiral chelate catalyst is certainly connected to the coordinating ability of the ligands and to the coordination geometry. This may explain, for example, the failure to observe optical induction if CH= in **199** is replaced by C(Me)=. It is known that this simple change of substituent leads to another preferred conformation of the phenylethyl group, and this altered geometrical situation may seriously affect the coordinating ability of the ligand to the copper atom.

Easily available copper(II) tartrate has also been used for an enantioselective cyclopropanation. From 3-methoxystyrene and 4-bromo-1-diazo-2-butanone, the cyclopropanes *cis*/*trans*-**204** were obtained; the mainly formed *trans*-isomer displayed an enantiomeric excess of 46% [199]. This reaction constituted the opening step of asymmetric total syntheses of equilenin and estrone.

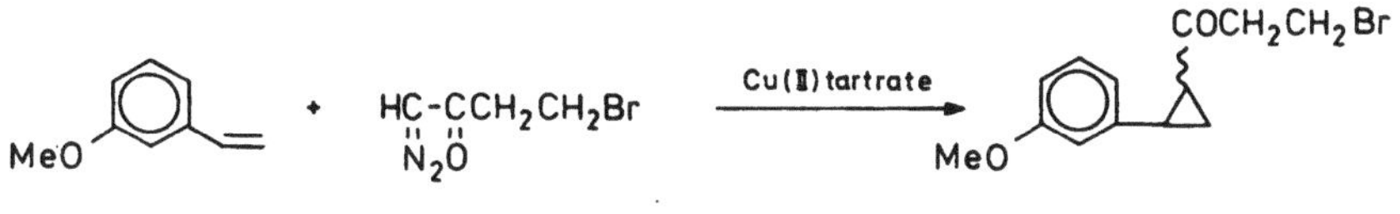

204

trans : *cis* = 12 : 1

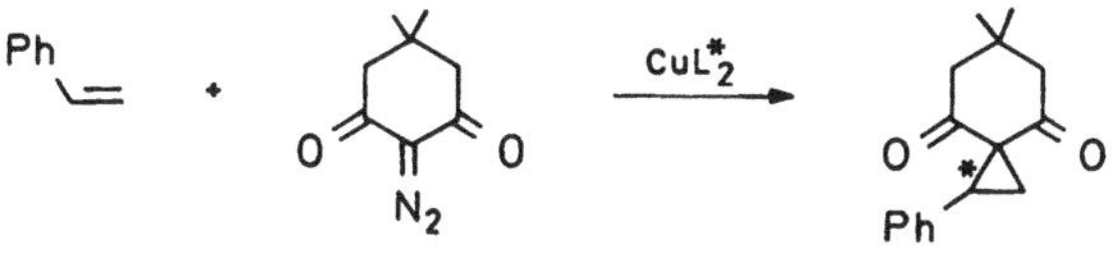

206

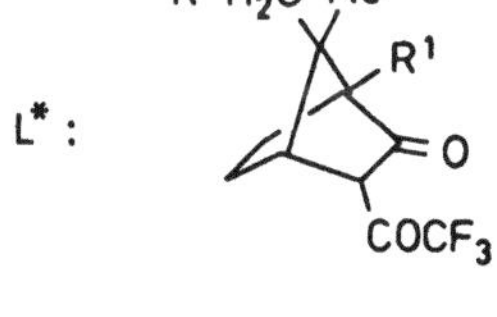

205

205	R^1	R^2	% *206*	% e.e. of *206*
a	Me	H	36	92
b	Me	I	21	73
c	$CH=CH_2$	H	48	100
d	$-(CH_2)_2-Si(MeO)(OMe)-OSi-H$	H	43	98

Copper chelates in which the ligands are rigid chiral β-diketonates of type **205** are responsible for the highest optical yields known in carbenoid cyclopropanation reactions [200]. The cyclopropane **206** was even obtained enantiomerically pure from 2-diazodimedone and styrene in the presence of CuL_2^* (L* = **205c**).

Even when the trifluoroacetyl-(+)-camphor ligand is linked to a solid support (Hypersil silica; **205d**), if retains its activity both in terms of yield and optical induction.

Nakamura and Otsuka have introduced chiral bis(1,2-dioximato)cobalt(II) complexes **207a–c**, which are conveniently available when starting from naturally occurring *d*-camphor or (—)-β-pinene, respectively [88,95,201].

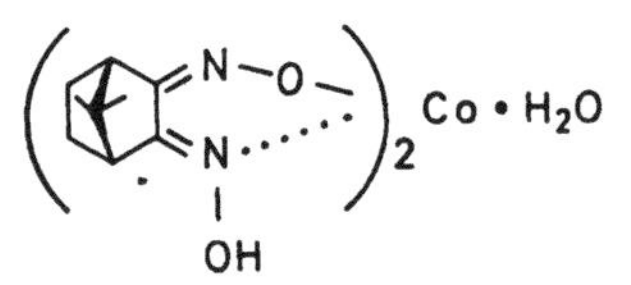

207a

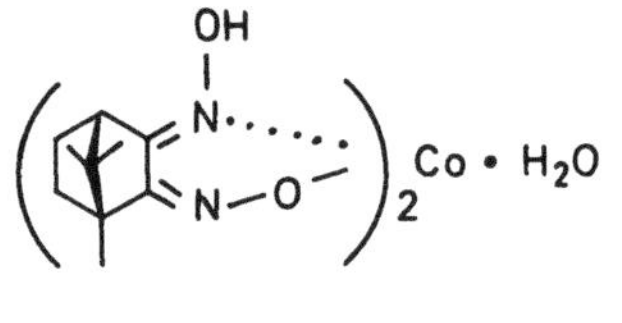

207b

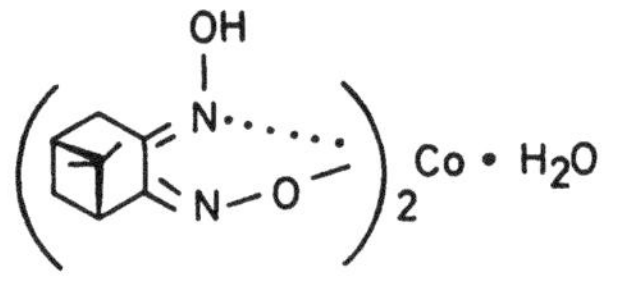

207c

In the presence of catalytic amounts of **207a** and at moderate temperatures (—15 to +30 °C), the cyclopropanes derived from styrene and various alkyl diazoacetates were obtained in good yields (80–95%) with remarkably high enantiomeric excess for both the *cis*(1*S*, 2*R*) and the *trans*(1*S*, 2*S*) isomer. With increasing steric bulk of the ester substituent (methyl → neopentyl), both the *trans*/*cis* ratio (0.69 → 2.34) and the optical yield (61 → 88% for the *trans*-cyclopropane at 0 °C) became higher [88,95].

Ethyl 2-phenylcyclopropanecarboxylate, obtained in the presence of **207a**, has *S* configuration at C-1 in both the *cis*- and *trans*-isomer. As that carbon has been furnished by the diazo ester, this result indicates enantiofacial selection at the carbenoid. In contrast, hardly any discrimination between the enantiofaces of the prochiral olefin occurs. Only when the ester substitutents become bulkier, does this additional stereochemical feature gain importance, and the *S* configuration at C-2 of the cyclopropane is favored.

As Table 13 shows, the opposite enantiomers of both *cis*- and *trans*-cyclopropanes are accessible if **207b** or **207c** are used instead of **207a** [88]. An analogous result was found for cyclopropanation of 1,1-diphenylethylene with ethyl diazoacetate (Table 14).

These facts are explained by assuming that **207b** and **207c**, which are equal within their first coordination sphere, are quasienantiomeric to **207a**. The reverse Cotton

Table 13. Asymmetric synthesis of ethyl 2-phenylcyclopropanecarboxylate with various Co(II) catalysts (0 °C, neat styrene, 3 mol % of catalyst)[a]

catalyst	yield [%]	*trans*/*cis*	*cis*-isomer		*trans*-isomer	
			configuration	optical yield [%]	configuration	optical yield [%]
207a	92	0.85	1*S*,2*R*	66	1*S*,2*S*	75
207b	62	0.65	1*R*,2*S*	63	1*R*,2*R*	64
207c	83	0.91	1*R*,2*S*	21	1*R*,2*R*	74

[a] From Ref. [88]; [b] Determined from the specific rotation of the acid.

Table 14. Asymmetric synthesis of ethyl 2,2-diphenylcyclopropanecarboxylate with various Co(II) catalysts (5 °C, neat styrene, 2.2–2.9 mol % of catalyst)[a]

catalyst	yield [%]	configuration	optical yield [%][b]
207a	95	1*S*	70
207b	77	1*R*	50
207c	87	1*R*	54

[a] From Ref. [88]; [b] Determined from the specific rotation of the acid.

effect with maxima of comparable intensities underlines this relationship. Structures **207aA** and **207bA**, in which the usual square-planar coordination around the metal is distorted towards a local C_{2v} geometry, represent a geometry which allows for enantiomeric first coordination spheres of both catalysts. From a practical point of view, the unimportance of the outer region of the ligand for stereochemical differentiation in asymmetric synthesis implies that for both **207a** and **207b**, readily available *d*-camphor can be chosen as starting material. It is by Co(II) complexation with two different geometric isomers of dioximes (*syn*/*anti* isomerism around C=N) that the quasienantiomeric chelates are constructed.

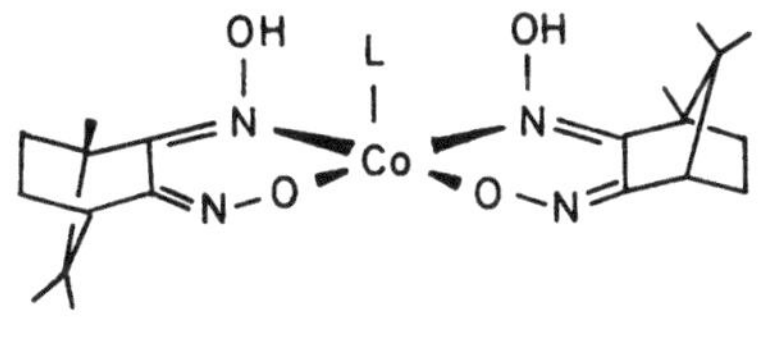

207aA

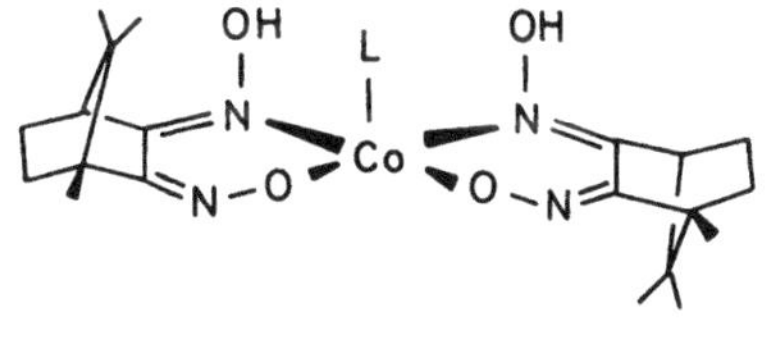

207bA

Some remarks concerning the scope of the cobalt chelate catalysts **207** seem appropriate. Terminal double bonds in conjugation with vinyl, aryl and alkoxycarbonyl groups are cyclopropanated selectively. No such reaction occurs with alkyl-substituted and cyclic olefins, cyclic and sterically hindered acyclic 1,3-dienes, vinyl ethers, allenes and phenylacetylene [95]. The cyclopropanation of electron-poor alkenes such as acrylonitrile and ethyl acrylate (optical yield in the presence of **207a**: 33%) with ethyl diazoacetate deserve notice, as these components usually undergo [3+2] cycloaddition to give pyrazolines and only a few catalysts are suited for the cyclopropanation process (see Sect. 2.3.2). In a mixture of styrene and methyl acrylate, only styrene is cyclopropanated, however [88]. Besides diazoacetates, diazomethane, ω-diazoacetophenone and dicyanodiazomethane have been used. Diazodiphenylmethane and diazofluorene were not decomposed even at 60 °C. Presumably, steric interactions between these sterically demanding diazo compounds and the cobalt chelate hinder a close approach of the two components.

It has already been mentioned that prochirality of the olefin is not necessary for successful enantioselective cyclopropanation with an alkyl diazoacetate in the presence of catalysts **207**. What happens if a prochiral olefin and a non-prochiral diazo compound are combined? Only one result provides an answer to date: The cyclopropane derived from styrene and dicyanodiazomethane shows only very low optical induction (4.6% e.e. of the (2*S*) enantiomer, catalyst **207a**) [95]. Thus, it can be concluded that with the cobalt chelate catalysts **207**, enantioface selectivity at the olefin is generally unimportant and that a prochiral diazo compound is needed for efficient optical induction. As the results with chiral copper 1,3-diketonates **205** and 2-diazodimedone show, such a statement can not be generalized, of course.

Preparation of chrysanthemic (**208**, $R^1=R^2=Me$) and permethric acid (**208**, $R^1=R^2=Cl$) derivatives is a very useful testing ground for enantioselective (and

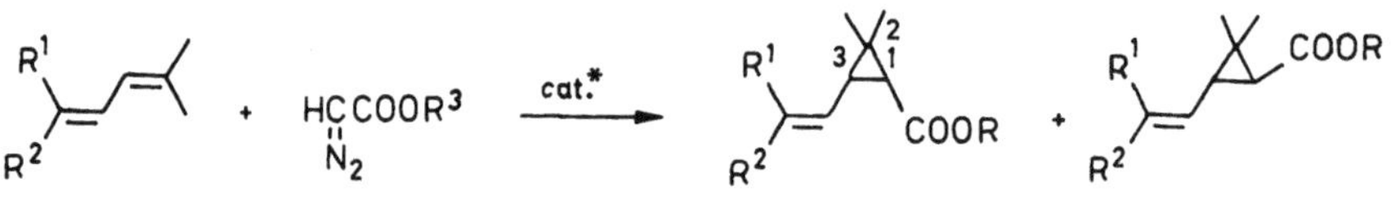

cis-*208* *trans*-*208*

cat*.:

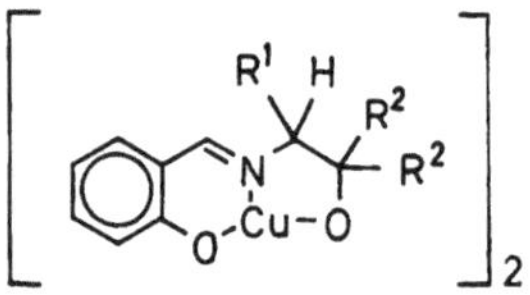

195a: R^1 = Me
R^2 = 5-*t*-butyl-2-octyloxyphenyl

195c: R^1 = $PhCH_2$
R^2 = 5-*t*-butyl-2-heptyloxyphenyl

195

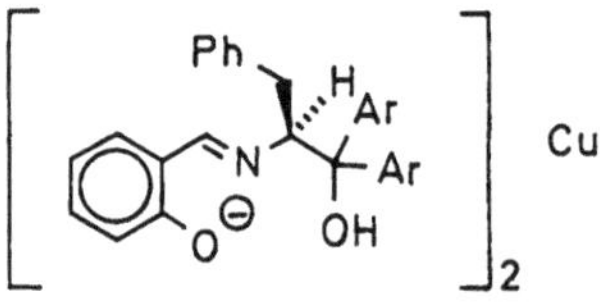

209

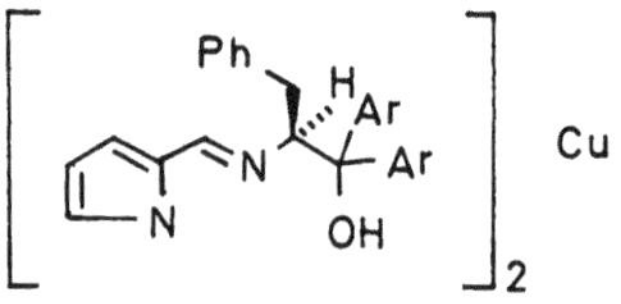

210

Ar = 2-anisyl

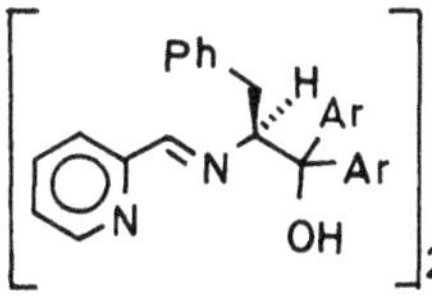

211

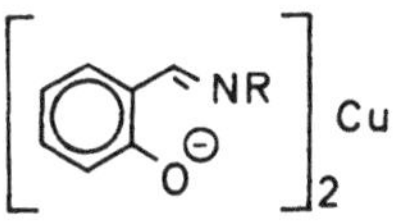

212

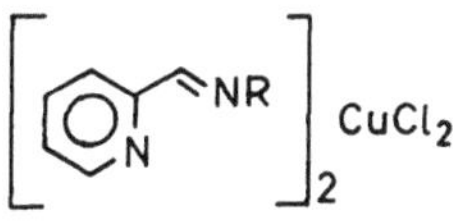

213

212a, 213a: =NR:

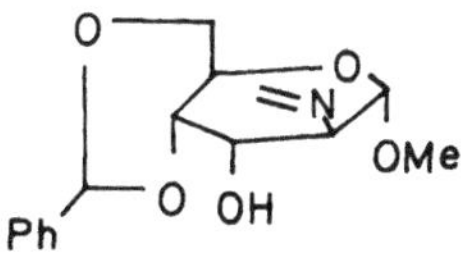

212b, 213b: =NR:

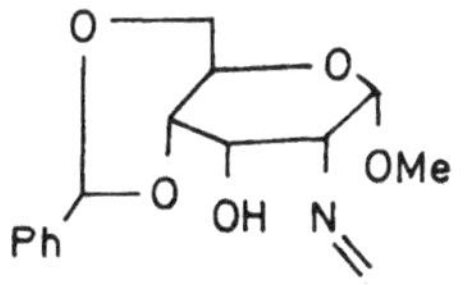

Scheme 29

stereoselective) cyclopropanation. Esters of these acids represent very powerful insecticides; as far as the cyclopropane ring is concerned, highest activity is found for the (1*R*) configuration [202, 203]. Several investigators have sought for an efficient solution to this problem by carbenoid reaction between the appropriate 1,3-diene and an alkyl diazoacetate in the presence of a chiral catalyst (Scheme 29).

The results are displayed in Table 15. Highest optical induction in the synthesis of alkyl chrysanthemates is furnished by the binuclear copper chelates **195**. (*R*)-configuration at the ligand's chiral center induced (1*R*)-selectivity in both the *cis*- and the *trans*-cyclopropane. Ligands derived from naturally occuring L-α-amino acids, thus having (*S*)-configuration, caused selectivity for the (1*S*)-enantiomer of both geometrical isomers. Chirality of the ester substituent of the diazo compound had no decisive influence on enantioselectivity. Only moderate optical induction was achieved upon cyclopropanation of 1,1-dichloro-4-methyl-1,3-pentadiene; with catalyst **210**, despite its (*S*)-configurated ligand, a small preference for the (1*R*)-*trans*-cyclopropane resulted. With Schiff bases derived from amino sugars as catalyst ligands [210], the direction of optical induction at C-1 of the cyclopropane was mainly determined by the configuration at C-2 of the sugar: (*S*)-configuration as in the 2-amino-D-altropyranoside-derived ligand (**212a**, **213a**) tended to furnish the (1*S*)-cyclopropane preferentially, whereas the (1*R*)-cyclopropane was favored when working with 2-amino-D-glucopyranoside- or 2-amino-D-allopyranoside-based ligands having (*R*) configuration at C-2 (**212b**, **213b**). Additional comparisons suggest that the degree of optical induction depends on the configuration at C-1 and C-3 of the sugar. It is clear that neither the stereoselectivity nor the enantioselectivity induced by these catalysts recommend them for preparative use, but they may have some value for

Table 15. Asymmetric synthesis of 2,2-dimethyl-3-vinyl-1-cyclopropanecarboxylates **208** according to Scheme 29

1,3-diene		diazoester	catalyst	yield of cyclopropane [%]	*cis/trans*	*cis*		*trans*		Ref.
R^1	R^2	R^3				% e.e.	config.	% e.e.	config.	
Me	Me	Et	(*R*)-**195a**	54	49/51	62	(1*R*,3*S*)	68	(1*R*,3*R*)[a]	92,204)
			(*S*)-**195a**	54	49/51	62	(1*S*,3*R*)	68	(1*S*,3*S*)[a]	204,205)
Me	Me	*d*-menthyl	(*R*)-**195a**	64	28/72	59	(1*R*,3*S*)	90	(1*R*,3*R*)[a]	92,204)
		dl-menthyl	(*R*)-**195a**	67	19/81	75	(1*R*,3*S*)	90	(1*R*,3*R*)[a]	92,204)
		l-menthyl	(*R*)-**195c**	42	9/91	22	(1*R*,3*S*)	86	(1*R*,3*R*)[a]	92,204)
Cl	Cl	*l*-menthyl	(*S*)-**195a**	52	36/64	31	(1*S*,3*R*)	51	(1*S*,3*S*)	91)
		Et	**209**	33	40/60	15	(1*S*,3*R*)	22	(1*S*,3*S*)	206)
			210	16	43/57	2	(1*S*,3*R*)	16	(1*R*, 3*R*)[b]	206)
Cl	CF_3	Et	**209**	9	51/49	0		0		206)
Cl	Cl	Et	**211**		40/60	17	(1*S*,3*R*)	5	(1*R*,3*R*)	207)
			212a	44	42/58	5	(1*S*,3*R*)	7	(1*S*,3*S*)	208,209)
			213a	22	40/60	5	(1*R*,3*S*)	17	(1*S*,3*S*)	208,209)
			212b	43	43/57	16	(1*R*,3*S*)	12	(1*R*,3*R*)	208,209)

[a] Determined after hydrolysis to give the acid and esterification with *d*-2-octanol; [b] Determined after transesterification with *d*-2-octanol/Ti(O-n-Bu)$_4$.

Table 16. Asymmetric synthesis of alkyl cyclopropanecarboxylates **215**[a]

entry	olefin **214**		diazoester $N_2CHCOOR$	catalyst	yield of cyclopropane **215** [%]	*cis/trans*	*cis*		*trans*		Ref.
	X	Y	R				% e.e.	config.	% e.e.	config.	
1	Cl	Cl	Et	(*S*)-**195a**	59	85/15	91	(1*R*,3*S*)	11	(1*R*,3*R*)	[91]
2	Cl	Cl	*l*-menthyl	(*S*)-**195a**	54	85/15	93	(1*R*,3*S*)	19	(1*R*,3*R*)	[91]
3	Cl	H	Et	(*S*)-**195a**	71	88/12	85	(1*R*,3*S*)	31	(1*R*,3*R*)	[91]
4	Cl	H	*l*-menthyl	(*S*)-**195a**	57	86/14	90	(1*R*,3*S*)	24	(1*R*,3*R*)	[91]
5	Br	H	*l*-menthyl	(*S*)-**195a**	56	83/17	95	(1*R*,3*S*)	23	(1*R*,3*R*)	[91]
6	H	H	*l*-menthyl	(*S*)-**195a**	59	83/17	[b]	(1*R*,3*S*)	23	(1*R*,3*R*)	[91]
7	CF_3	Cl	Et	**209**	12	62/38	26	(1*R*,3*S*)	26	(1*R*,3*R*)	[206]
8	CF_3	Cl	Et	**211**[c]	17	69/31	25	(1*R*,3*S*)	38	(1*S*,3*S*)	[206]
9	CF_3	Cl	Et	**210**	14	54/46	7	(1*R*,3*S*)	13	(1*S*,3*S*)	[206]
10	CF_3	Cl	Et	**212b**	13	58/42	41	(1*R*,3*S*)	5	(1*S*,3*S*)	[208,211]
11	CF_3	Cl	Et	**213b**	8	80/20	44	(1*R*,3*S*)	38	(1*R*,3*R*)	[208,211]

[a] Reaction temperature: 30 °C (entries 1–6); 70–75 °C (entries 7–9); 80 °C (entries 10–11); [b] Not completely resolved; [c] For a similar catalyst, see Ref. [212].

model studies of enantioface selectivity in the cyclopropanation reactions [208]. No enantioselectivity at all resulted from cyclopropanation of 2-chloro-5-methyl-1,1,1-trifluoro-2,4-hexadiene with chiral copper chelates. Thus, the degree of enantioselectivity parallels the nucleophilicity of the cyclopropanated double bond.

Enantioselective cyclopropanation of monoolefins **214** has also been performed. With the already mentioned chiral catalysts **195a** and **209–213** rather high enantiomeric excess was achieved in some cases (Table 16), and the vinylcyclopropane structure was obtained in a subsequent dehydrohalogenation step.

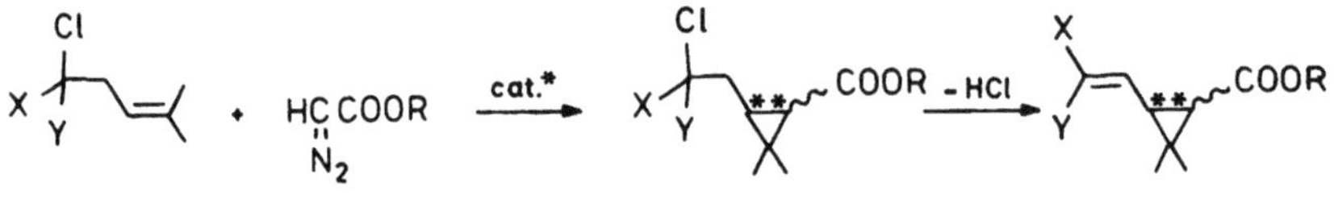

214 *215*

Most remarkably, the homoallylic halides **214** not only yield the thermodynamically unfavored *cis*-cyclopropanes **215** preferentially (see Sect. 2.2.3), but also give rise to enantioselective formation of the (1*R*) configuration, in contrast to the cyclopropanation of 1,3-butadienes with the same catalysts (see Table 15). Only in the case of olefin **214** (X = CF_3, Y = Cl), may the (1*S*)-*trans* isomer be obtained enantioselectively, depending on the catalyst (Table 16, entries 8–11). In these few cases, optical induction occurs at C(3) of the cyclopropane rather than at C(1).

The asymmetric synthesis of dihydrochrysanthemolactone **217** by intramolecular cyclopropanation of diazoacetate **216** in the presence of chiral salicylaldimine/

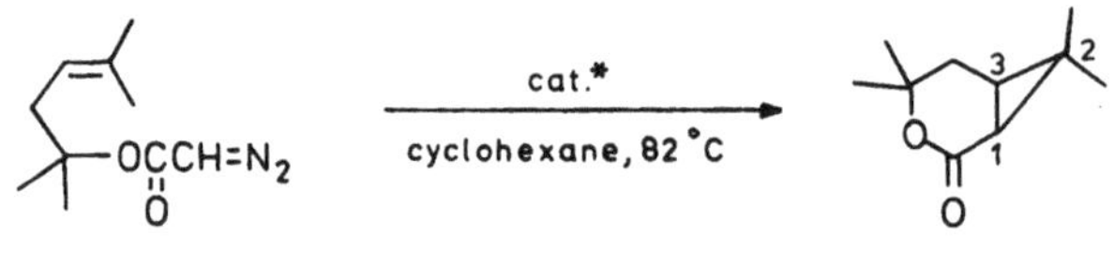

216 *217*

catalyst	% *217*	% e.e. of *217*
(*S*)-*218*	59	6.4 (1*R*)
(*R*)-*195e*	47	23 (1*S*)

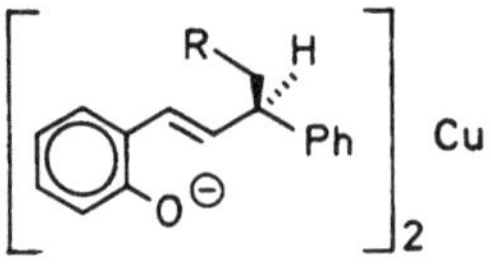

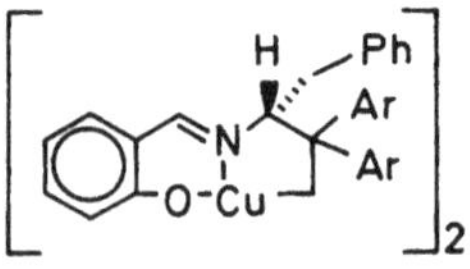

(*S*) - *218*　　(*R*) - *195e*

R = p-tolyl　　Ar = 2-(2-propyloxy)-phenyl

copper complexes **218** or **195e** is another example where the sense of optical induction is different from that found in the synthesis of chrysanthemic esters with the same type of catalysts (see Scheme 29): From (*S*)-**218**, the (1*R*, 3*S*)-cyclopropane is obtained, whereas with (*R*)-**195e**, the (1*S*) configuration prevails [213].

Use of a chiral diazo ester proved less rewarding in terms of enantioselective cyclopropanation. Only very low enantiomeric excesses were obtained when styrene was cyclopropanated with the carbenoid derived from diazoacetic esters **219** bearing a chiral ester residue [214].

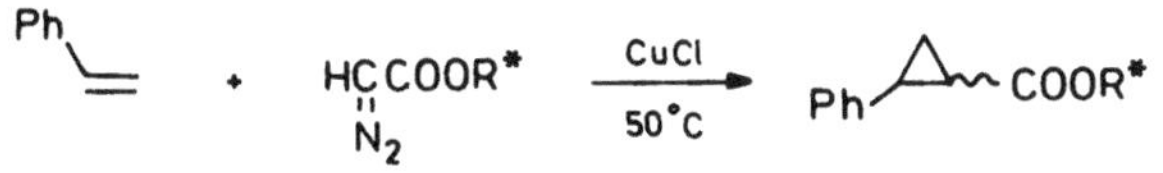

219

R*	*trans*:*cis* ratio	cyclopropane % optical purity (sign of rotation) *cis*	*trans*
(−)-bornyl	2.9	4.6(−)	1.2(+)
(+)-bornyl	2.8	4.5(+)	1.4(−)
(−)-menthyl	2.3	11.7(−)	0.3(−)
(−)-2-methylbutyl	2.7	inactive	0.4(−)

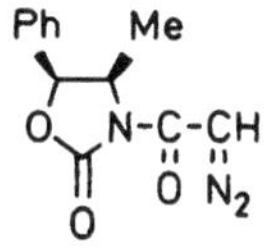

220

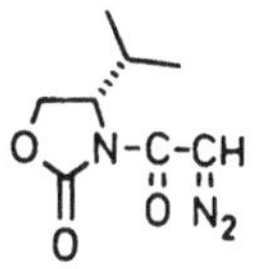

221

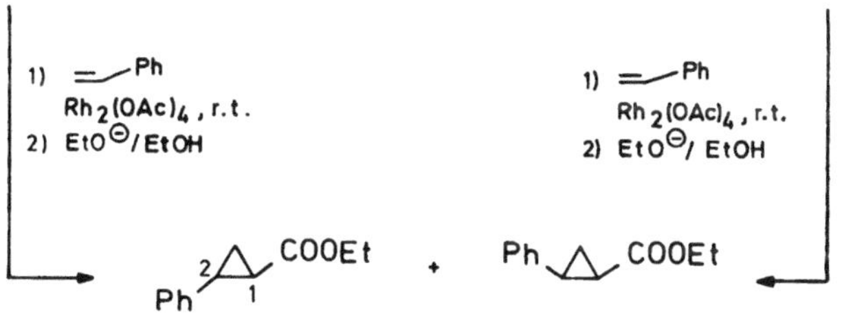

trans-222 *cis-222*

	from *220*	from *221*
total yield	35–40%	20–24%
trans:*cis*	1.8	1.8
% e.e	*trans*: 14(1*R*)	13(1*R*)
	cis: 13(1*R*)	not determined

Optical induction emanating from a chiral diazoacetamide is apparently not much higher. The 2-phenylcyclopropanecarboxylates *cis*-**222** and *trans*-**222**, obtained in low yield from (N-diazoacetyl)oxazolidones **220**, **221** and styrene in the presence of $Rh_2(OAc)_4$ followed by ethanolysis, showed only small enantiomeric excesses [215]. Starting with either diazo compound, the (1*R*) enantiomer was predominant in both *cis*- and *trans*-**222**.

Diastereoface-differentiating reactions of a carbenoid with an alkene bearing an easily removable, chiral substituent have been used only ocassionally for the enantioselective production of a cyclopropane [216]. A recent example is given by the cyclopropanation of the (—)-ephedrine-derived olefin **223** with $CH_2N_2/Pd(OAc)_2$; after removal of the protecting group, (1*R*, 2*R*)-2-phenylcyclopropane carbaldehyde was isolated with at least 90% e.e. [37].

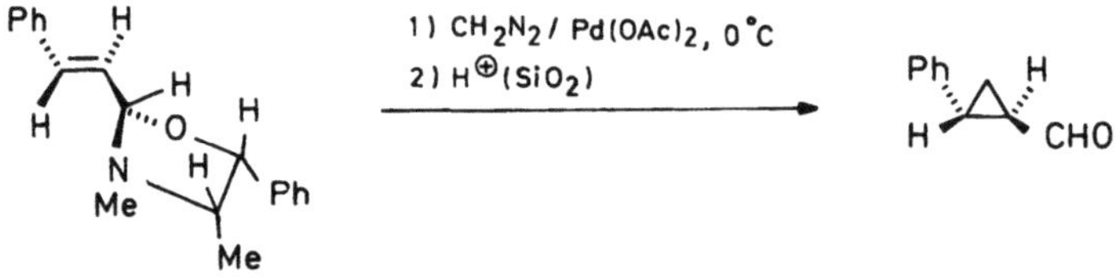

223 (1*R*, 2*R*)

3 Reaction with Acetylenes

Transition-metal catalyzed decomposition of alkyl diazoacetates in the presence of acetylenes offers direct access to cyclopropene carboxylates **224**; in some cases, the bicyclobutane derivatives **225** were isolated as minor by-products. It seems justified to state that the traditional copper catalysts have been superseded meanwhile by $Rh_2(OAc)_4$, because of higher yields and milder reaction conditions [217, 218] (Table 17). $[(\eta^3\text{-}C_3H_5)PdCl]_2$ has been shown to promote cyclopropenation of 2-butyne with ethyl diazoacetate under very mild conditions, too [219], but obviously, this variant did not achieve general usage. Moreover, $Rh_2(OAc)_4$ proved to be the much more efficient catalyst in this special case (see Table 17).

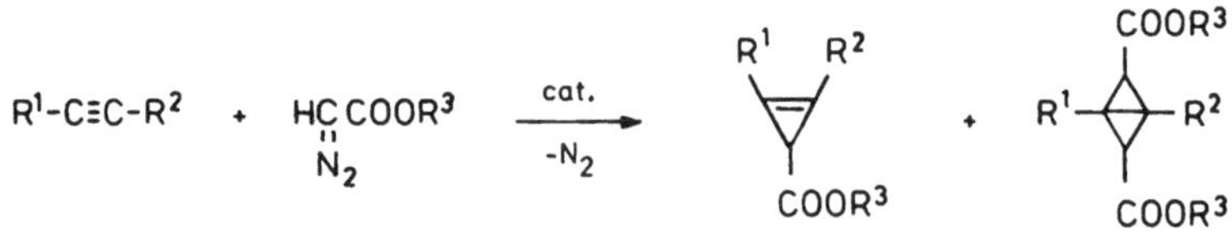

224 *225*

As has already been mentioned for cyclopropanation of olefins, the diazoester should be added slowly to the mixture of alkyne and $Rh_2(OAc)_4$, in order to minimize formation of carbene dimers. The reaction works well with mono- and

Table 17. Cyclopropenation with alkyl diazoacetates according to: $R^1C{\equiv}CR^2 + N_2CHCOOR^3 \rightarrow$ **224** + **225**

R^1	R^2	R^3	conditions	[alkyne]:[diazoester]:[catalyst]	%**224**	%**225**	Ref.
Alkyl, $SiMe_3$	$SiMe_3$	Me	$CuSO_4$	8-fold excess of alkyne	8–54		220)
c-C_3H_5	$SiMe_3$	Me	$CuSO_4$, 115 °C	1.5-fold excess of diazoester	37	4	221)
$SiMe_3$	$SiMe_3$	Me	CuBr, Δ		18		222)
Ph	*t*-Bu	Et	$CuSO_4$, 120 °C		27		223)
4-anisyl	Ph	Et	Cu powder, 110 °C	2-fold excess of alkyne	64[a]		223)
n-Bu, *t*-Bu, C_6H_{11}, *i*-Pr	H	Me	$Rh_2(OAc)_4$, 25 °C	200:80:1	63–86		218, 223)
$MeOOCCH_2$	H	Me	$Rh_2(OAc)_4$, 25 °C	200:80:1	40		218)
$MeOCH_2$	H	Me	$Rh_2(OAc)_4$, 25 °C	200:80:1	46[b]		218)
$MeOCH_2$	$MeOCH_2$	Me	$Rh_2(OAc)_4$, 25 °C	200:80:1	38		218)
$AcOCH_2$	$AcOCH_2$	Et	$Rh_2(OAc)_4$, 26–36 °C	1364:424:1	30		166)
$AcOCH_2$	$AcOCH_2$	*t*-Bu	$Rh_2(OAc)_4$, 26–36 °C	670:165:1	65		166)
$ClCH_2$	$ClCH_2$	*t*-Bu	25 °C	12750:1950:1	59		167)
R-C_6H_4	Me	Me	$Rh_2(OAc)_4$, 25 °C	R=H: 2:0.8:0.01	66		217)
				4-Me: 1.8:1:0.01	41		
				4-Cl: 2:0.8:0.01	34		
				3-Cl: 2:1.2:0.01	41		
				4-*t*-Bu: 2:0.8:0.005	50		
				4-OMe: 1270:600:1	18[a]		223)
Me	Me	Et	$Rh_2(OAc)_4$ 0–20 °C	2.3–3:1:1–3 · 10^{-3} 0–20 °C	60–70		224)
n-Pr	*n*-Pr	Me	$Rh_2(OAc)_4$, 25 °C	~1200:600:1	16		223)
n-Pr	*n*-Pr	CH_2CH_2Br	$Rh_2(OAc)_4$, 25 °C	30:10:0.03	55		225)
Me_3Si	Me	CH_2CH_2Br	$Rh_2(OAc)_4$, 25 °C	30:10:0.03	50		225)

[a] After saponification; [b] The rearranged product, methyl 2-methoxymethyl-1-cyclopropenecarboxylate, was obtained.

disubstituted triple bonds; no C/H insertion was observed with terminal acetylenes [218]. Problems arise when either the alkyne or the cyclopropene are not inert towards the catalyst. For example, ethoxyacetylene and phenylacetylene are polymerized by $Rh_2(OAc)_4$ and cyclopropene carboxylates may be polymerized by $Rh_2(CF_3COO)_4$ [217]. The competition of a triple bond and a double bond for the ketocarbenoid is poorly selective [217], similar to the situation in copper-catalyzed reactions [14]. Alternatively, the carbenoid derived from 2-bromoethyl diazoacetate and $Rh_2(OAc)_4$ (25 °C) or $Cu/CuSO_4$ (100 °C) reacted with 2-methyl-1,5-hexadien-3-yne at the double bonds exclusively [224]. The same holds true for the synthesis of bis(1-methylcyclopropyl)acetylene from 2,5-dimethyl-1,5-hexadien-3-yne and diazomethane; copper(II) bis (N-α-phenylethylsalicylaldiminate) has been recommended there as an efficient catalyst [226].

Ylide-derived products may be formed as minor by-products from propargylic chlorides or ethers; this contrasts with the inertness of allylic chloride or ethers

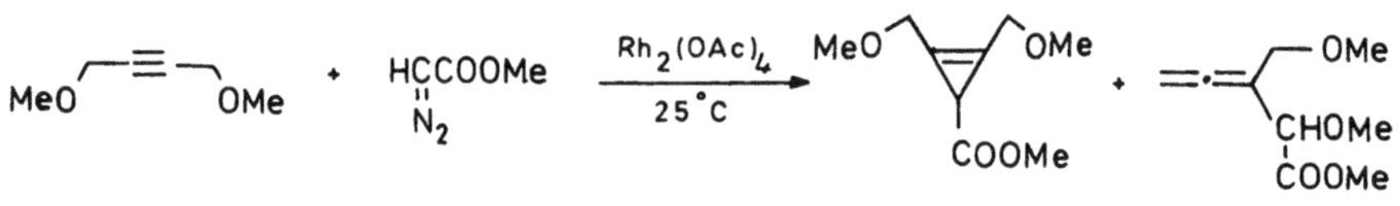

226

(38%) (8%)

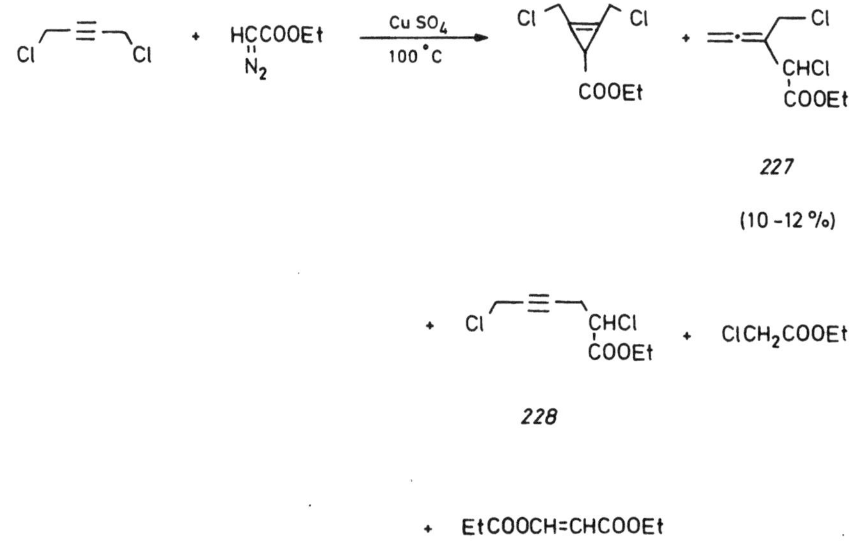

227

(10–12%)

228

towards such transformations. Thus, 1,4-dimethoxy-2-butyne gave allene **226**, probably resulting from a [2,3]-sigmatropic rearrangement of an oxonium ylide, besides the expected cyclopropene in the $Rh_2(OAc)_4$-catalyted reaction with methyl diazoacetate [218]. From the $CuSO_4$-catalyzed reaction between 1,4-dichloro-2-butyne and ethyl diazoacetate, a complex product mixture resulted which also contained [2,3]- and [1,2]-ylide rearrangement products (**227** and **228**) [227]; no such compounds were reported for the $Rh_2(OAc)_4$-catalyzed reaction [167].

Reaction of propargylic alcohols **229** with alkyl diazoacetates entails competition between O/H insertion and cyclopropenation.

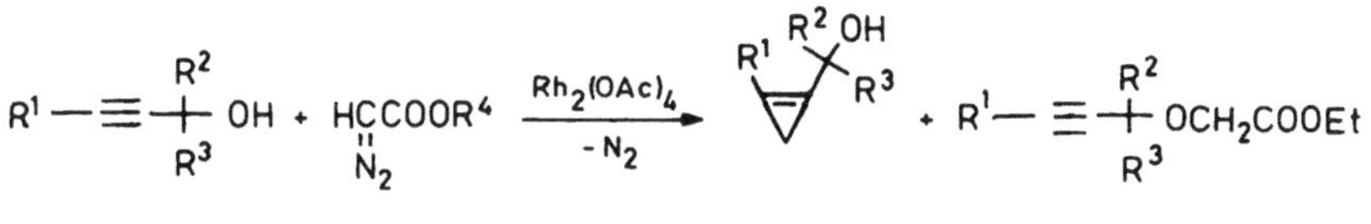

229

Under the catalytic action of $Rh_2(OAc)_4$, formation of a propargylic ether from a terminal alkyne (**229**, $R^1 = H$) is preferred as long as no steric hindrance by the adjacent group is felt [162, 218]. Otherwise, cyclopropenation may become the dominant reaction path [e.g. **229** (R^1 = H, R^2 = R^3 = Me) and methyl diazoacetate: 56% of cyclopropene, 36% of propargylic ether [162]], in contrast to the situation with allylic alcohols, where O/H insertion is rather insensitive to steric influences.

Ethyl diazopyruvate, under copper catalysis, reacts with alkynes to give furane-2-carboxylates rather than cyclopropenes [113] (Scheme 30). What looks like a [3 + 2] cycloaddition product of a ketocarbenoid, may actually have arisen from a primarily formed cyclopropene by subsequent copper-catalyzed ring enlargement. Such a sequence has been established for the reaction of diazoacetic esters with acetylenes in the presence of certain copper catalysts, but metallic copper, in these cases, was not able to bring about the ring enlargement [14]. Conversely, no cyclopropene derivative was detected in the diazopyruvate reaction.

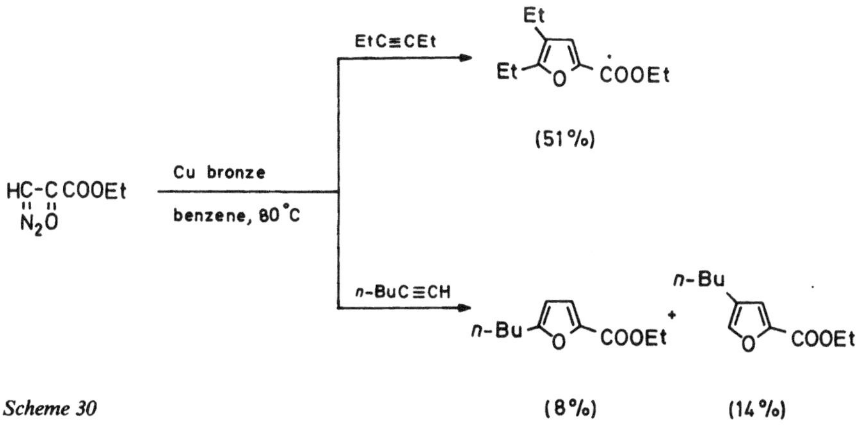

Scheme 30

4 Reaction with Aromatic and Heteroaromatic Compounds

4.1 Benzene and its Derivatives

Copper-catalyzed cyclopropanation of benzene and its derivatives by a diazoacetic ester yields a norcaradiene **230** which undergoes spontaneous ring opening to cycloheptariene **231**. At the temperatures needed for successful cyclopropanation, sigmatropic H-shifts leading to conjugated isomers of cycloheptatriene carboxylates cannot be avoided. The situation is complicated by the formation of regioisomers upon cyclopropanation of substituted benzenes, and separation of the cycloheptatriene isomers may became tedious if not impossible.

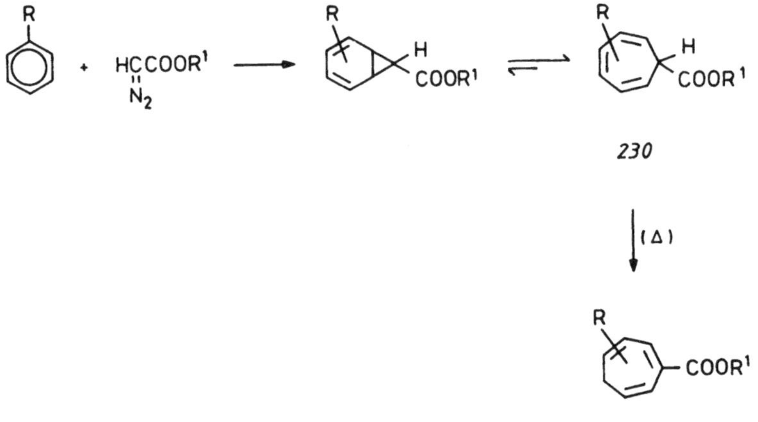

230

231

A major improvement came with the introduction of rhodium(II) trifluoroacetate [[$Rh_2(CF_3COO)_4$]] as cyclopropanation catalyst, which allows efficient and fast production of the unrearranged cycloheptatriene carboxylates **230** at room temperature [228]. Cyclopropanation with methyl diazoacetate at 22 °C gave the following total yields ([substrate]/[EDA]/[catalyst] = 5000/250/1): benzene, 100%; toluene, 95%; *o*-xylene, 80%; *m*- or *p*-xylene, 90%; mesitylene, 60%; indan, 53%; anisole, 73%; chlorobenzene, 72%; fluorobenzene, 46%; ethyl benzoate, 10%; hexafluorobenzene, ~5%. Increasingly lower yields are obtained with $Rh_2(C_6F_5COO)_4$, $Rh_2(CH_3OCH_2COO)_4$, $Rh_2(CH_3COO)_4$ and $Rh_2[(CH_3)_3(COO)]_4$ as catalyst. This findings support the assumption of an electrophilic metal carbene intermediate, which is further corroborated by relative reactivity studies comparing benzene and some substituted derivatives. However, the reaction seems more sensitive to steric than electronic effects, as the reduced yields in the series toluene, xylene, mesitylene on the substrate side or methyl, ethyl, *t*-butyl on the diazoester side show. Not unexpectedly, steric effects also contribute to the regioselectivity of cyclopropanation, but the results presented in Scheme 31 [228] do not allow one to generally speak of steric effects as being the major factor.

$Rh_2(CF_3COO)_4$-catalyzed cyclopropanation of aromatic substrates is not confined

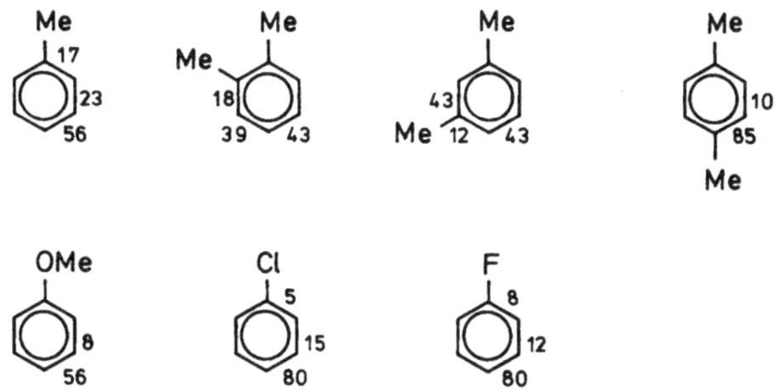

Scheme 31. Isomer distribution [%] of $Rh_2(CF_3COO)_4$-catalyzed cyclopropanation of substituted benzenes with methyl diazoacetate at 22 °C. The numbers refer to the percentage of 1,3,5-cycloheptatriene-7-carboxylate from the total cycloheptatriene isomers.

to diazoacetic esters; both inter- and intramolecular addition of ketocarbenoids derived from α-diazoketones has been promoted under mild reaction conditions and usually in high yield.

The synthesis of cycloheptatrienyl ketones **233** from benzene and α-diazoketones **232** proceeds in "essentially quantitative yield"; the products were not isolated but directly transformed into benzyl ketones **234** through the action of trifluoroacetic acid [229].

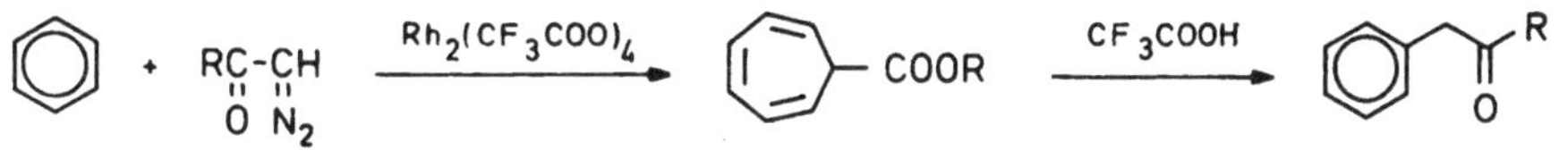

232 233 234

R = Me, Me_2CH, 4-Me-C_6H_4, Ph, $Cl(CH_2)_3$, $Cl(CH_2)_2$, $MeOCH_2$, cyclopropyl, $ClCH_2$, MeCHBr

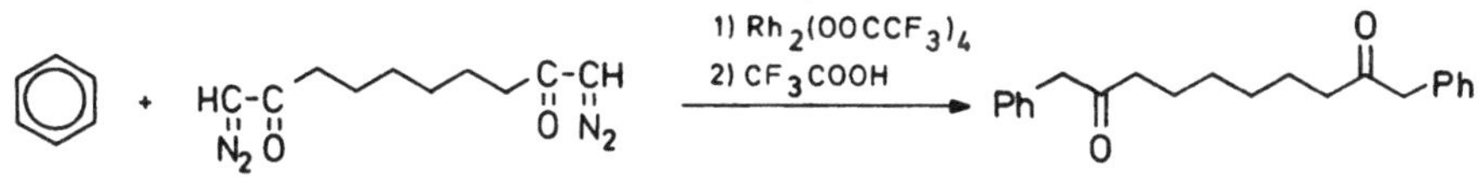

235

236
(70%)

1,10-Bis(diazo)-2,9-decanedione **235** undergoes a twofold reaction to give the diketone **236** [229]. If so desired, the cycloheptatrienyl ketones **233** can be isolated, except for those with R = CH_2Cl or CHBrMe which isomerize spontaneously. In the case of 1-diazo-3-phenyl-2-propanone (**232**, R = $PhCH_2$), cyclopropanation competes with intramolecular cyclization leading to 2-indanone in 49% yield [229].

Anisole has been cyclopropanated with benzhydryl 6-diazopenicillanate **237** in the presence of $Rh_2(CF_3COO)_4$, but again, cycloheptatrienes (**238** and **239**) rather than norcaradienes were isolated [230].

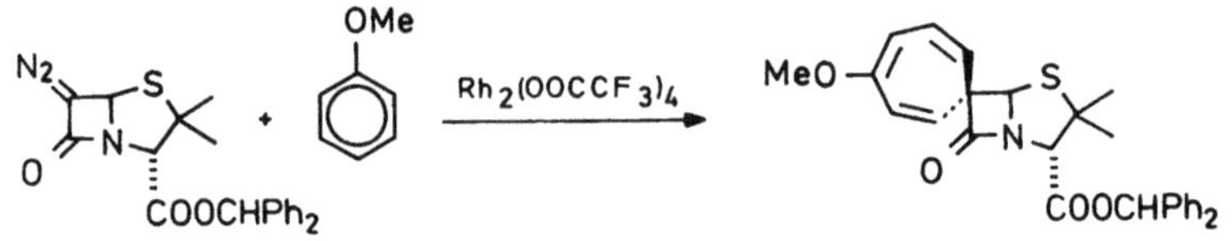

237 *238* (13%)

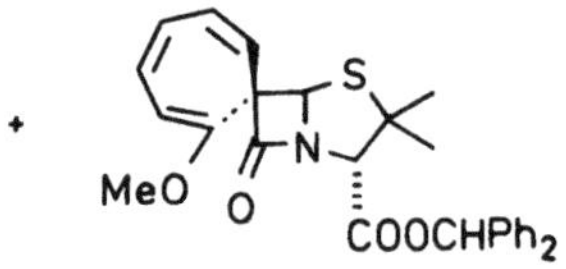

239 (2%)

Intramolecular cyclopropanation of 4-aryl-1-diazo-2-butanones **240** allows construction of the bicyclo[5.3.0]decane framework [12]. In a reaction sequence analogous to that described above for the intermolecular ketocarbenoid reaction, bicyclo-[5.3.0]deca-1,3,5-trien-8-ones **241** are formed. They rearrange to the conjugated isomers **242** at the high temperatures needed if the reaction is catalyzed by copper [231] or CuCl [232], but can be isolated in excellent yield from the $Rh_2(OAc)_4$-promoted reaction which occurs at lower temperature [233].

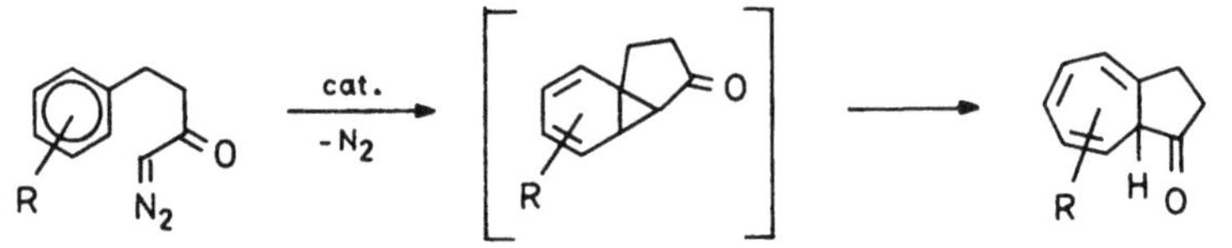

240 *241*

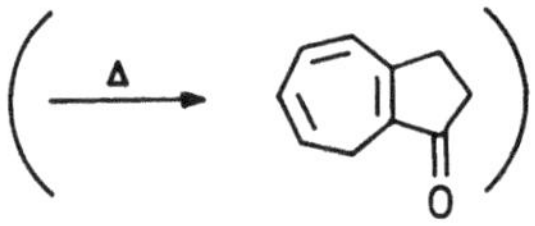

242

By analogy with the acid-induced ring contraction of cycloheptatrienyl ketones **233**, high-yield rearrangement to 2-tetralones is possible [233]. As Scheme 32 shows, substituents on the aromatic nucleus determine the regioselectivity of cyclopropanation. Remarkably enough, the sterically most hindered site of diazoketone **243** is cyclopropanated regioselectively; only trace amounts of such products resulted from the intermolecular reaction between anisole and diazopenicillanate **237** [230] or ethyl diazoacetate [234] (for the reaction between anisole and methyl diazoacetate, this regioisomer was not reported, see Scheme 31).

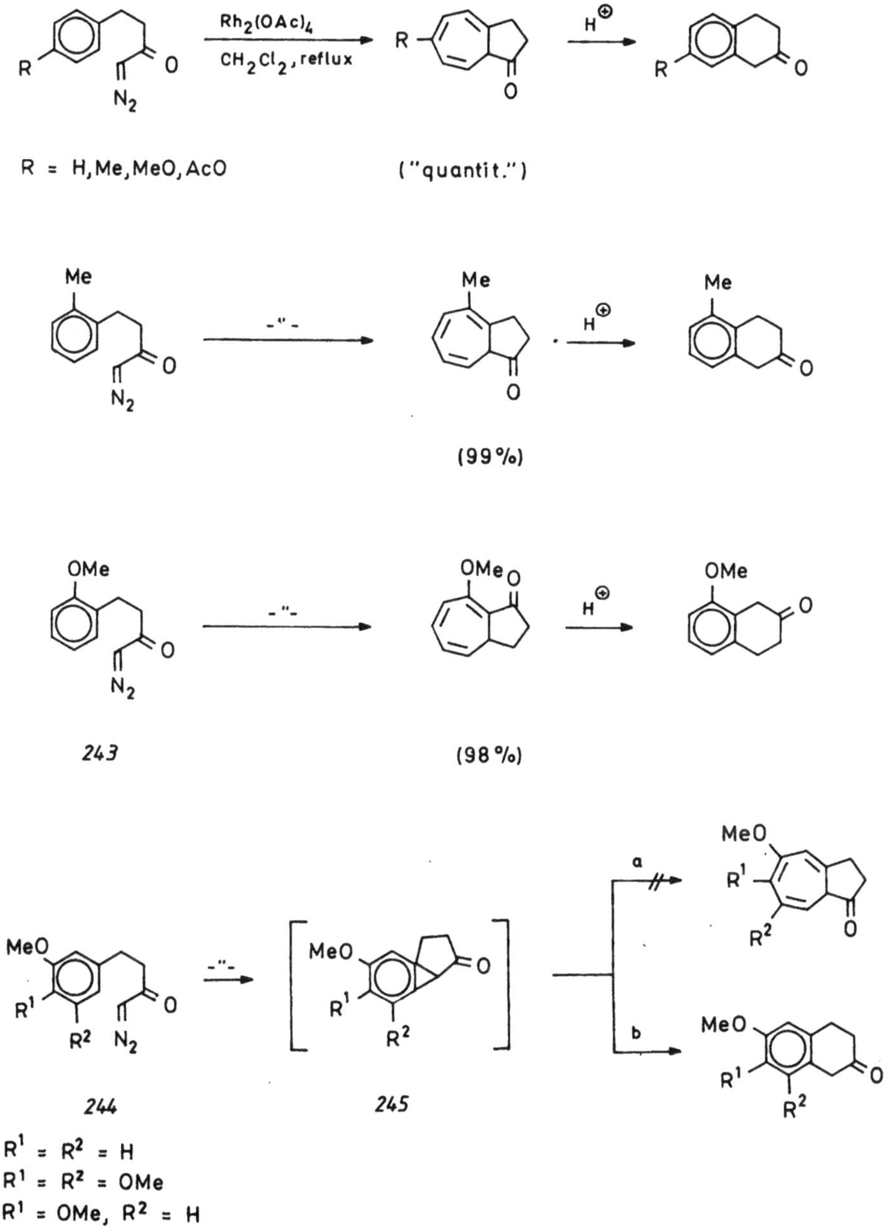

Scheme 32

$Rh_2(OAc)_4$-catalyzed decomposition of *m*-methoxy-substituted diazoketones **244** furnishes 6-methoxy-2-tetralones rather than the expected bicyclo[5.3.0]decatrienones. This demonstrates that the direction of ring opening in the norcaradienone intermediate **245** may well be influenced by the nature and position of a substituent.

The cyclization of phenolic diazoketones **246** to spirodienones **247** was reinvestigated with $Rh_2(Me_3CCOO)_4$, $Rh_2(OAc)_4$ and $Pd(OAc)_2$ as catalysts [235]; the yields were found to be better than with CuCl used in earlier studies.

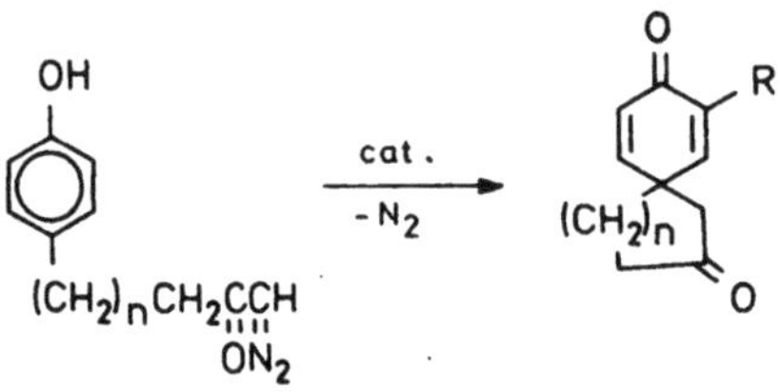

246 *247*

Two novel example of intramolecular aromatic C/H insertion are still to be mentioned here: 4-Aryl-2-hydroxy-1-naphthoates **249** have been obtained from the $Rh_2(OAc)_4$-catalyzed decomposition of ethyl 5-aryl-2-diazo-3-oxopent-4-enoates **248** [236]. If the *cis*-aryl group is missing, Wolff rearrangement rather than cyclization occurs. Similarly, 3-oxo-3,4-dihydro-2*H*-1-benzopyrans **253–255** are available by copper(II) hexafluoroacetylacetonate promoted decomposition of 1-diazo-3-aryloxy-2-propanones **250–252** [237]. In both cases, it is unknown whether a direct C/H insertion process or an intramolecular cyclopropanation/rearrangement sequence accounts for the cyclization product. The close structural relationship between **250–252** and 4-aryl-1-diazo-2-butanones **240** which are known to undergo intramolecular cyclopropanation may suggest analogous reactivity, but electrophilic attack of the metal carbene intermediate on the aromatic nucleus, followed by proton transfer, explains the formation of **253–255** also.

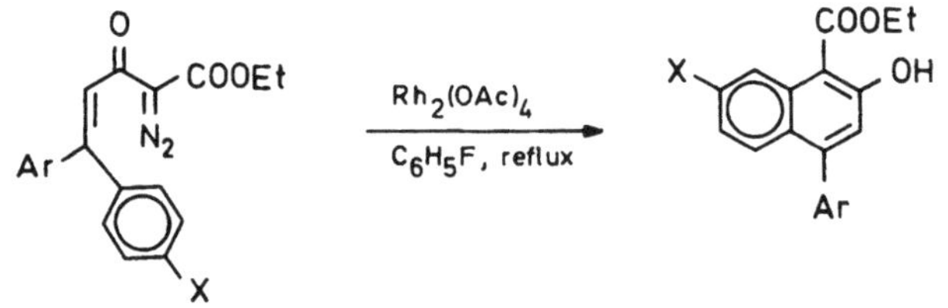

248 *249*

Ar	X	yield [%]
Ph	H	90
4-Me-C_6H_4	Me	76
4-Cl-C_6H_4	Cl	97

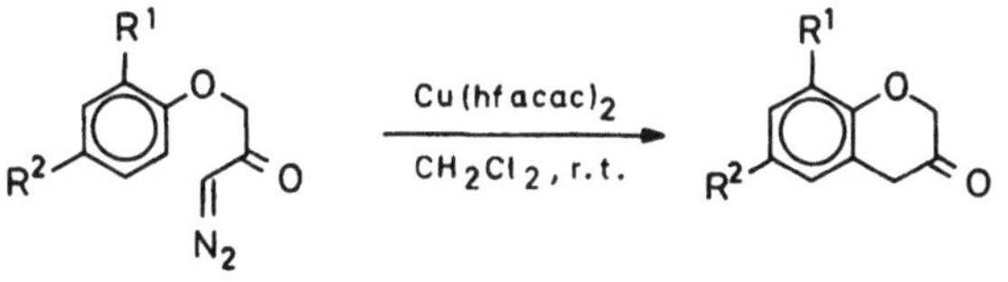

250 253

R^1, R^2 = H, Me, *t*-Bu (73 - 86%)

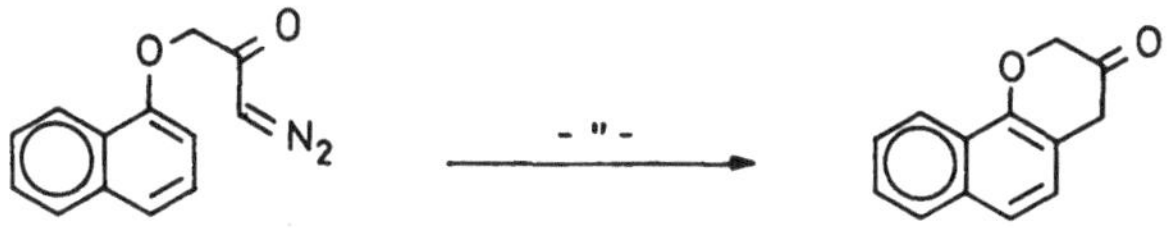

251 254 (84%)

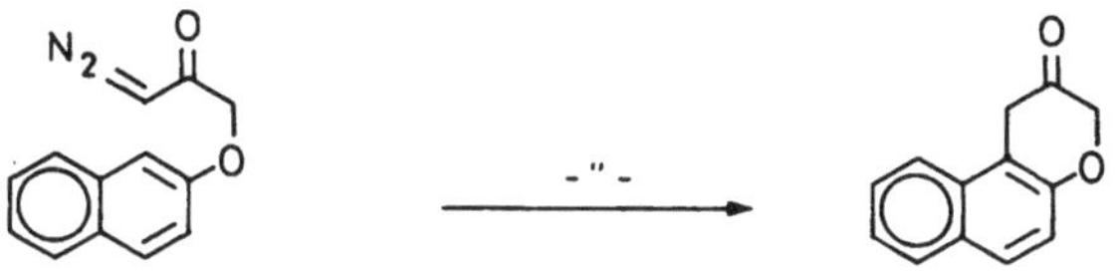

252 255 (87%)

4.2 Heteroaromatic Compounds

Reactions of π-excessive heteroaromatic compounds such as pyrroles, thiophenes and furans with carbenoids have been known for several years [6,10,14]. Recent activities were directed towards further synthetic applications of already known reactions, evaluation of the efficacy of novel catalysts and towards mechanistic insights.

Pyrroles

Reaction of ketocarbenoids with pyrrole and N-alkylpyrroles yields the product of formal insertion into the α-C—H bond (**256**); in many cases the β-insertion product **257** is formed concomitantly, but generally in lower yield [238-241]. The regioselectivity varies according to the catalyst, the diazo compound and the N-alkyl substituent. Some examples concerning the former two variables are given in Table 18 [239,240].

Table 18. Regioselectivity of ketocarbenoid insertion into C—H bonds of N-methylpyrrole.

R^1	R^2	catalyst (mol %)	temperature [°C]	total yield [%] (**256** + **257**)	ratio **256**/**257**
H	Et	Cu bronze (10)	100	40	5.2
		$Cu(acac)_2$ (0.5)	50	34	14.2
		$Cu(hfacac)_2^a$ (1.5)	55	57	3.5
		$Cu(C_{10}H_{12}NO)_2^b$ (1.5)	50	50	16.9
		$Cu(OTf)_2$ (1.5)	40	63	1.4
		$Rh_2(OAc)_4$ (1.5)	80	c	c
COOMe	COOMe	Cu bronze (10)	110	~30	6.5
		$Cu(acac)_2$ (1)	80	65	9.0
		$Cu(hfacac)_2^a$ (1)	70	83	7.3
		$Cu(C_{10}H_{12}NO)_2^b$ (1)	90	76	12.0
		$Cu(OTf)_2$ (1)	70	70	5
		$Rh_2(OAc)_4$ (0.5)	70	92	9.8
MeCO	COOEt	$Cu(hfacac)_2^a$ (1)	75	68	—
		$Cu(C_{10}H_{12}NO)_2^b$ (1)	90	62	—
		$Rh_2(OAc)_4$ (0.5)	70	72	—

[a] hfacac = hexafluoroacetylacetonate; [b] $C_{10}H_{12}NO$ = N-isopropylsalicylaldiminate; [c] Complex product mixture with little **256** and **257**.

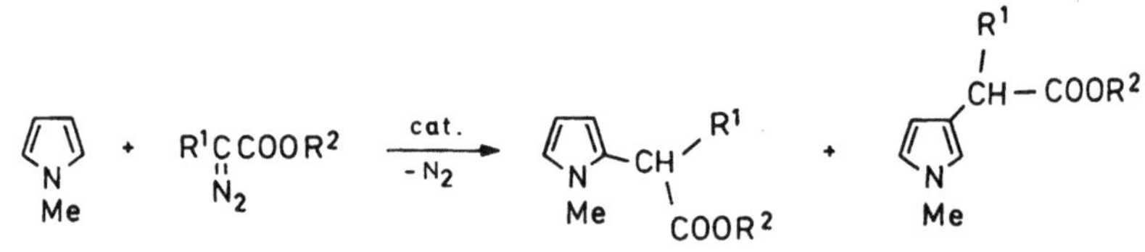

256 *257*

Concerning the influence of the N-alkyl substituent, it was found that for the reaction with ethyl diazoacetate, the amount of β-isomer increased in the series H < Me < *i*-Pr < *t*-Bu (exclusive formation of the β-isomer in the *t*-Bu case), i.e. with steric bulk of this group.

It was proposed that electrophilic addition of a metal carbene rather than direct insertion or fragmentation of an initially formed 2-azabicyclo[3.1.0]hex-3-ene accounts for the formal insertion products [239] (Scheme 33).

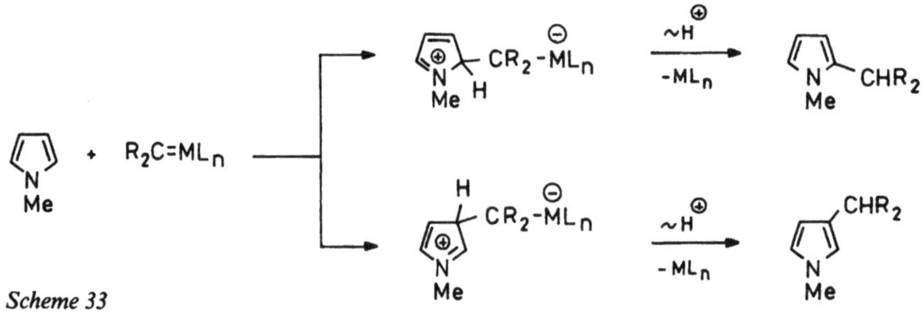

Scheme 33

In an intramolecular version of ketocarbenoid α-C/H insertion, copper-promoted decomposition of 1-diazo-3-(pyrrol-1-yl)-2-propanone (**258a**) or 1-diazo-4-(pyrrol-1-yl)-2-butanone (**258b**) resulted in quantitative formation of the respective cyclization product **259** [242]. The cyclization **260** → **261**, on the other hand, is a low-yield reaction which is accompanied by olefin formation. The product ratio was found to vary with the copper catalyst used, but the total yield never exceeded 35% [243].

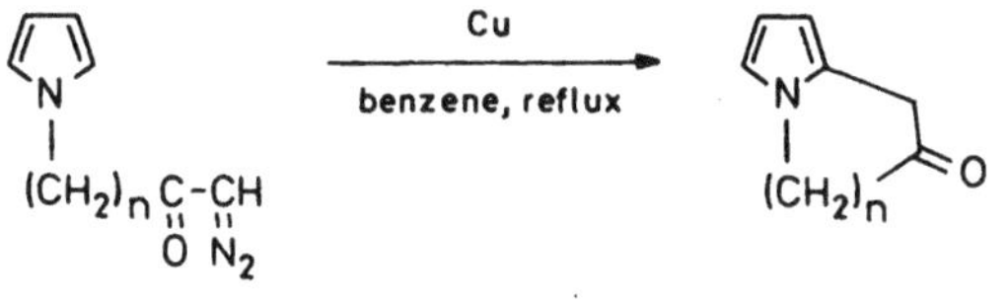

258a (n = 1)
258b (n = 2)

259a,b
(100%)

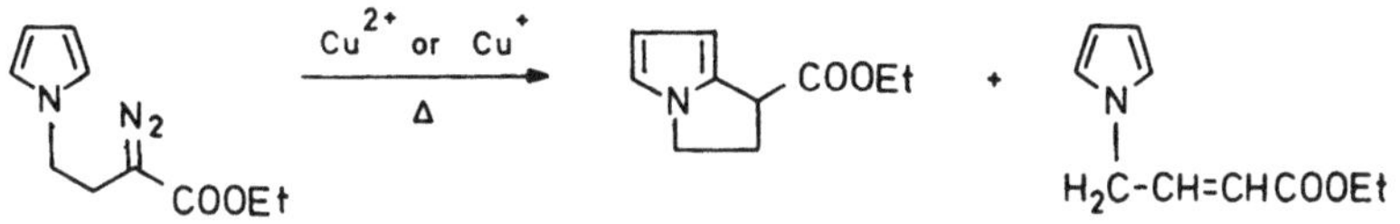

260 *261* (*Z* + *E*)

Whereas pyrrole was reported not to give N/H insertion by ketocarbenoids, such a reaction mode does occur with imidazole: Copper-catalyzed decomposition of ethyl diazoacetate at 80 °C in the presence of imidazole gives ethyl imidazol-1-ylacetate exclusively (93%); small amounts of a C-alkylated imidazole were obtained additionally under purely thermal conditions [244]. N/H insertion also takes place at benzimidazole [245a]. The reaction is thought to begin with formation of an N^3-ylide, followed by N^1 → C proton transfer leading to the formal N/H insertion product. Diazomalonic esters behave analogously; however, they suffer complete or partial dealkoxycarbonylation under the reaction conditions [244] (Scheme 34). N-alkylation of imidazole and benzimidazole by the carbenoids derived from ω-diazoacetophenone and 2-(diazoacetyl)naphthalene has also been reported [245b].

Thiophenes

Rhodium(II) acetate was found to be much more superior to copper catalysts in catalyzing reactions between thiophenes and diazoesters or diazoketones [246]. The outcome of the reaction depends on the particular diazo compound [246]: With *t*-butyl diazoacetate, high-yield cyclopropanation takes place, yielding 6-*exo*-substituted thiabicyclohexene **262**. Dimethyl or diethyl diazomalonate, upon $Rh_2(OAc)_4$-catalysis at room temperature, furnish stable thiophenium bis(alkoxycarbonyl)methanides **263**, but exclusively the corresponding carbene dimer upon heating. In contrast, only 2-thienylmalonate (36%) and carbene dimer were obtained upon heating the reactants for 8 days in the presence of CuI · $P(OEt)_3$. The Rh(II)-promoted ylide formation

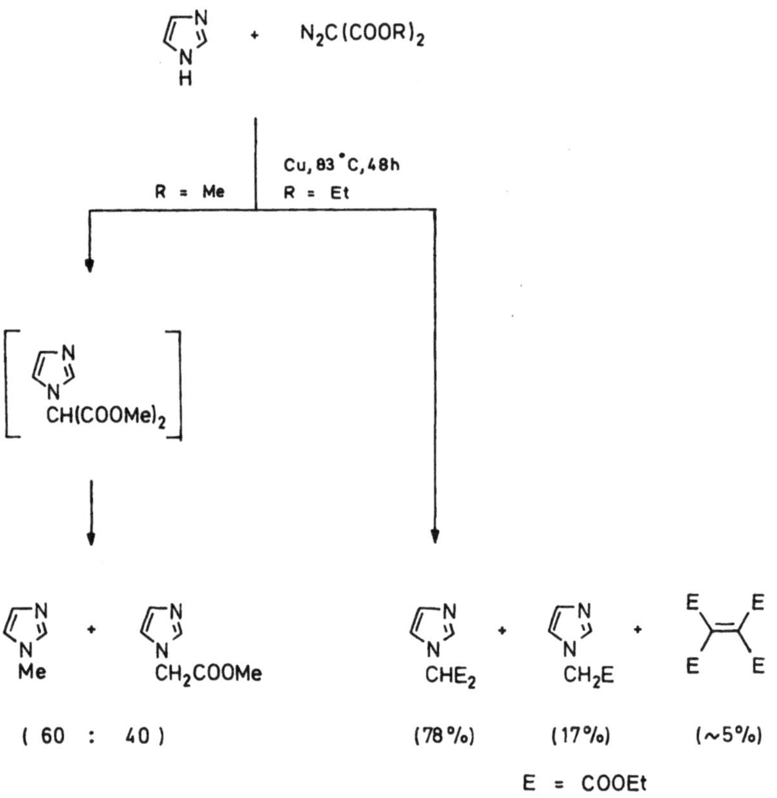

Scheme 34

could also be realized with 2- and 2,5-substituted thiophenes (2,5-dichloro-, 2-methyl-, 2-bromo-, 2-hydroxymethyl-, 2-bromo-3-methyl) as well as with benzo[*b*]thiophene and dibenzothiophene [246,247]. Diazo Meldrum's acid, a cyclic diazomalonate, proved to be resistant towards $Rh_2(OAc)_4$ at moderate temperatures, rendering ylide formation impossible. Ethyl diazoacetacetate gave both the formal C/H insertion product **264** and the thiabicyclohexene **265** [246,248]. An insertion product analogous to **264** was also isolated in low yield from the reaction between thiophene and ω-diazoacetophenone [246].

The view has been expressed that a primarily formed ylide may be responsible for both the insertion and the cyclopropanation products [230,246,249]. In fact, ylide **263** rearranges intramolecularly to the 2-thienylmalonate at the temperature applied for the CuI · $P(OEt)_3$ catalyzed reaction between thiophene and the diazomalonic ester [250]; this readily accounts for the different outcome of the latter reaction and the $Rh_2(OAc)_4$-catalyzed reaction at room temperature. Alternatively, it was found that 2,5-dichlorothiophenium bis(methoxycarbonyl)methanide, in the presence of copper or rhodium catalysts, undergoes typical carben(oid) reactions intermolecularly [251,252]; whether this has any bearing on the formation of **262** or **265**, is not known, however.

Another possible isomerization of a primarily formed S-ylide would be a thiophene

→ 2*H*-thiopyran ring enlargement by a Stevens rearrangement. Two examples of this reaction mode have been published meanwhile: Benzhydryl 6-diazopenicillanate **237** gives spiro-connected 2*H*-thiopyran **266** with thiophene besides the cyclopropanation product **267** [230].

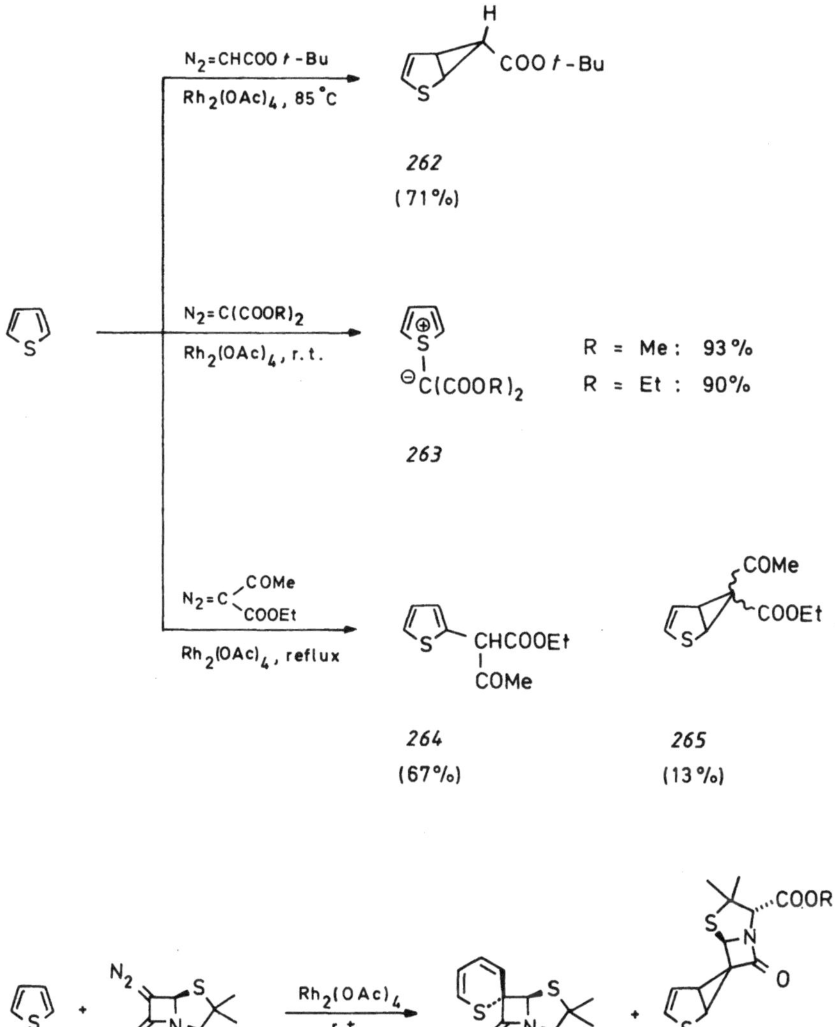

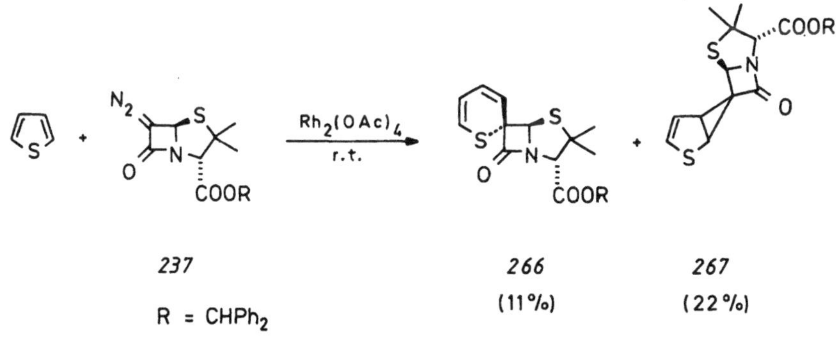

An intramolecularly formed S-ylide, formed by $Rh_2(OAc)_4$-catalyzed decomposition of methyl [α-(2-thienyl)benzoyl]diazoacetate **268**, is thought to furnish **269** by a

[1,2] rearrangement[253]. A formal C/H insertion product, **270**, occurs as a minor by-product; we do not speculate on its mode of formation here.

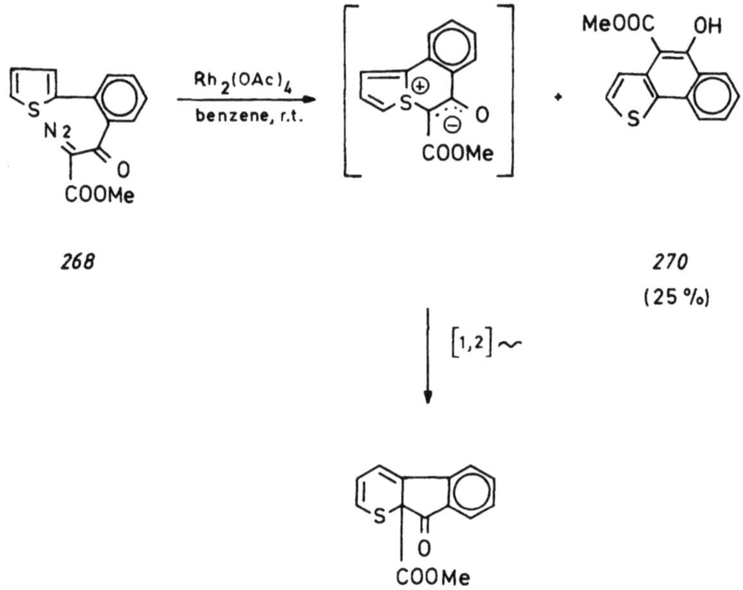

268 *270* (25%) *269* (68%)

Some functionalized thiophenes have been investigated in order to assess the scope of ylide-derived chemistry. As already mentioned, 2-(hydroxymethyl)thiophene still gives the S-ylide upon $Rh_2(OAc)_4$-catalyzed reaction with dimethyl diazomalonate [246], but O/H insertion instead of ylide formation seems to have been observed by other workers (Footnote 4 in Ref. [254]). From the room temperature reaction of 2-(aminomethyl)thiophene and dimethyl diazomalonate, however, salt **271** was isolated quite unexpectedly[254]. $Rh_2(OAc)_4$, perhaps deactivated by the substrate, is useless in terms of the anticipated carbenoid reactions. Formation of a diazomalonic ester amide and amine-catalyzed cyclization to a 5-hydroxytriazole seem to take place instead.

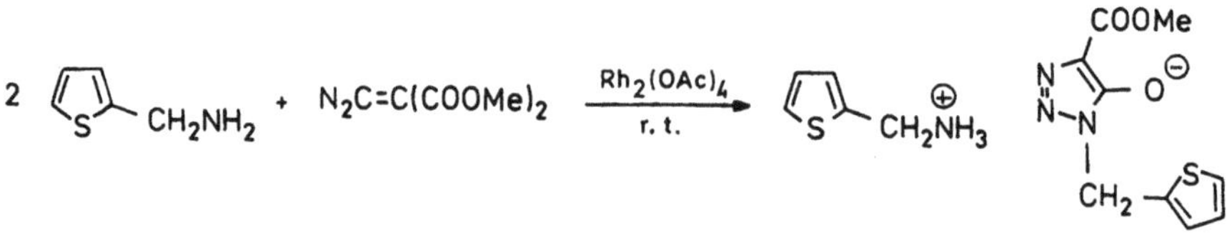

271 (67%)

2-Alkenylthiophenes offer both the olefinic bond and the heterocycle as possible reaction sites to a ketocarbenoid [249]. The transition-metal catalyzed reaction between dimethyl diazomalonate and thiophenes having a $-CR=CH_2$ (R=H, Me) side chain leads exclusively to the (2-thienyl)cyclopropane derivative, irrespective of whether CuCl (80 °C) or $Rh_2(OAc)_4$ (20 °C) serves as catalyst. Replacement of the terminal $=CH_2$ group by =CHMe gives rise to both cyclopropane **272** and formal C/H insertion product **273** upon CuCl catalysis. In the $Rh_2(OAc)_4$-catalyzed reaction at room temperature, S-ylide **274** is formed additionally, which rearranges to **272** and **273** at 100 °C. Finally, if the 5-position of the alkenylthiophene is blocked by an Et substituent, products analogous to **273** and **274** are no longer formed, and conversely no cyclopropane corresponding to **272** is accessible from 2-isobutenylthiophene. Similar dependencies on the side-chain substitution pattern have already been described for copper-catalyzed reactions between 2-alkenylfurans and ethyl diazoacetate [255]; in those cases, both steric and electronic effects of methyl substitution in the side chain were presented as an explanation [256].

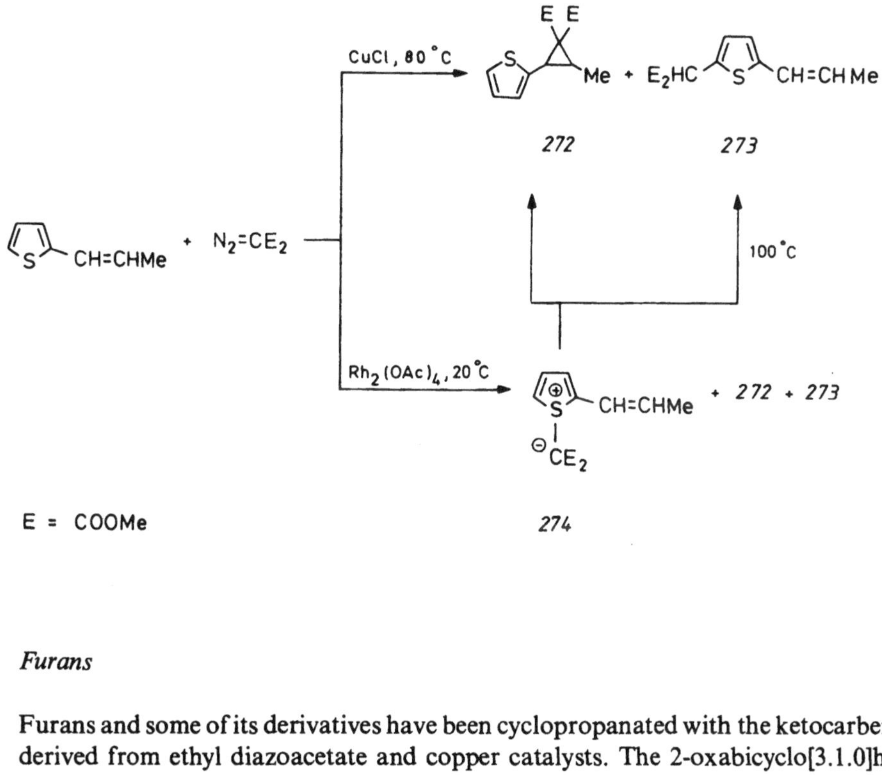

Furans

Furans and some of its derivatives have been cyclopropanated with the ketocarbenoids derived from ethyl diazoacetate and copper catalysts. The 2-oxabicyclo[3.1.0]hex-3-enes thus formed are easily ring-opened to 1,4-diacylbutadienes thermally, thermocatalytically or by proton catalysis [14,136]. The method has been put to good use by $Rh_2(OAc)_4$-catalyzed cyclopropanation of furan with diazoketones **275** to bicyclic products **276**. Even at room temperature, they undergo electrocyclic ring-opening and *cis, trans*-dienes **277 a** are obtained with fair selectivity [257,258]. These compounds served as starting materials in the total syntheses [257-259] of some HETE's (monohydroxy eicosatetraenoic acids) which constitute oxidative metabolites of arachidonic acid.

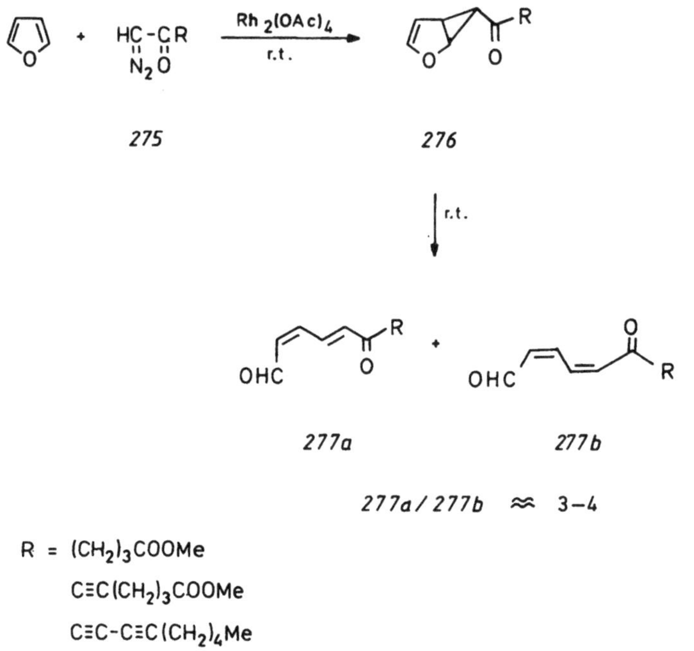

5 Reaction with C=X Groups (X = N, O)

Few efforts have been devoted to carbenoid reactions with C=N and C=O groups since the last two reviews [14,260] covering this field were written.

5.1 Reaction with C=N

Aziridines have been synthesized, albeit in low yield, by copper-catalyzed decomposition of ethyl diazoacetate in the presence of an imine [260]. It seems that such a carbenoid cyclopropanation reaction has not been realized with other diazo compounds. The recently described preparation of 1,2,3-trisubstituted aziridines by reaction of phenyldiazomethane with N-alkyl aldimines or ketimines in the presence of zinc iodide [261] most certainly does not proceed through carbenoid intermediates; rather, the metal salt serves to activate the imine to nucleophilic attack from the diazo carbon. Replacement of ZnI_2 by one of the traditional copper catalysts resulted in formation of imidazoline derivatives via an intermediate azomethine ylide [261].

Reaction of the imine moiety of **278** with excess ethyl diazoacetate in the presence of $Cu(acac)_2$ led to the cyclopentane-annulated product **279** the structure of which was confirmed by an X-ray analysis [262]. It is assumed that **279** results from reaction between a carbene dimer (diethyl fumarate) and an intermediate N-ylide or the isomeric aziridine. Indeed, a three-component reaction between ethyl diazoacetate, **278** and dimethyl fumarate produced the tricyclic system **280**.

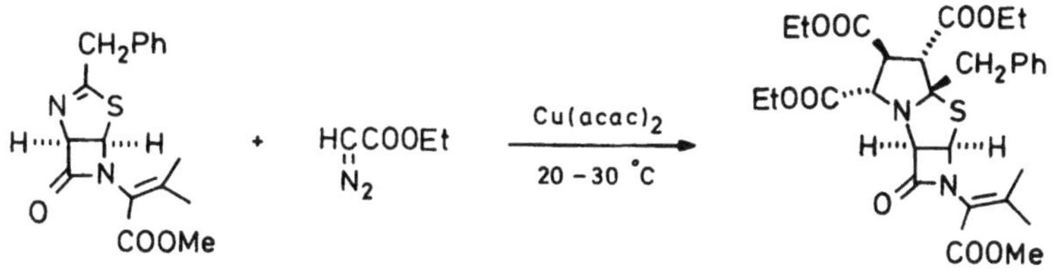

278 *279*

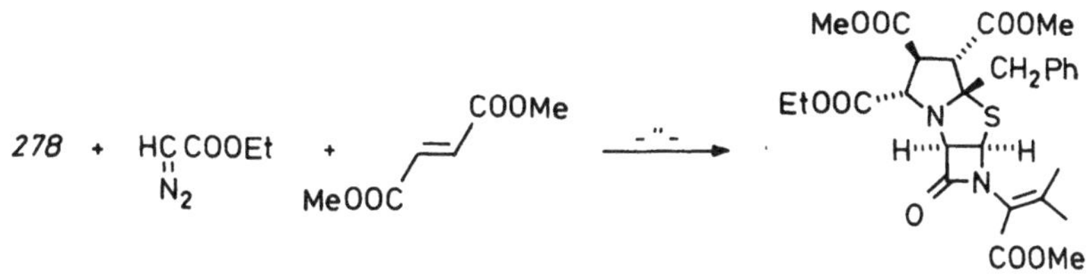

280

The outcome of the copper-catalyzed decomposition of a diazo compound in the presence of a 1,1-diarylmethanimine depends on the nature of the diazo compound. With diazodiphenylmethane, the N/H insertion product **281** and the isomeric imine

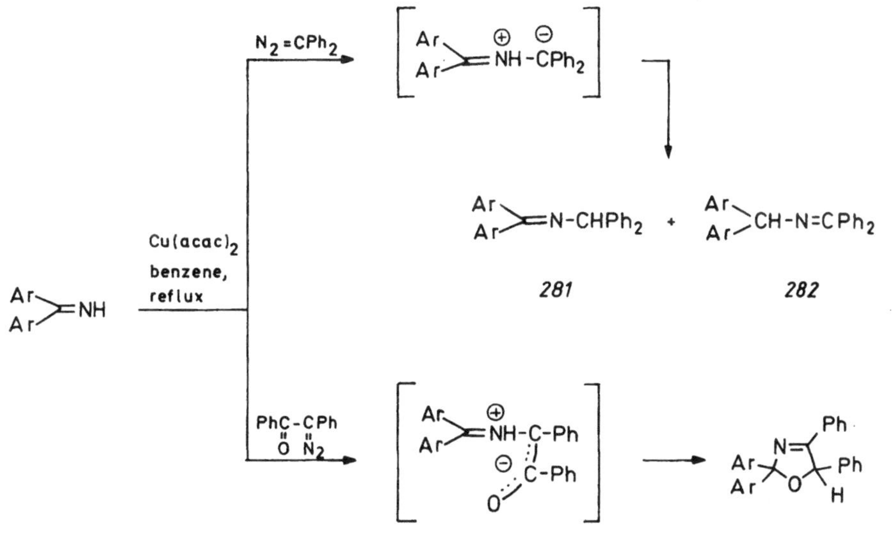

281 *282* *283*

Ar = C_6H_5, 4-Me-C_6H_4 etc.

282 are formed [263]. Both compounds which are not interconvertible under the reaction conditions, are likely to result from a common N-ylide intermediate. With azibenzil, however, 3-oxazolines **283** are the reaction products, although they are obtained in low yield only [264]. The latter result contrasts with that of the purely thermal reaction [265]. There, of course, Wolff rearrangement-derived products appear, but copper catalysts are known to prevent this rearrangement.

Alkyl diazoacetates react with N,N'-diisopropylcarbodiimide in the presence of $Cu(OTf)_2$ or $Rh_2(OAc)_4$ to give 5-alkoxy-4-oxazolines **284** rather than iminoaziridines [266,267].

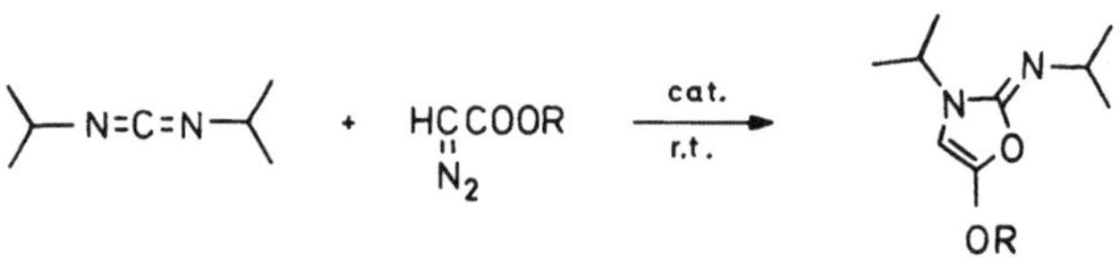

284
(~70 %)

R = Me, Et, *n*-Bu
cat. = $Cu(OTf)_2$, $Rh_2(OAc)_4$

5.2 Reaction with C=O

Interaction of a carbonyl group with an electrophilic metal carbene would be expected to lead to a carbonyl ylide. In fact, such compounds have been isolated in recent years [14]; the strategy comprises intramolecular generation of a carbonyl ylide whose substituent pattern guarantees efficient stabilization of the dipolar electronic structure. The highly reactive 1,3-dipolar species are usually characterized by [3 + 2] cycloaddition to alkynes and activated alkenes. Furthermore, cycloaddition to ketones and aldehydes has been reported for 1-methoxy-2-benzopyrylium-4-olate **286**, which was generated by $Cu(acac)_2$-catalyzed decomposition of *o*-methoxycarbonyl-ω-diazoacetophenone **285** [268].

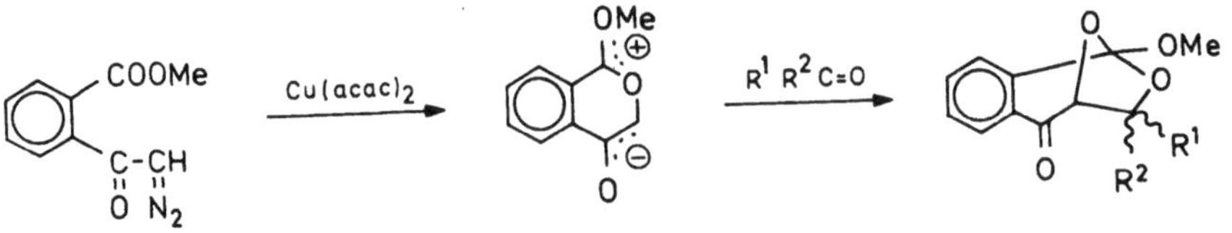

285 *286*

Very recently, the first synthesis of the mesoionic 1,3-dioxolium-4-oxide **288a** by palladium-promoted decomposition of the aryldiazoacetic benzoic anhydride **287a** has been realized. Being too unstable for isolation, **288a** was trapped by [3 + 2] cycloaddition to several acetylenes, ultimately giving furans **289** in good yield [269]. $Cu(acac)_2$ proved to be inefficient for decomposition of the diazoacid anhydride, but served well for ylide generation from **287b**. Once again, the dipolar species **288b** was trapped by an acetylenic dipolarophile [269]. Similar reactions with acenaphthylene and N-methylmaleimide as dipolarophiles have also been reported [269].

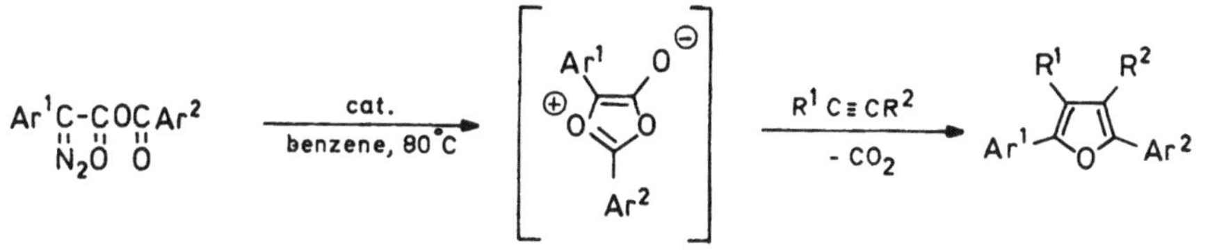

287 *288* *289*

a: Ar^1 = 4-NO_2-C_6H_4; Ar^2 = Ph; cat. = $[(\eta^3\text{-}C_3H_5)PdCl]_2$

b: Ar^1 = Ar^2 = 4-Cl-C_6H_4; cat. = $Cu(acac)_2$

Intramolecular carbonyl ylide formation was also invoked to explain the formation of the 4*H*-1,3-oxazin-5(6*H*)-ones **291a, b** upon copper-catalyzed decomposition of diazoketones **290a, b** [270]. Oxapenam **292**, obtained from **290b** as a minor product, originates from an intermediary attack of the carbenic carbon at the sulfur atom. In fact, this pathway is followed exclusively if the C(Me, COOMe) group in **290b** is replaced by a CH_2 function (see Sect. 7.2).

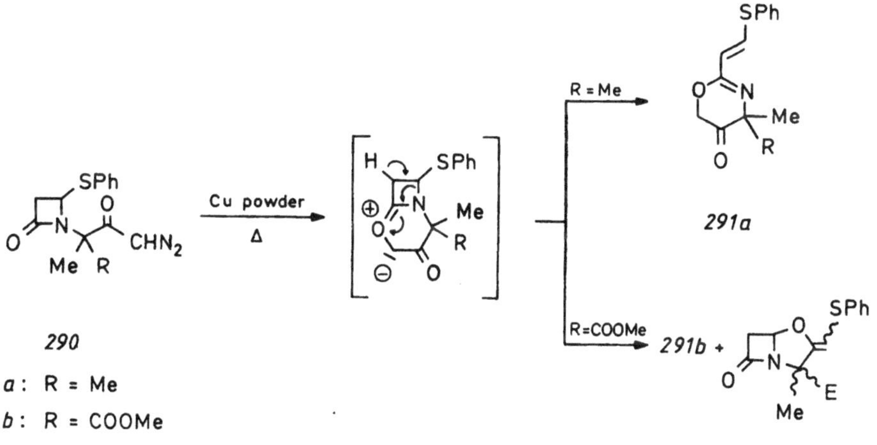

290

a: R = Me

b: R = COOMe

291a

292

(3 diastereomers)

Interaction between a carbonyl oxygen and a metal carbene leading to a transient carbonyl ylide may also be considered to be involved in the production of a vinyl ether by CuCl-catalyzed decomposition of ethyl diazoacetate in the presence of enolizable ketones such as acetone and cyclohexanone [271,272]. The carbonyl ylide derived from benzaldehyde and bis(methoxycarbonylcarbene) was trapped by 1,3-dipolar cycloaddition either to excess aldehyde [273] or to an electron-poor olefin [274]. In both cases, improved yields distinguish the metal-catalyzed decomposition of the diazo precursor over the purely thermal reaction. As has been shown for the first mentioned reaction, the product yield and distribution (1,3-dioxolanes **293a, b** and oxiranes **294**) depends on the catalyst, among other parameters [273].

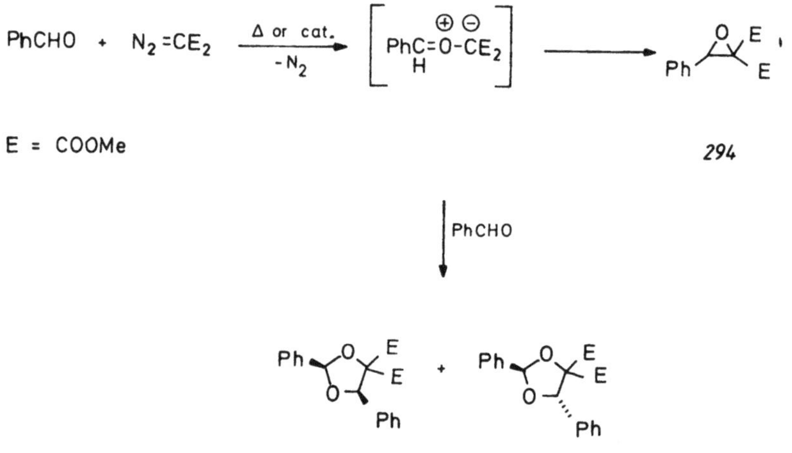

E = COOMe

294

293a *293b*

conditions[a]	% (293a + 293b)	203a:293b	% 294
Δ(125 °C)	56	58:42	10
$Cu(acac)_2$ (125 °C)	82	55:45	7
$Rh_2(OAc)_4$ (75 °C)	72	55:45	—
CuOTf (25 °C)	87	71:29	—

[a] An excess of PhCHO was used. In the metal-catalyzed reactions, the diazomalonate was added gradually.

The copper-catalyzed decomposition of methyl 2-diazo-3-oxobutyrate **57a** in the presence of aldehydes gives ready excess to 1,3-dioxole-4-carboxylates [275]. Copper(II)

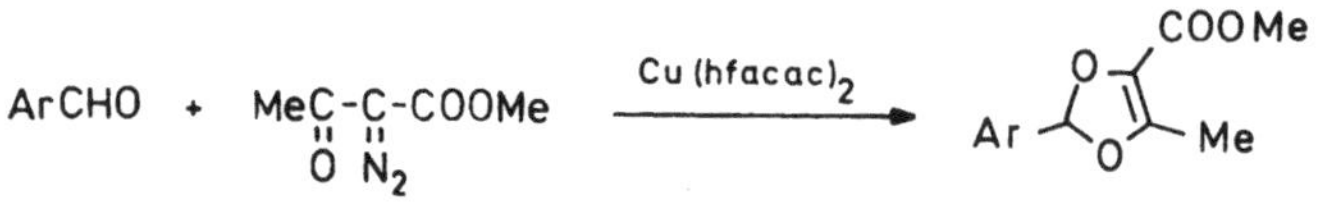

57a 76 - 83%

Ar = *n*-Pr, *i*-Pr, Ph, 2-furyl, *trans*-CH=CHMe

[*57a*] : [catalyst] = 34 : 1

hexafluoroacetylacetonate seems to be the catalyst of choice, but the yields are dependent on the catalyst: diazoketone ratio.

The reaction, formally speaking a [3 + 2] cycloaddition between the aldehyde and a ketocarbene, resembles the dihydrofuran formation from **57a** or similar α-diazoketones and alkenes (see Sect. 2.3.1). For that reaction type, 2-diazo-1,3-dicarbonyl compounds and ethyl diazopyruvate **56** were found to be suited equally well. This similarity pertains also to the reactivity towards carbonyl functions; 1,3-dioxole-4-carboxylates are also obtained by copper chelate catalyzed decomposition of **56** in the presence of aliphatic and aromatic aldehydes as well as enolizable ketones [276]. No such products were reported for the catalyzed decomposition of ethyl diazoacetate in the presence of the same ketones [271,272]. The reasons for the different reactivity of ethoxycarbonylcarbene and α-ketocarbenes (or the respective metal carbenes) have only been speculated upon so far [276].

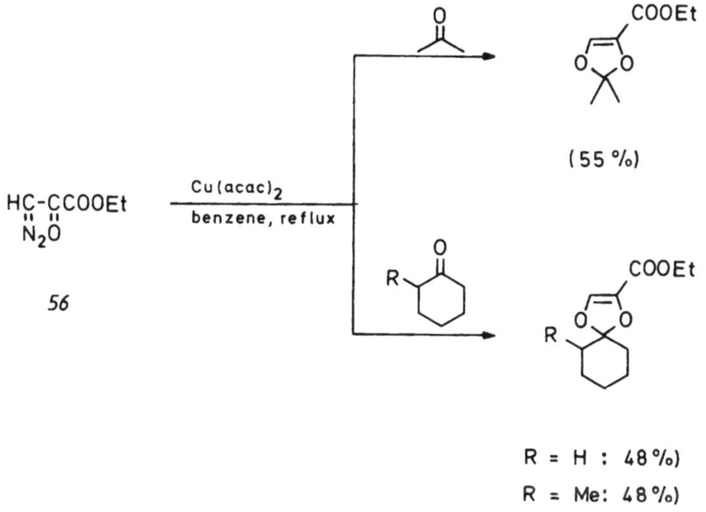

6 Insertion into X—H Bonds

6.1 Insertion into Aliphatic C—H Bonds[3]

Intramolecular C/H insertion by copper-catalyzed decomposition of α-diazoketones provides a convenient cyclization procedure which is limited, however, to diazo compounds which allow energetically favorable realization of the transition state leading to the cyclized product.

This prerequisite is well documented by quite a few examples [12] and has been underlined again by a recently published study on the $CuSO_4$-catalyzed decompo-

[3] Recent results dealing with insertion reactions into allylic (Sect. 2.3.3), aromatic (Sect. 4.1) and heteroaromatic (Sect. 4.2) C—H bonds have already been discussed.

sition of 1-methylcycloalkyl diazomethyl ketones **295** [277]. On heating in cyclohexane, cyclopentanones **296**, cyclobutanones **297** and solvent insertion products **298** are formed; whereas C/H insertion leading to the four-membered ring remains a minor side-reaction in each case, cyclopentanone formation is markedly dependent on the ring size of the cycloalkane residue. This points to the importance of ring conformation for supplying the required proximity of the functional groups involved in the insertion process. In the chain-extended α-diazoketone **299**, entropy factors are likely to prevent productive intramolecular C/H insertion. The formation of a solvent insertion product can be suppressed when the reaction is run in Freon TF instead of cyclohexane, but the yield of the cyclization products is not improved thereby [277].

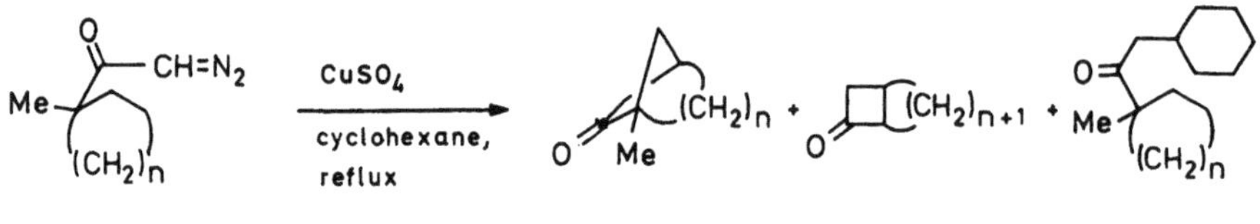

295 *296* *297* *298*

	296	297	298
n = 4	72%	9%	>5%
n = 3	58%	4%	18%
n = 2	19%	3%	9%
n = 1	1%	—	29%

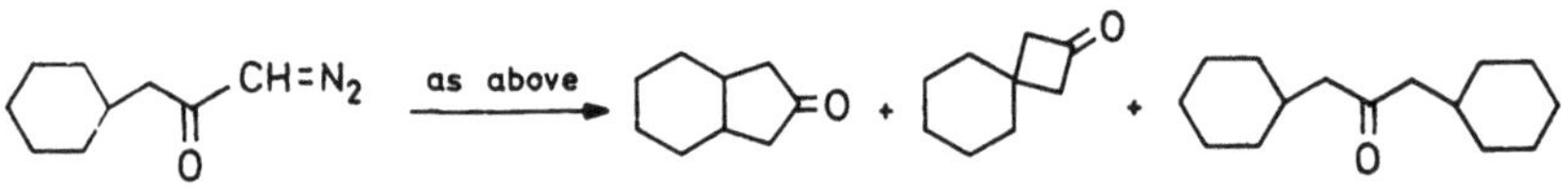

299 (*cis* : 9%) (*trans* : 5%) (1%) (23%)

Nickel(II) acetylacetonate turned out to be a highly efficient, homogeneous catalyst not only for intramolecular cyclopropanation (see Sect. 2.4) but also for intramolecular insertion reactions. For the synthesis of bicyclo[3.2.1]octan-6-ones **301** and **302** from diazoacetyl-cyclohexanes **300**, it proved superior to heterogeneous (Cu_2O, CuO, $CuSO_4$) and homogeneous (CuOTf) copper catalysts as well as to palladium(II) and cobalt (III) acetylacetonate [278]. Similarly, the yield of the $Ni(acac)_2$-catalyzed regioselective C/H insertion reaction **303** → **304** is about twice as high as that of the copper-catalyzed reaction [278]. The regioselectivity of the insertion process for **300** (R^1 = H, R^2 = aryl) corresponds to expectations [279]; it remains unaltered by using $Ni(acac)_2$ instead of copper catalysts.

The use of rhodium(II) acetate in carbenoid chemistry has also been extended to promoting intramolecular C/H insertion reactions of ketocarbenoids [277,280,280a]. From the α-diazo-β-ketoester **305**, highly functionalized cyclopentane **306** could thus be constructed in acceptable yields by regiospecific insertion into an unactivated

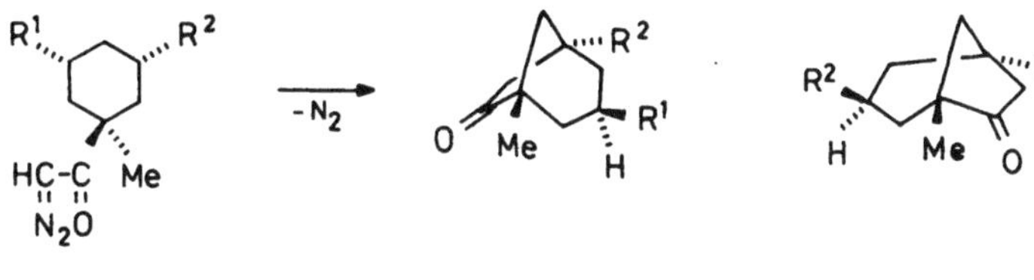

300 *301* *302*

R^1	R^2	conditions	catalyst	% (*301* + *302*)	*301*:*302*
H	Ph	cyclohexane, reflux, hv (tungsten lamp)	Cu_2O, CuO or $CuSO_4$	60–66	85:15 to 75:25
		hv (tungsten lamp)	CuOTf	60	85:15
		hv (tungsten lamp)	$Ni(acac)_2$	86	90:10
H	4—OMe—C_6H_4	hv (tungsten lamp)	Cu_2O, CuO or $CuSO_4$	60–75	83:17 to 70:30
		hv (tungsten lamp)	CuOTf	60	85:15
		hv (tungsten lamp)	$Ni(acac)_2$	85	85:15
Me	Ph	hv(tungsten lamp)	Cu_2O, CuO or $CuSO_4$	60–70	30:70 to 48:52
			CuOTf	65	48:52
			$Ni(acac)_2$	92	45:55

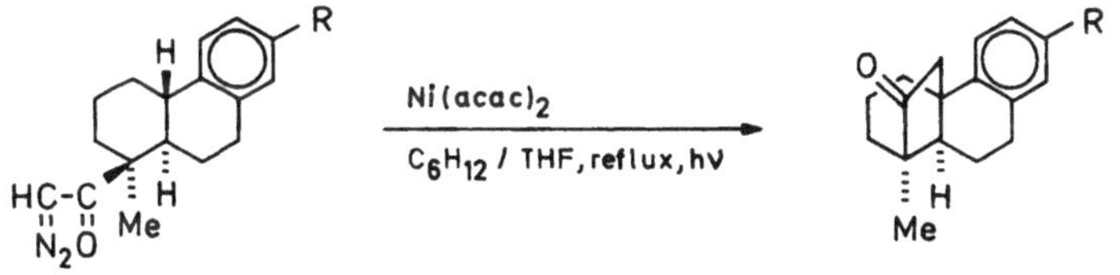

303 *304*

R = H : 82 %
R = OMe: 80 %

C—H bond [174,280,281]. For comparison, only trace amounts of cyclopentane resulted from the $CuSO_4$-catalyzed decomposition of 1-diazo-2-octanone or 1-diazo-4,4-dimethyl-2-pentanone [277]. It is obvious that the use of $Rh_2(OAc)_4$ considerably extends the scope of transition-metal catalyzed intramolecular C/H insertion, as it allows for the first time, *efficient* cyclization of ketocarbenoids derived from freely rotating, acyclic diazoketones. This cyclization reaction can also be highly diastereoselective, as the exclusive formation of a *trans*-2,3-disubstituted cyclopentane carboxylate from **307** shows [281a]. The stereoselection has been rationalized by assuming a chair-like six-membered transition state for the insertion process, with both Ph and Me in equatorial position.

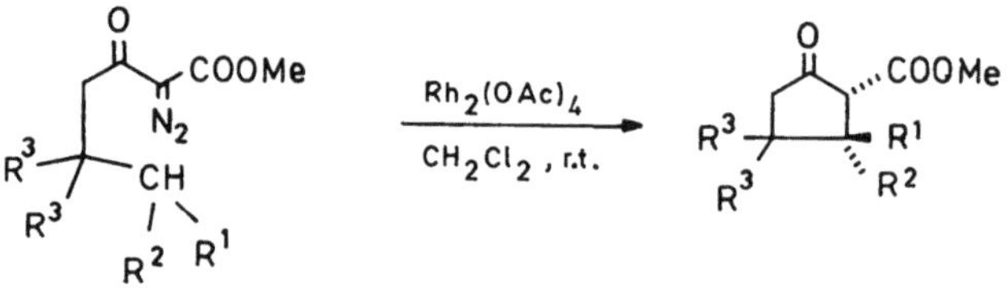

305 306

R^1	R^2	R^3	% 306
H	n-C_8H_{17}	H	68
i-Pr	H	H	55
Me	Me	H	64
$CH=CH_2$	H	H	48 (+ cyclopropane, 10%)
$CH_2-CH=CMe_2$	H	Me	77
CH(Me)⌒⌒=⟨	H	H	58
$CH_2-\equiv-(CH_2)_5-O-THP$	H	H	>74[a]
$-(CH_2)_8-O-THP$	H	H	>87[a]

[a] Relative stereochemistry not reported

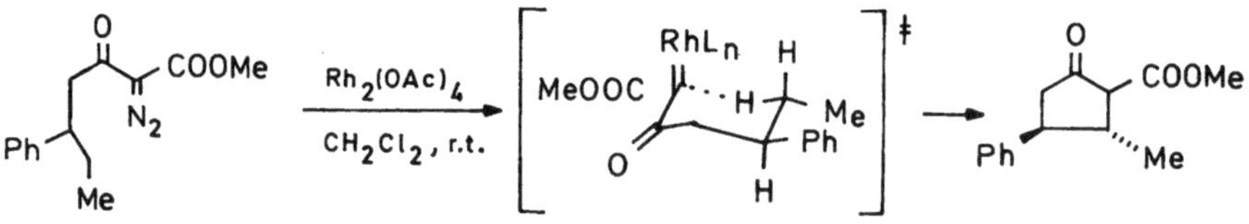

307

Copper-catalyzed ketocarbenoid C/H insertion has been shown to occur with retention of configuration [282], and the same is true for the $Rh_2(OAc)_4$-promoted reaction. Advantage has been taken of this fact for a synthesis of (+)-α-cuparenone,

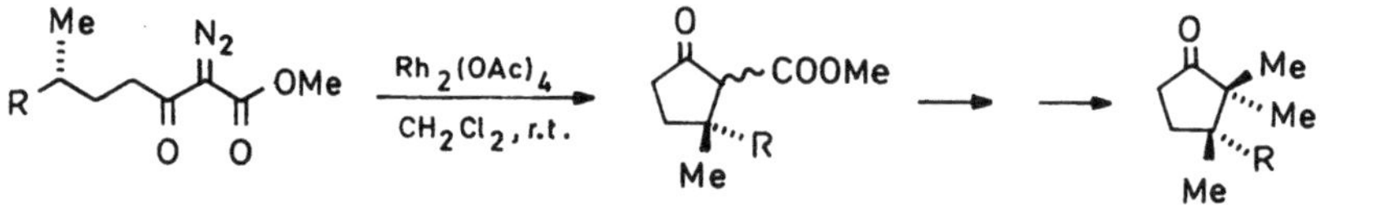

308 309 (67%) (+)-α-cuparenone

R = —⟨◯⟩—Me

the key step of which consists of intramolecular C/H insertion at the chiral C-6 of methyl 2-diazo-6-(4-tolyl)-3-oxoheptanoate **308** leading to a diastereomeric mixture of cyclopentanes **309** [283].

Chiral induction at the γ-CH_2 group of diazoesters **310** could be realized when a chiral ester residue was present [284]. The best results were obtained with esters derived from 2-aryl-3-hydroxybornanes. In order to explain the observed diastereoselectivities (**312a:312b**), it has been proposed again that the transition state geometry of the intermediate metal carbene is that of a chair-like six-membered ring which includes the hydrogen atom to be transferred. Provided that the C/H insertion occurs with retention of configuration (see above) and that the substituent R occupies an equatorial position, the two transition states **311a** and **311b** can be imagined which are enantiomeric to each other (leaving aside the chiral substituent R*). The front face of the prochiral carbenoid carbon being shielded by the naphthyl group, **311b** becomes energetically disfavored compared to **311a**, so that insertion product **312a** is formed preferentially.

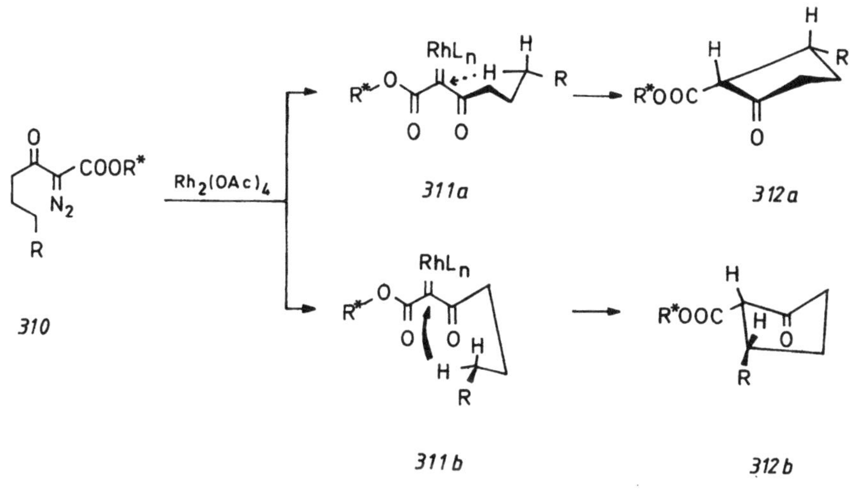

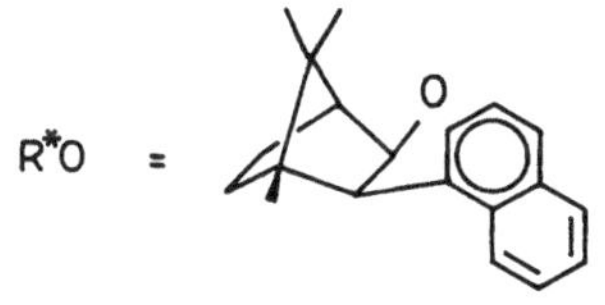

R	*312a*:*312b*
n-C_5H_{11}	87:13
CH=CH_2	92: 8
i-Pr	83:17
t-Bu	83:17
Ph	85:15

For 5-(2-diazo-1,3-dioxobutyl)-1-oxa-5-azaspiro[5,5]undecane (**313**), intramolecular carbenoid insertion into a (N)C—H bond represents quite an unusual way of constructing a β-lactam ring [285].

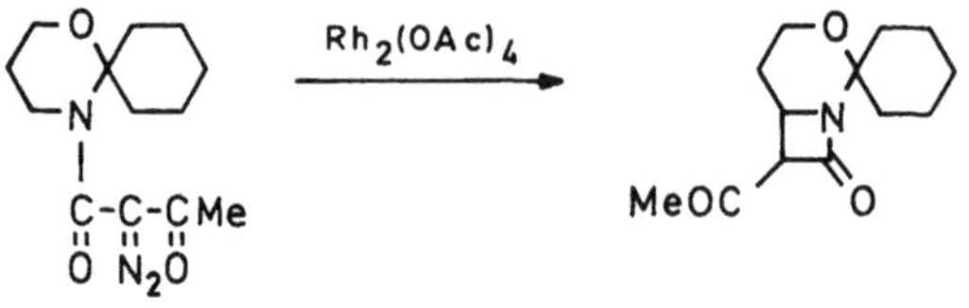

313

Similar to the *intra*molecular insertion into an unactivated C—H bond, the *inter*molecular version of this reaction meets with greatly improved yields when rhodium carbenes are involved. For the insertion of an alkoxycarbonylcarbene fragment into C—H bonds of acyclic alkanes and cycloalkanes, rhodium(II) perfluorocarboxylates [286], rhodium(II) pivalate or some other carboxylates [287,288], and rhodium(III) porphyrins [287] proved to be well suited (Tables 19 and 20). In the era of copper catalysts, this reaction type ranked as a quite uncommon process [14], mainly because the yields were low, even in the absence of other functional groups in the substrate which would be more susceptible to carbenoid attack. For example, $CuSO_4$(CuCl)-catalyzed decomposition of ethyl diazoacetate in a large excess of cyclohexane was reported to give 24% (15%) of C/H insertion, but 40% (61%) of the two "carbene dimers" [289].

As Table 20 shows, the yields of the Rh(II)-promoted reaction are temperature-dependent. Furthermore, competitive experiments between pairs of alkanes revealed a marked dependence on the alkoxy group of the diazoester and on the perfluoroalkyl carboxylate part of the catalyst. The observed relative selectivities have been taken as evidence for the importance of lipophilic interactions between carbenoid and alkane.

In the presence of rhodium(II) acetate and trifluoracetate as well as iodorhodium(III) *meso*-tetraphenylporphyrin (**47a**), insertion into the terminal CH_3 group of an *n*-alkane is highly disfavored as compared to insertion into the various CH_2 groups (Tables 19 and 20). In contrast, photochemically generated alkoxycarbonylcarbenes usually display a lower selectivity in competitive methyl/methylene insertion reactions [7,282,290]. If, however, iodorhodium(III) *meso*-tetramesitylporphyrin (**47c**) serves as catalyst, insertion into the terminal CH_3 group becomes more favorable and increases steadily as the chain length of the alkane is increased (Table 19). Similarly, replacing Rh(II) acetate by Rh(II) 9-triptycenecarboxylate led to enhanced insertion into primary C—H bonds: Relative yields of insertion into C_1—H, C_2—H, C_3—H of pentane changed from 4/63/33 to 30/61/9, and for 2,3-dimethylbutane, the 3°/1° insertion ratio decreased from 115/1 to 12/1 (corrected for number of hydrogens) [228]. These results underline the crucial role of bulky groups close to the reaction site in determining the regioselectivity of the C/H insertion process; the remarkable ability of iodorhodium(III) *meso*-tetramesitylporphyrin to alter the stereochemical course of intermolecular cyclopropanation reactions has already been mentioned (see Sect. 2.2.3). With rhodium(III) *meso*-tetraarylporphyrins as catalysts,

Table 19. Yields of C/H monoinsertion products from rhodium-catalyzed decomposition of ethyl diazoacetate in n-alkanes (60 °C, molar ratio $4 \cdot 10^3$–$1 \cdot 10^4$ (alkane)/10^3 (diazoester)/1 (catalyst)[a]

n-alkane	catalyst[b]	total yield [%]	relative yields [%] of attack at						1°/2°[c]
			C-1	C-2	C-3	C-4	C-5	C-6	
n-hexane	Rh piv	50	5	62	33				0.07
	Rh TPPI	46	8	71	21				0.12
	Rh TMPI	36	25	61	14				0.44
n-octane	Rh piv	33	3	49	27	21			0.06
	Rh TPPI	35	6	52	22	20			0.13
	Rh TMPI	80	21	52	14	13			0.53
n-decane	Rh piv	29	2	42	20	18	18		0.05
	Rh TPPI	40	5	40	19	18	18		0.14
	Rh TMPI	24	20	48	11	10	11		0.67
n-dodecane	Rh piv	13	1	32	17	15	18	17	0.03
	Rh TPPI	17	4	33	20	13	15	15	0.14
	Rh TMPI	21	20	45	10	9	8	8	0.83

[a] From Ref. [287]; [b] Rh piv = $Rh_2(OOCCMe_3)_4$; Rh TPPI = iodorhodium(III) *meso*-tetraphenylporphyrin (**47a**); Rh TMPI = iodorhodium(III) *meso*-tetramesitylporphyrin (**47c**); [c] Ratio of $CH_3:CH_2$ insertion, corrected for number of H atoms.

Table 20. Yields of C/H insertion products in the $Rh_2(CF_3COO)_4$-catalyzed decomposition of ethyl diazoacetate in alkanes (22 °C; 100 mmol of cycloalkane or 200 mmol of acyclic alkane, 3 mmol of diazoester, 2.0–2.2 · 10^{-3} mmol of catalyst)[a]

alkane	yield [%][b]	alkane	yield [%]	rel. yield of insertion at			
				C-1	C-2	C-3	C-4
cyclopentane	50(68)	n-pentane	65(92)	7	66	27	
cyclohexane	78(90)		46	12	88		
cycloheptane	43(62)	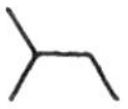	71	5	25	66	4
cyclooctane	64	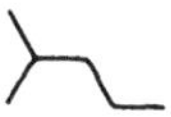	68				

[a] From Refs. [286] and [288]; [b] Yields at boiling point of alkane are given in parentheses.

it is also possible to obtain considerable amounts of CHCOOEt insertion at the methyl groups of toluene, *p*-xylene and mesitylene [287b]. This happens at the expense of cyclopropanation of the aromatic ring, which occurs exclusively or nearly so in the presence of $Rh_2(OAc)_4$ [287b] or $Rh_2(CF_3COO)_4$ [228]. For example, the insertion: cyclopropanation ratio for the system mesitylene/ethyl diazoacetate changes from 1:99 with $Rh_2(OAc)_4$ to 78:22 with **47c** as catalyst. If one pictures the geometry of approach between rhodium carbenoid and substrate as given in **314** (for C/H insertion) and **315** (for cyclopropanation), one recognizes that the latter is disfavored because of higher steric interaction between the two reactants.

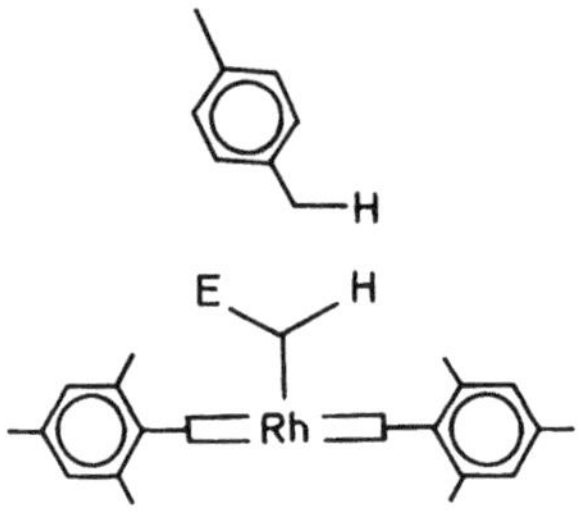

314

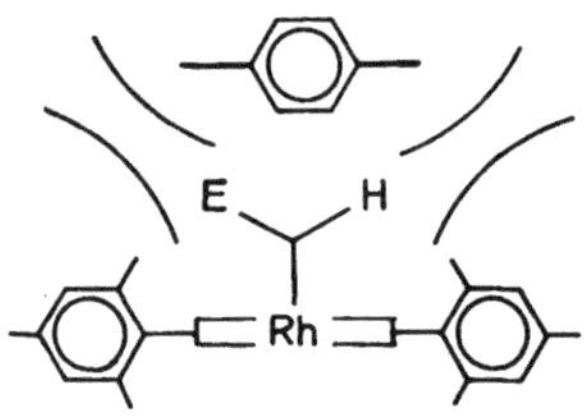

315

6.2 N/H Insertion[4]

Insertion of a carbene unit into the N—H bond of primary or secondary amines by copper salt catalyzed decomposition of diazo compounds has been known for a number of years [14)]. The copper chelate promoted reaction of diazodiphenylmethane [291)] or 2-diazo-1,2-diphenyl-1-ethanone [292)] with primary benzylamines or

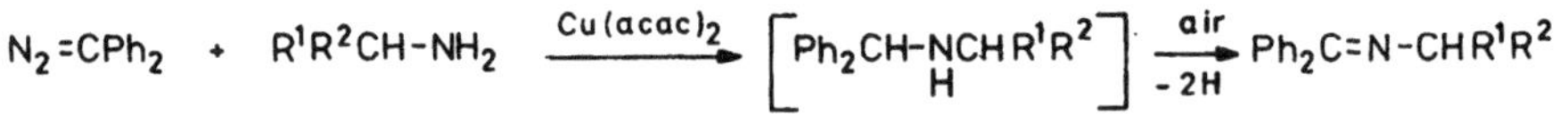

316 (54 - 68%)

R^1	Me	Ph	Me
R^2	Ph	Ph	4-Me-C_6H_4

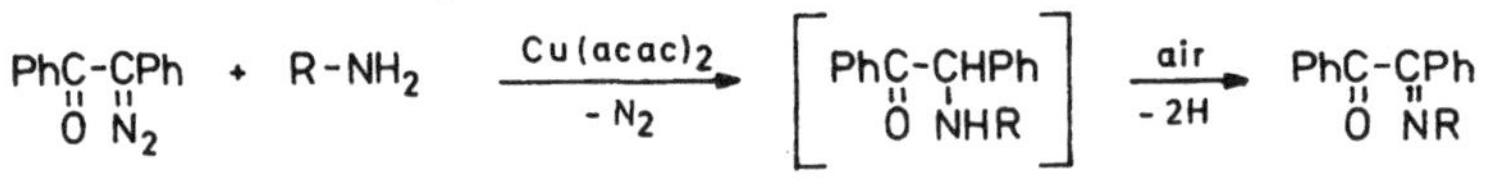

317

R = $CHPh_2$: 43%
R = CHMePh : 28%
R = Ph : 53%

[4] Formal insertion reactions into the N—H bond of imidazoles (Sect. 4.2) and imines (Sect. 5.1) have already been discussed.

aniline does not stop at the stage of the respective insertion products (**316** and **317**), however, since these compounds are readily dehydrogenated to produce Schiff bases.

A similar reaction sequence allows the preparation of symmetrical and unsymmetrical ketazines **318** from hydrazones and diazodiphenylmethane or 2-diazo-1,2-diphenyl-1-ethanone [293].

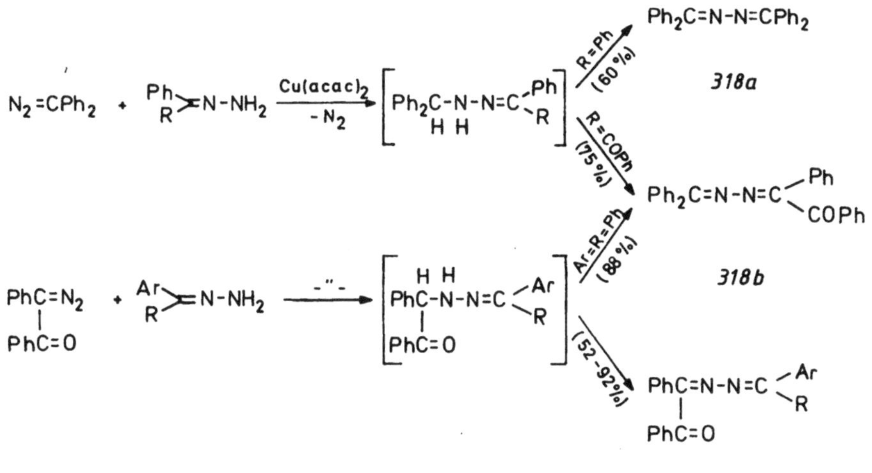

318a *318b* *318c*

Ar = Ph, 4-X-C_6H_4 (X = OMe, Cl, Me)

R = aryl, COAr

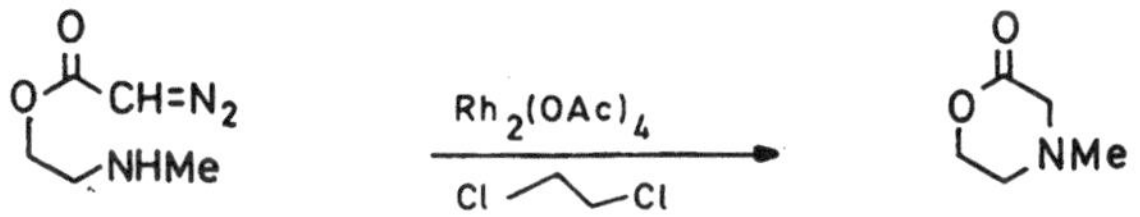

319a (~18%)

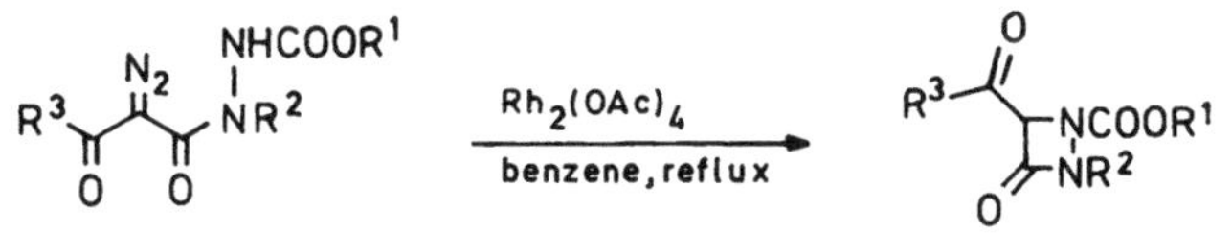

319b (75 - 100%)

R^1 = *t*-Bu, $PhCH_2$

R^2 = $PhCH_2$, CH_2COOEt

R^3 = OEt, Me

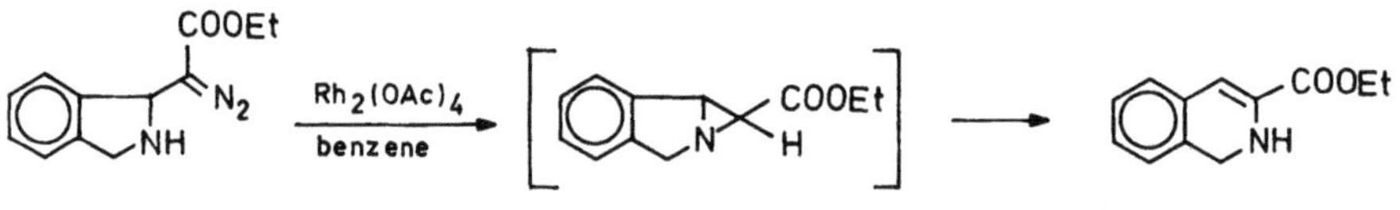

320 321

Rhodium(II) acetate has proven its versatility also in the field of ketocarbenoid N/H insertion. With diazocarbonyl compounds **319a** [294a)] and **319b** [294b)], intramolecular versions of this reaction have been realized. Furthermore, it was suggested that the transformation of diazo ester **320** into dihydroisoquinoline **321** begins with an intramolecular ketocarbenoid N/H insertion, followed by spontaneous ring-opening of the aziridine intermediate [295)]; the possibility of direct ring expansion at the carbene or carbenoid stage has not been considered.

$Rh_2(OAc)_4$ has become the catalyst of choice for insertion of carbene moieties into the N—H bond of β-lactams. Two cases of intermolecular reaction have been reported. The carbene unit derived from alkyl aryldiazoacetates **322** seems to be inserted only into the ring N—H bond of **323** [246)]. Similarly, N-malonyl-β-lactams **327** are available from diazomalonic esters **325** and β-lactams **326** [297)]. If, however, the acetate function in **326** is replaced by an alkylthio or arylthio group, C/S insertion rather than N/H insertion takes place (see Sect. 7.2). Reaction of ethyl diazoacetoacetate **57b** with **328** also yields an N/H insertion product (**329**) [298)], rather than ethyl 1-aza-4-oxa-3-methyl-7-oxabicyclo[3.2.0]hex-2-ene-2-carboxylate, as had been claimed before [299)].

For intramolecular N/H insertion involving a β-lactam, $Rh_2(OAc)_4$ was found to be superior to other catalysts and to the photochemical route [300)]. Therefore, this procedure has been appraised to be the most efficient one for constructing a bicyclic β-lactam and, consequently, has become a standard method for synthesizing

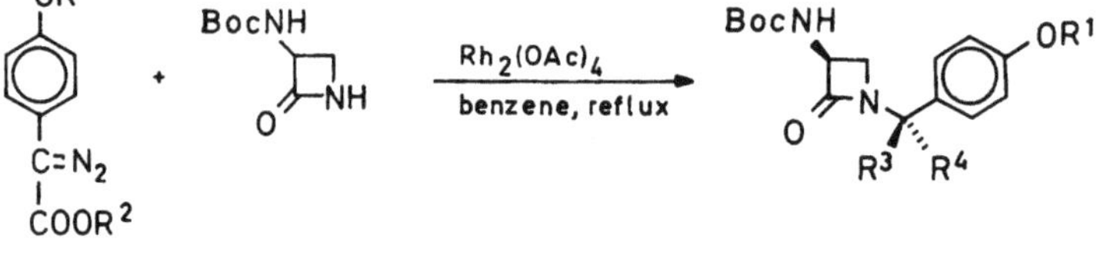

322 323 324a : R^3 = H, R^4 = $COOR^2$

324b : R^3 = $COOR^2$, R^4 = H

Boc = *t*-BuOOC

R^1	R^2	% *324*	*324a*:*324b*
Me	*t*-Bu	26	4:3
Me	Et	60	4:3
CH_2Ph	CH_2Ph	67	1:1.2

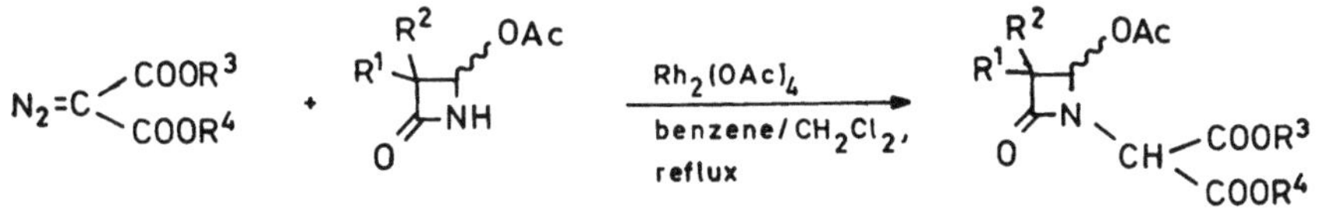

325 *326* *327*

(22 - 28%)

R^1 = H, Me, Et, phthalimido; R^2 = H, Me
R^3 = Me, 4-nitrobenzyl; R^4 = 4-nitrobenzyl

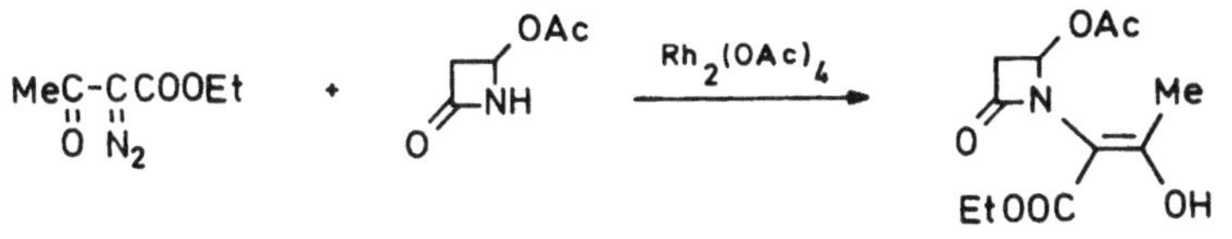

57b *328* *329*

carbapenam, oxapenam, carbacephem and oxacephem systems from 2-azetidinones bearing an appropriate diazocarbonyl-containing side chain at C-4. Examples are given in Table 21. It will be noted that even highly functionalized azetidinones undergo the cyclization reaction in high yield.

Table 21. $Rh_2(OAc)_4$-catalyzed intramolecular carbenoid insertion into the N—H bond of β-lactams.

Diazo compound[a]	Reaction conditions[c]	Product[a]	Yield [%]	Ref.
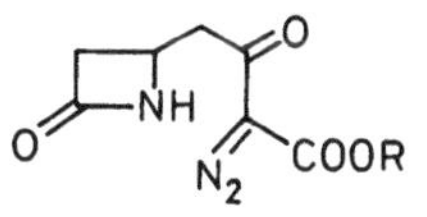	Benzene, 80°C	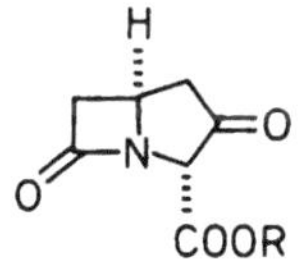		
		R = Et	93	301)
		R = CH_2Ph	77-85	300)
		R = PNB	81	302)
	CH_2Cl_2, r.t.	R = ONB	43	303)
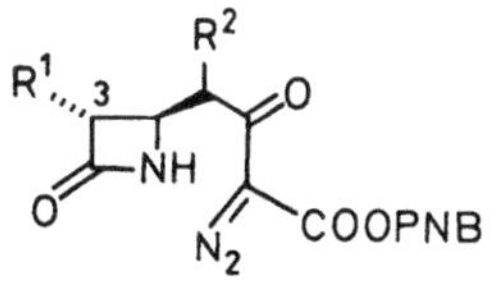	Benzene, 80°C	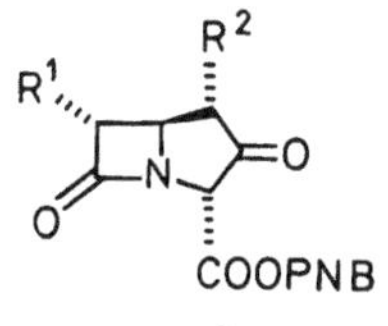		
		R^1 = 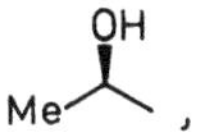, R^2 = H	100	304)

Diazo compound[a]	Reaction conditions[c]	Product[a]	Yield [%]	Ref.
		R^1 = 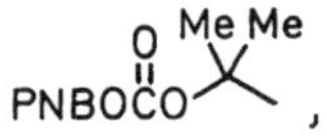 , R^2 = H	95	305)
		ring closure of C-3 epimer	100	306)
		R^1 = 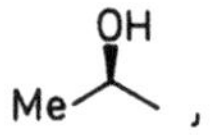 , R^2 = Me	not given	307)
		R^1 = 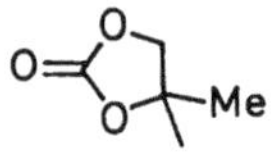 , R^2 = H	not given	308)
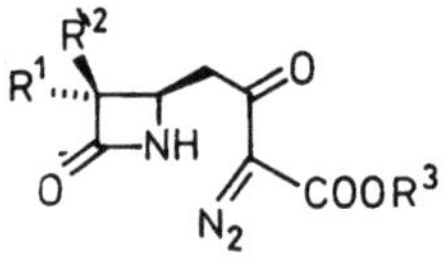	Benzene, 80°C	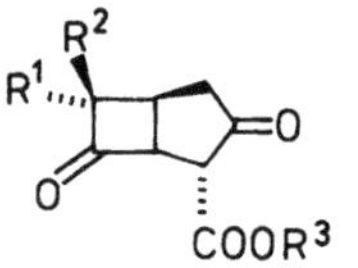 R^1 / R^2 / R^3: Et, H, t-Bu i-Pr, H, t-Bu Me, Me, t-Bu Et, H, CH_2Ph		309)
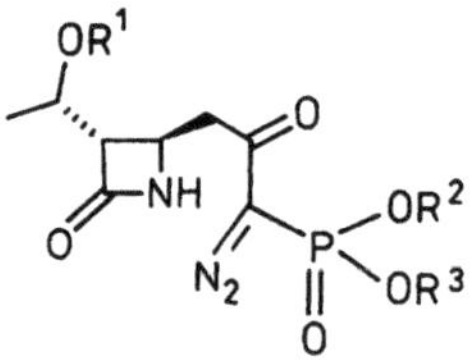	Benzene, 80°C	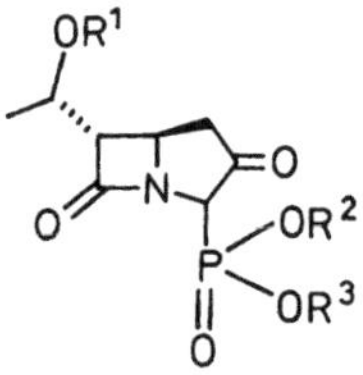 R^1 = t-$BuMe_2Si$, R^2 = R^3 = Me R^1 = t-$BuMe_2Si$, R^2 = R^3 = CH_2Ph R^1 = H, R^2 = R^3 = CH_2Ph R^1 = H, R^2 = Me, R^3 = Bz	60-90	310)
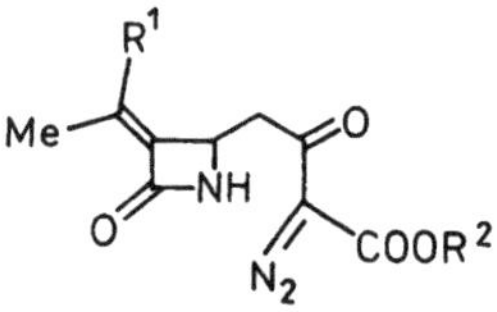	Benzene, 80°C	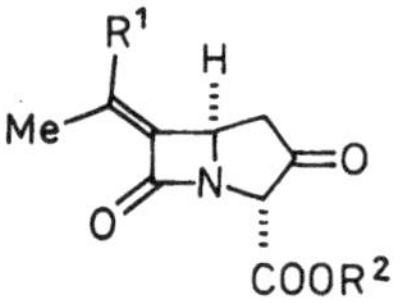		
		R^1 = $CH_2OCOOPNB$, R^2 = PNB	63-99	311)
		R^1 = Me, R^2 = CH_2Ph	not given	312a)
		R^1 = H, R^2 = PNB	86	312b)
		R^1 = CH_2Br, R^2 = PNB	81	312b)
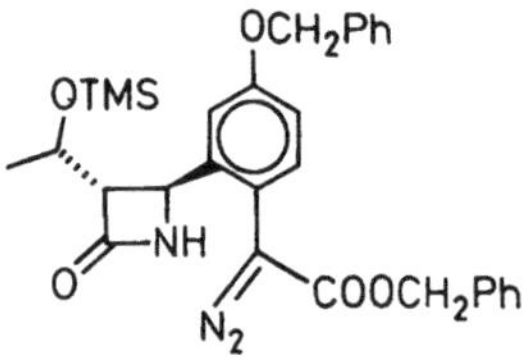	Toluene, 80°C	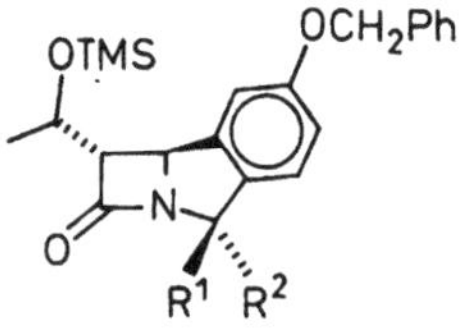 a : R^1 = H, R^2 = $COOCH_2Rh$ b : R^1 = $COOCH_2Ph$, R^2 = H	not given a/b = 2	313)

Diazo compound[a]	Reaction conditions[c]	Product[a]	Yield [%]	Ref.
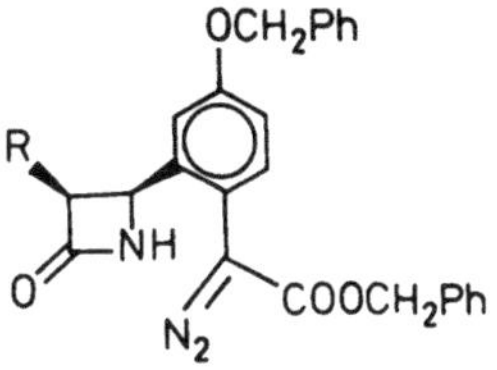 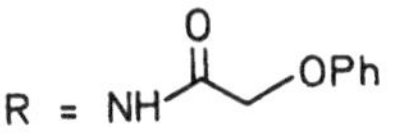	Toluene, 80 °C	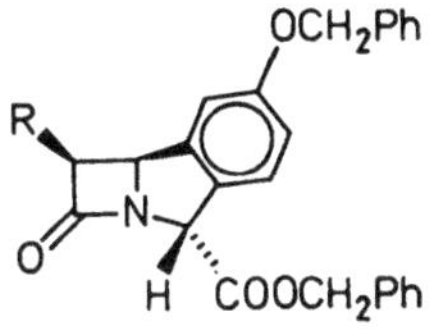	not given	314)
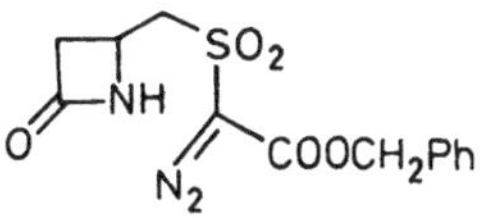	Benzene, 80 °C	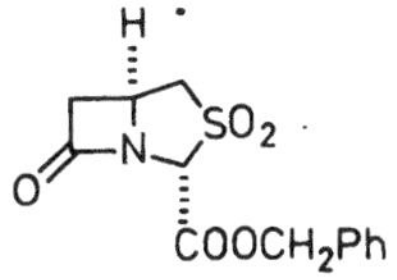	95	315)
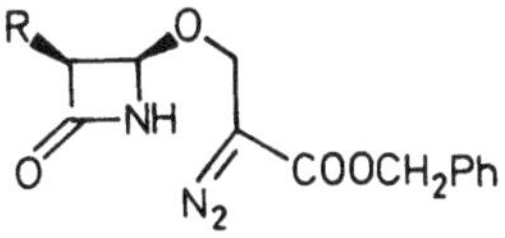 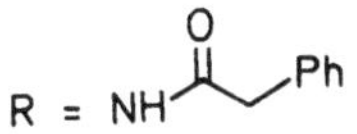	Benzene, r.t.	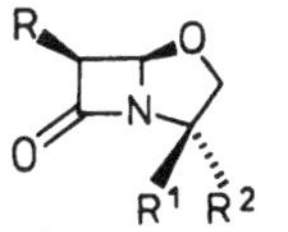 R^1 = H, R^2 = $COOCH_2Ph$	10	316)
		R^1 = $COOCH_2Ph$, R^2 = H	45	
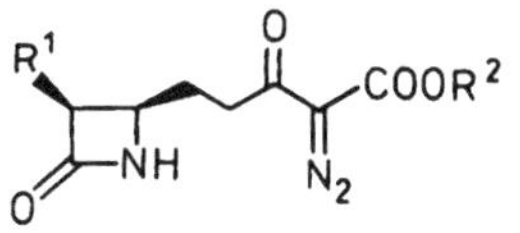		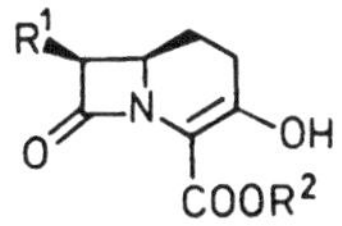		
	Benzene, 75 °C	R^1 = H, R^2 = *t*-Bu, $CHPh_2$, CH_2Ph	>70	317)
	$CHCl_3$, 60 °C	R^1 = NHBoc, R^2 = Me, CH_2Ph	>75	318)
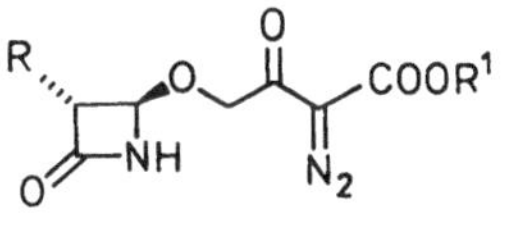 R = NHCOPh	Benzene, 80 °C	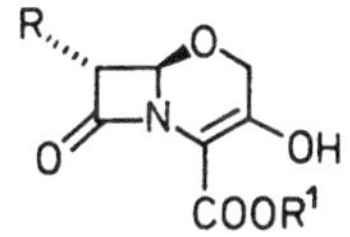 R^1 = PNB, *t*-Bu, $CHPh_2$	53–85	319, 320)
		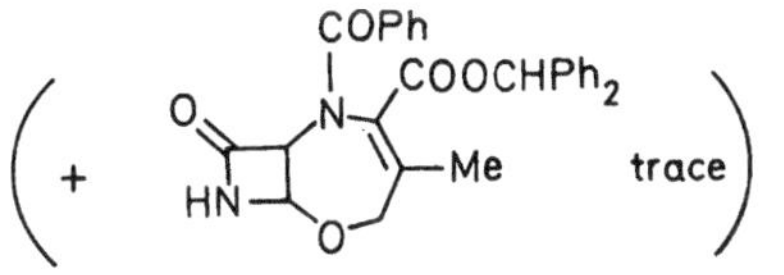	trace	320)

[a] PNB = 4-nitrobenzyl; ONB = 2-nitrobenzyl.
[b] Typically, the diazo compound: catalyst molar ratio is (300–1000):1.

Failure to obtain the desired azacarbacephem **331** had to be accepted with the diazetidinone **330**. Instead of the hoped-for N/H insertion, the ketocarbenoid derived from **330** attacked the more nucleophilic N-1 atom to give an intermediate ammonium ylide which then went on to the products **332** and **334** as suggested in the formula scheme [321]. Evidence for the intermediacy of **333** was provided by the isolation of **335**, when the reaction was carried out in the presence of norbornene.

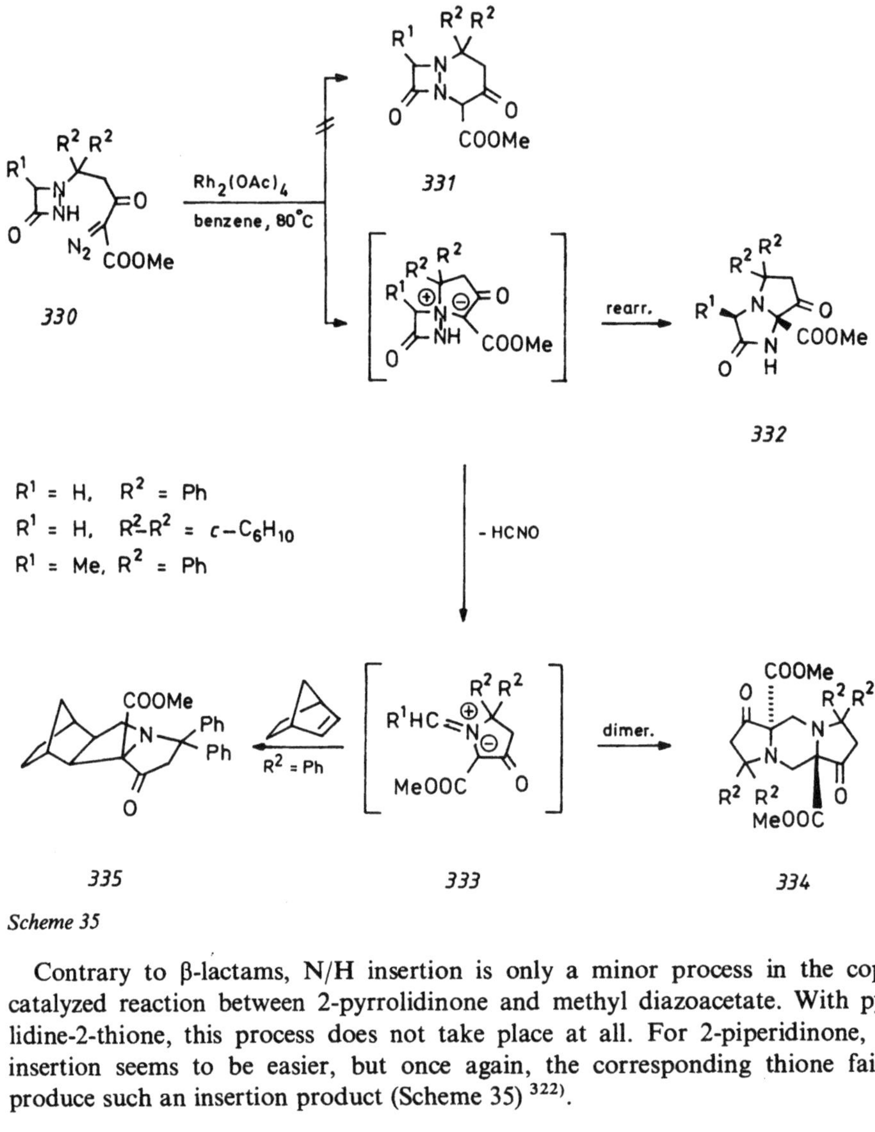

Scheme 35

Contrary to β-lactams, N/H insertion is only a minor process in the copper-catalyzed reaction between 2-pyrrolidinone and methyl diazoacetate. With pyrrolidine-2-thione, this process does not take place at all. For 2-piperidinone, N/H insertion seems to be easier, but once again, the corresponding thione fails to produce such an insertion product (Scheme 35) [322].

6.3 O/H and S/H Insertion

Synthesis of α-alkoxyketones from α-diazocarbonyl compounds and alcohols under the influence of copper or rhodium catalysts is well established as an alternative to the Lewis or proton acid catalyzed variant of this synthetic transformation. The sole recent contribution to the aspect of general reactivity deals with the competition between O/H insertion and cyclopropanation of unsaturated alcohols [162]. The results

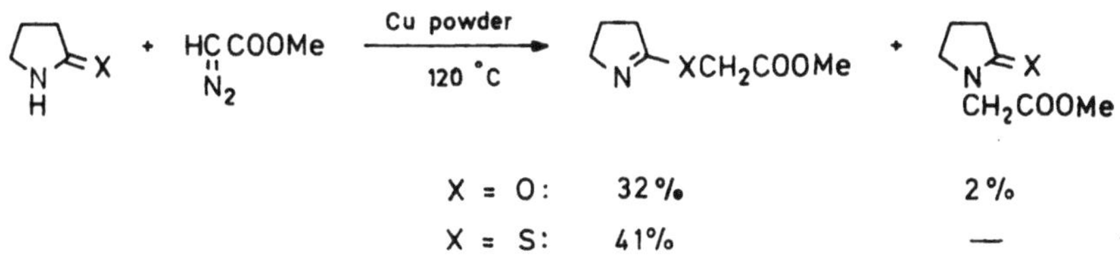

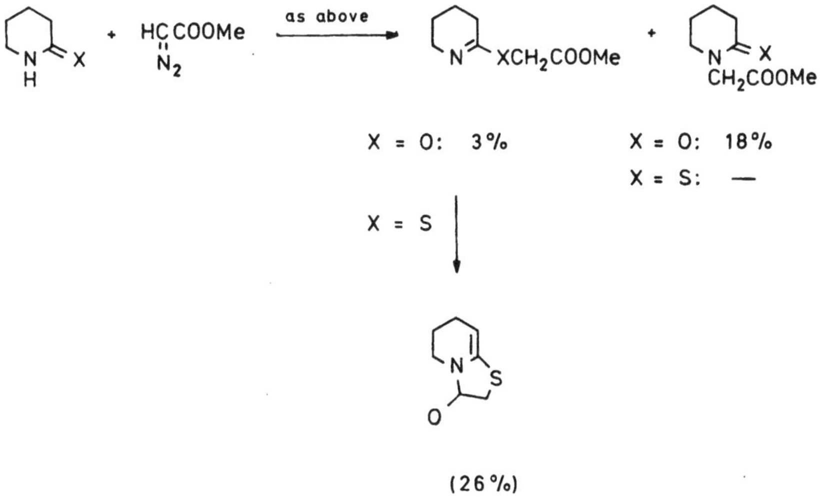

have already been discussed in preceding sections of this review (Sects. 2.3.5 and 3). It shall only be summarized here that, in the rhodium(II) carboxylate catalyzed reaction between alkyl diazoacetate and allylic or other olefinic alcohols, O/H insertion always prevails over cyclopropanation, whereas with acetylenic alcohols, steric hindrance at the hydroxy site may render O/H insertion less favorable than carbenoid addition to the triple bond.

An interesting application of carbenoid O/H insertion is the synthesis of macrocyclic oxacrown ethers **337** from α,ω-diazoketones **336** and oligoethylene glycols [323].

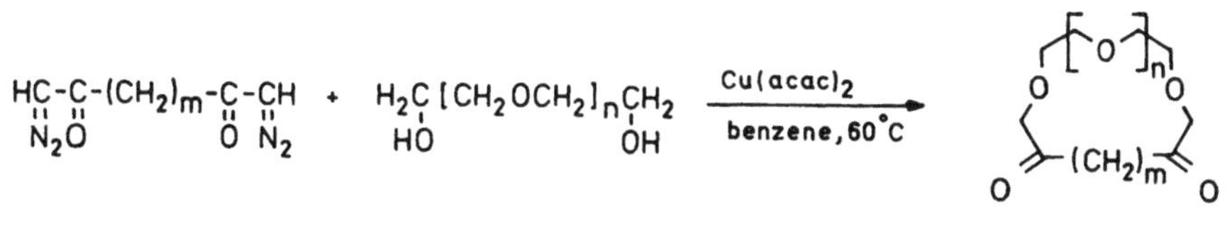

336 *337*

m = 6, 8, 9; n = 2, 3 (7-26%)

Concerning the mechanism of O/H insertion, "direct" carbenoid insertion, oxonium ylide and proton transfer processes have been discussed [7]. A recent contribution to this issue is furnished by the $Cu(acac)_2$- or $Rh_2(OAc)_4$-catalyzed reaction of benzhydryl 6-diazopenicillanate [237] with various alcohols, from which 6α-alkoxypenicillanates **339** and tetrahydro-1,4-thiazepines **340** resulted [324]. Formation of **340** is rationalized best by assuming an oxonium ylide intermediate **338** which then rearranges as shown in the formula scheme. Such an assumption is justified by the observation of thiazepine derivatives in reactions which involved deprotonation at C-6 of 6β-aminopenicillanates [325, 326]. It is possible that the oxonium ylide is the common intermediate for both **339** and **340**.

The assumption that **339** arises from the oxonium ylide by a proton transfer process is supported by the reversed product ratio obtained in the reaction with ethanol in the presence of diazabicyclo[4.3.0]non-5-ene (DBN).

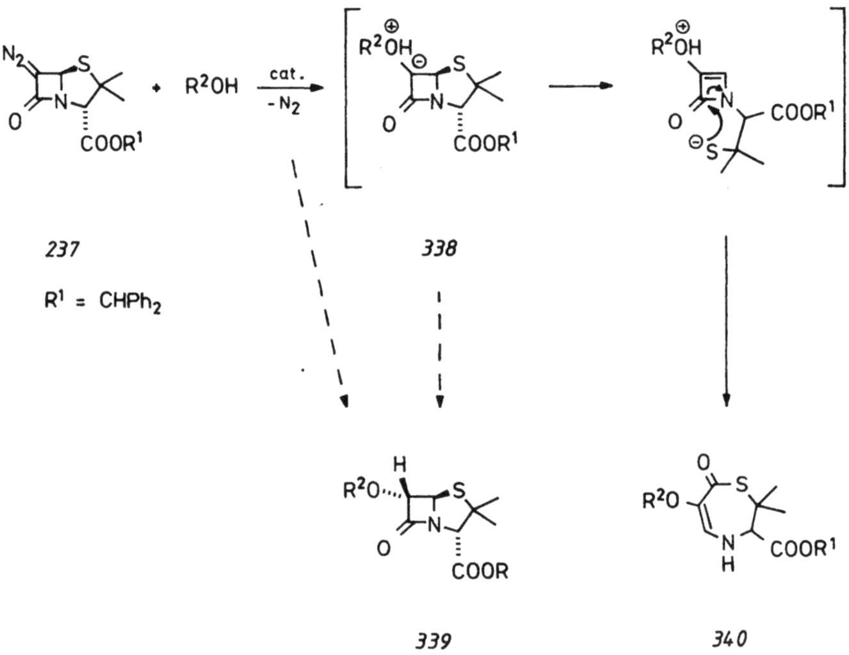

catalyst:	$Rh_2(OAc)_4$		$Cu(acac)_2$	
R^2OH	% *339*	% *340*	% *339*	% *340*
MeOH	55	19	56	23
EtOH	12	75	20	29
t-BuOH	6	72		
$PhCH_2OH$	<5	67		
$CH_2=CHCH_2OH$	<5	70	9	56
EtOH/DBN	55	20		

The known examples of carbenoid insertion into an S—H bond have been supplemented by the $Rh_2(OAc)_4$-catalyzed synthesis of α-phenylthioketones from α-diazoketones and thiophenol [327]. By this method, a number of primary and secondary acyclic α-diazoketones, ethyl diazoacetate and cyclic diazoketones such as 2-diazocyclopentanone, 2-diazo-6-methylcyclohexanone and 2-diazocycloheptanone were converted at room temperature in good to high yield.

7 Further Reactions Involving Hetero Atoms

7.1 Ethers, Acetals, Epoxides and Orthoesters

Interaction of a metal carbene with an unshared electron pair of an ether oxygen atom leads to an oxonium ylide which is able to rearrange via a 1,2-carbon shift to a formal C—O insertion product. It seems, however, that this is a rather unfavorable pathway, since reports on this reaction mode are scarce [7,10,14,154]. One may expect that allylic ethers are better substrates for stimulating the reaction, as an oxonium ylide can now be trapped by a symmetry-allowed [2,3]-sigmatropic rearrangement. Alternatively, competitive carbenoid cycloaddition to the double bond may occur. In fact, it was recently reported that the $Rh_2(OAc)_4$-catalyzed reaction of ethyl diazoacetate with allylic ethers gave no or only trace amounts of ylide-derived products, whereas the ylide route competes successfully with cyclopropanation in the case of allylic acetals [154] (see Sect. 2.3.4). According to earlier reports, the copper-salt catalyzed decomposition of diazomalonates in the presence of allylic ethers gave the oxonium ylide derived product preferentially [328], whereas mainly cyclopropanation took place with $CuCl/CH_2N_2$ [329].

Kirmse has reported that the ring enlargement of optically active 2-methyloxetane to 3-methyltetrahydrofuran occured with racemization (e.e. $<2\%$) when $Rh_2(OAc)_4/CH_2N_2$ was used, but with partial retention of configuration upon photolysis of diazomethane [330]. Furthermore, the photochemical procedure gives a complex mixture of products (C/H insertion, 2-methyl- and 3-methyltetrahydrofuran, ring-opened products from intramolecular β-elimination at the oxonium ylide stage). These differences indicate that no free oxonium ylide occurs in the rhodium-catalyzed reaction; rather, a C-metalated species must be involved in the product-forming step.

Epoxides are easily deoxygenated by the action of dimethyl diazomalonate or ethyl trimethylsilyldiazoacetate in the presence of catalytic amounts of rhodium(II) acetate or pivalate according to Scheme 36 [331]. The corresponding olefins are formed with retention of configuration and are not cyclopropanated by the carbenoid under the reaction conditions. The latter feature makes this reagent combination superior to the ethyl diazoacetate/copper catalyst system [332] which produced complex reaction mixtures. Keto functions as well as Br, OAc and OMe substituents are tolerated, but aldehyde groups are not, because of competing carbonyl insertion reactions. The sterically shielded epoxide of adamantylidene adamantane could not be deoxygenated.

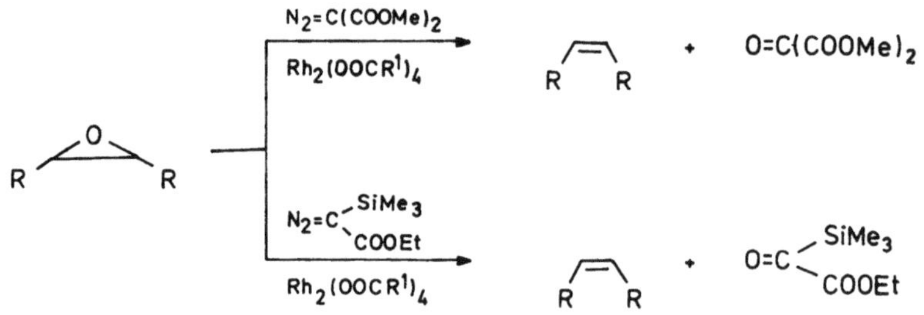

Scheme 36

Intramolecular oxonium ylide formation is assumed to initialize the copper-catalyzed transformation of α,β-epoxy diazomethyl ketones **341** to olefins **342** in the presence of an alcohol [333]. The reaction may be described as an intramolecular oxygen transfer from the epoxide ring to the carbenoid carbon atom, yielding a β,γ-unsaturated α-ketoaldehyde which is then acetalized. A detailed reaction mechanism has been proposed. In some cases, the oxonium-ylide pathway gives rise to additional products when the reaction is catalyzed by copper powder. If, on the other hand, diazoketones of type **341** are heated in the presence of olefins (e.g. styrene, cyclohexene, cyclopentene, but not isopropenyl acetate or 2,3-dimethyl-2-butene) and palladium(II) acetate, intermolecular cyclopropanation rather than oxonium ylide derived chemistry takes place [334].

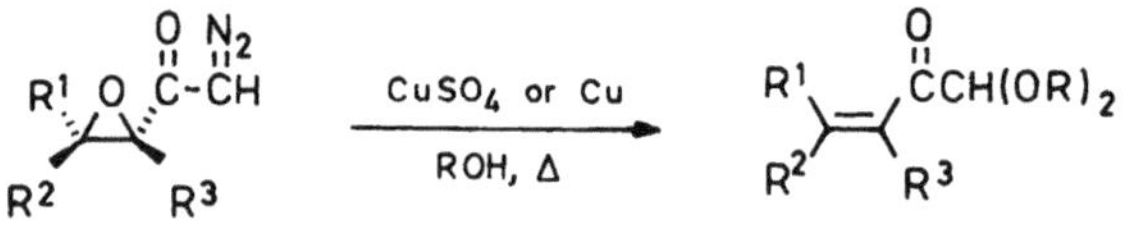

341 *342*

R^1	R^2	R^3
H	Ph	H
H	Ph	Ph
Ph	H	Ph
$-(CH_2)_5-$		H
$-(CH_2)_4-$		H
Me	Me	H
1-adamantyl	1-adamantyl	H

Insertion of a ketocarbene moiety into a C—O bond of orthoesters is normally performed with catalysis by $BF_3 \cdot Et_2O$. Copper(II) triflouromethanesulfonate was found to be a similarly efficient catalyst also, at least in some cases, whereas $Rh_2(OAc)_4$ was much less suited to promote this transformation [160]. Besides the C/O insertion product **343**, the alcohol insertion product **344** and, in reactions with ethyl diazoacetate, the formal carbene dimers were obtained. In agreement with $BF_3 \cdot Et_2O$, $Cu(OTf)_2$ did not bring about insertion into a C—O bond of trimethyl orthobenzoate. Moreover, changing the time of addition of the diazo compound or changing the catalyst concentration — parameters which are often so crucial for expedient cyclopropanation and other carbenoid reactions — had little influence on the yield of the C/O insertion products. All these facts were taken to suggest that $Cu(OTf)_2$ (or CuOTf, which may result from reduction of Cu(II) by the diazo compound) acts as a Lewis acid which generates a dialkoxycarbenium ion from the orthoester. Product formation then proceeds via attack if this cationic intermediate at the diazo carbon.

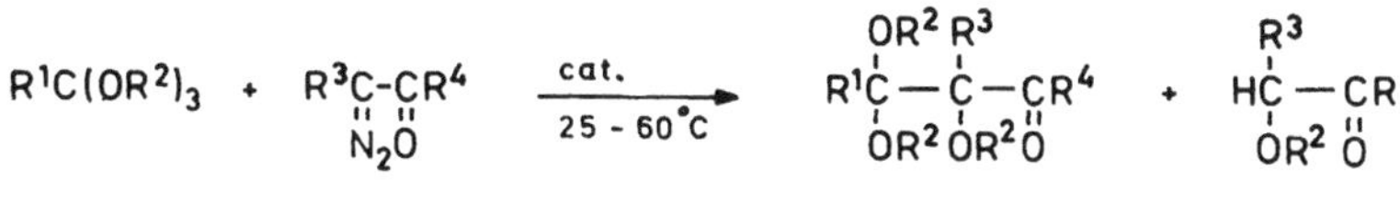

343 *344*

R^1	R^2	R^3	R^4	% *343*[a]	% *344*[a]
H	Me	H	OEt	45, 20, 83	6, 27, 5
		COOMe	OMe	46, 13	20, 20,
		H	COPh	34[b], ,58	34[b], ,4
Me	Et	H	OEt	38[c], 3, 50	15, 11,
		COOMe	OMe	<1, ,	20, ,
Et	Et	H	Et	<2, ,46	8, ,

[a] Yields for catalysis by $Cu(OTf)_2$, $Rh_2(OAc)_4$ and $BF_3 \cdot Et_2O$.
[b] 15% of α,α-dimethoxyacetophenone are also formed.
[c] Mixture of $Me—C(OEt)_2—CHOEt—COOEt$ and MeCO—CHOEt—COOEt.

7.2 Thioethers, Disulfides, Diselenides, Selenoesters

Interaction of an electrophilic carbene or carbenoid with R—S—R compounds often results in the formation of sulfonium ylides. If the carbene substituents are suited to effectively stabilize a negative charge, these ylides are likely to be isolable; otherwiese, their intermediary occurence may become evident from products of further transformation. Ando [152 b)] has given an informative review on sulfonium ylide chemistry, including their formation by photochemical or copper-catalyzed decomposition of diazocarbonyl compounds. More recent examples, including the generation and reactions of ylides obtained by metal-catalyzed decomposition of diazo compounds in the presence of thiophenes (Sect. 4.2), allyl sulfides and allyl dithioketals (Sect. 2.3.4) have already been presented.

Recently, the sulfonium-ylide pathway has been used repeatedly to perform one-carbon ring expansion by formal insertion of a carbene moiety into a C—S or N—S bond of a cyclic sulfur compound. Examples are compiled in Scheme 37. It can be seen that ring enlargement was successful with thiochroman-4-ones **345** and **348**, 1,3-dithianes **351** and 3-isothiazolone **352** (also with COOMe instead of Et). Thermally induced Stevens rearrangement of a primarily formed S-ylide may explain the ring-expanded products in all cases. However, triethylamine-induced rearrangement **346** → **347** shows that other mechanisms might also be operating [335)]. Attempts to achieve ring enlargement of 6*H*-dibenzo[b,d]thiopyrans **349** furnished mixed results [336]. The desired transformation took place when diazodiphenylmethane was decomposed purely thermally, whereas at best trace amounts of the 6,7-dihydrodibenzo[b,d]thiepin were obtained from ethyl diazoacetate/$CuSO_4$. Dimethyl diazomalonate, when decomposed by $CuSO_4$ in the presence of **349**, gave a stable ylide **350** which could be transformed thermally into the ring-enlarged product only when R = Ph; **350** (R = H) rearranged to a dibenzothiopyran exclusively. Copper-induced decomposition of ethyl diazoacetate in the presence of isothiochroman also failed to furnish a C/S insertion product [338].

A thermally stable sulfonium ylide is also obtained from the $CuSO_4$-catalyzed reaction between dimethyl diazomalonate and thioxanthene or its 9-alkyl derivatives [339]; rearrangement to the thioxanthen-9-ylmalonate occurs only with base catalysis.

From the results with the isolable ylides **350**, it can be concluded that the fate of less stable, non-isolated sulfonium ylides depends dramatically on their respective substituents [336,338]. Thus, the outcome of these reactions is programmed at the ylide stage and not during interaction of a presumed metal carbene with the sulfur-containing substrate.

$Rh_2(OAc)_4$-catalyzed decomposition of diazoester **352a** results in intramolecular C/S insertion, whereby a quaternary benzylic carbon atom without a heterosubstituent is generated. This transformation was used in a synthesis of (±)-cuparene [339a].

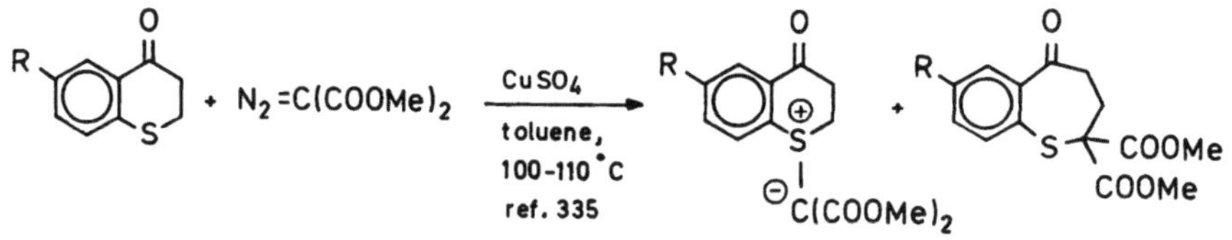

345 *346* *347*

R = H, Me, Cl

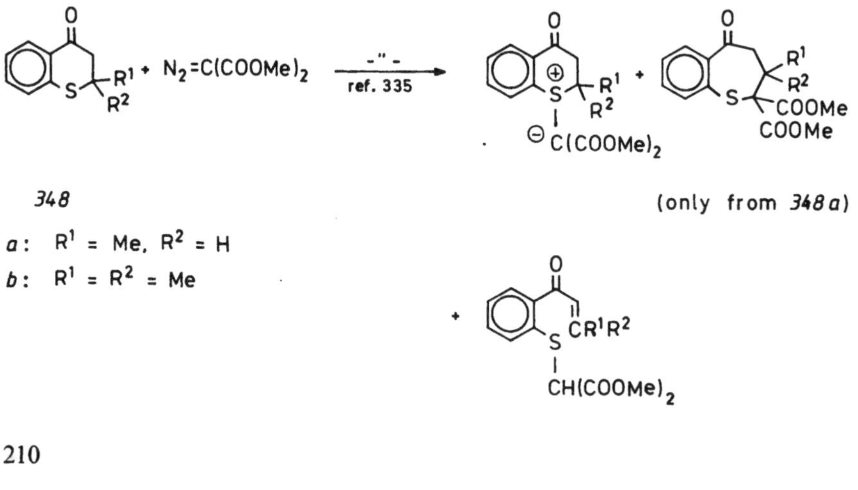

348 (only from *348a*)

a: R¹ = Me, R² = H

b: R¹ = R² = Me

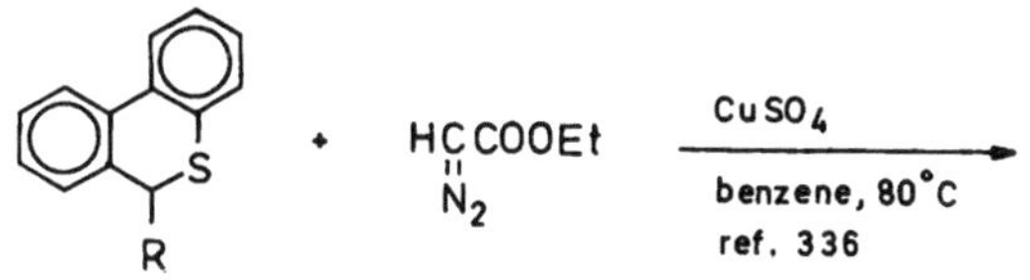

349

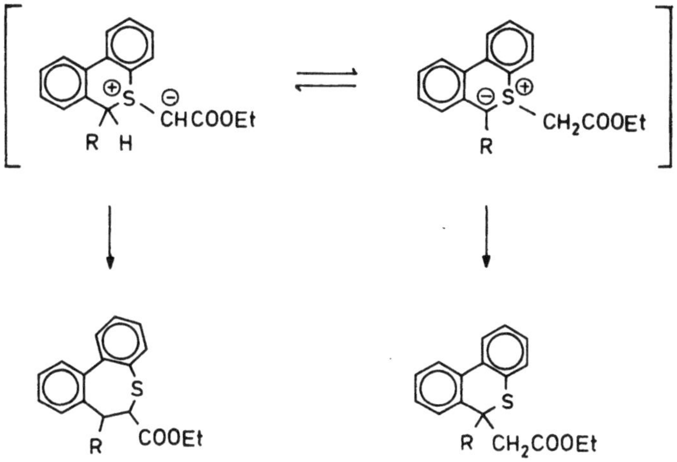

R = H : — 44%
R = Ph: trace 38%

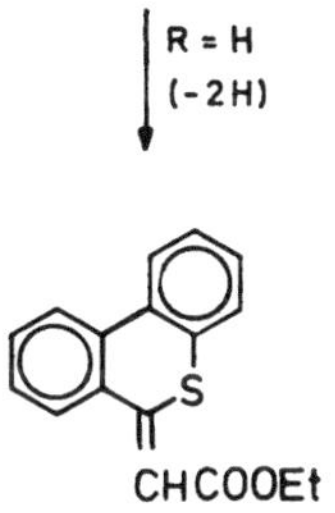

(7%)

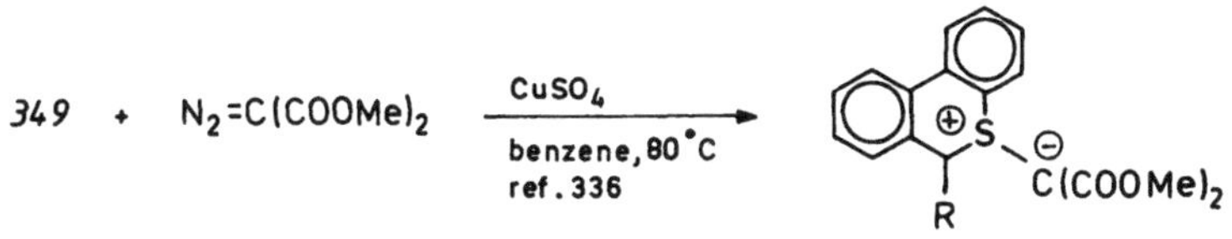

350

R = H : 85%
R = Ph: 90%

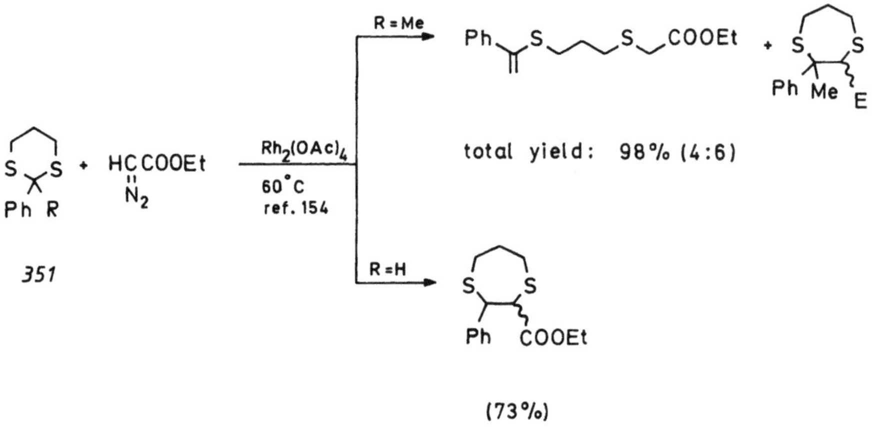

351

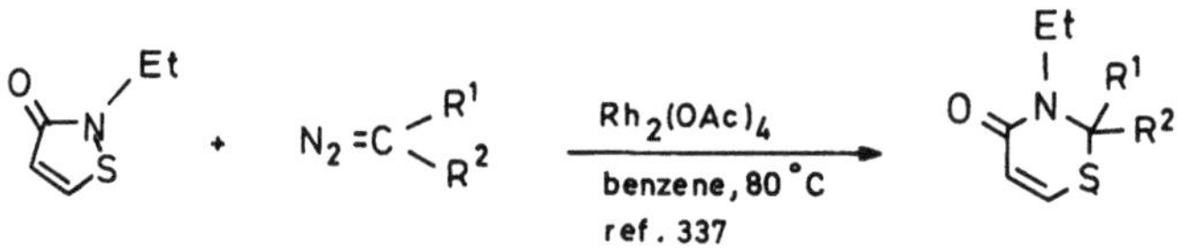

352

R^1	R^2	yield [%]
COOMe	COOMe	70
COMe	COOEt	74
COMe	COMe	58
-C(=O)-O-C(Me)(Me)-O-C(=O)-		74
-C(=O)-CH₂-C(Me)(Me)-CH₂-C(=O)-		65
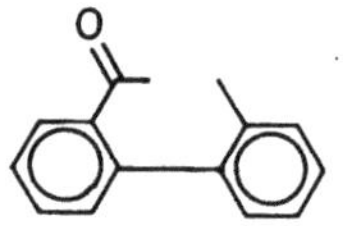		91

Scheme 37

Reactions of carbenoids with 4-thio-substituted 2-azetidinones have attracted much interest recently. Insertion of the carbene unit derived from diazomalonic esters [297,340] or ethyl diazo(diethoxyphosphoryl)acetate [340] into the C_4—S bond of simple β-lactams **353** and **354** took place irrespective of whether a N—H or a N—R

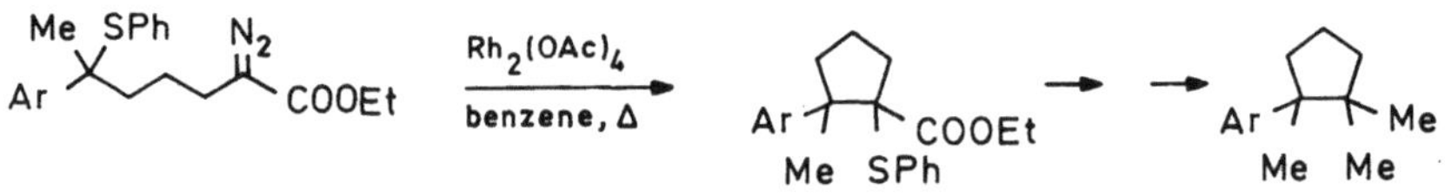

352a

function was present. Alternatively, $Rh_2(OAc)_4$-catalyzed decomposition of *p*-nitrobenzyl α-diazoacetoacetate in the presence of **355** resulted in the three-atom insertion product **357** the formation of which is easily understood as arising via the intermediary ylide **356** [297].

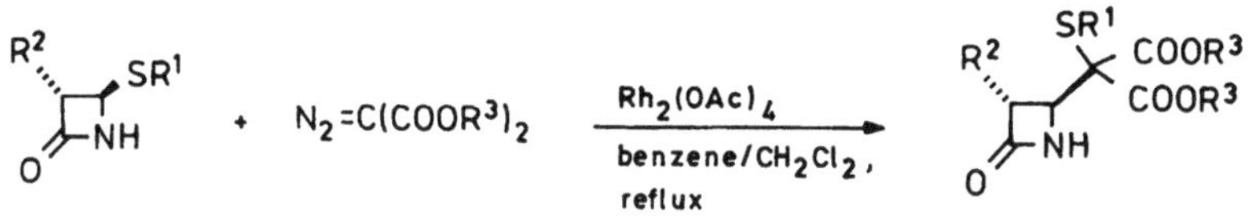

353

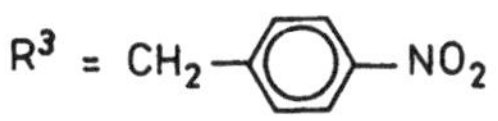

R¹	Me	Ph	Ph	Ph
R²	Et	Et	H	*i*-Pr
Yield [%]	30	51	59	45

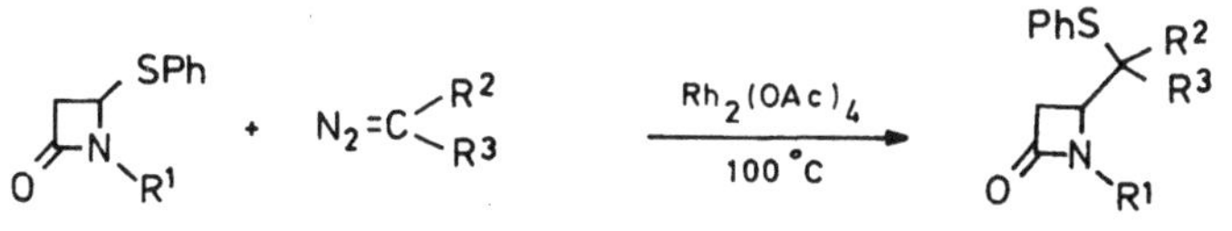

354

R¹	R²	R³	Yield [%]
$SiMe_2t$-Bu	COOMe	COOMe	60
CH_2COOt-Bu	COOEt	$PO(OEt)_2$	58

355

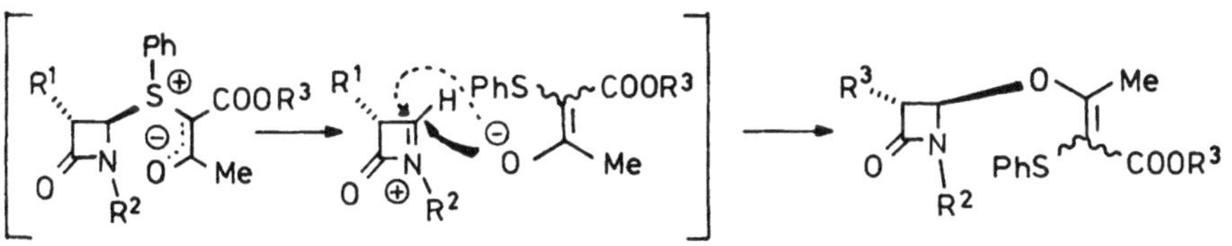

356 *357 (E + Z)*

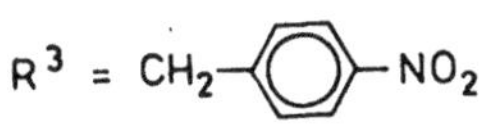

R¹ = Et, R² = THP: 79% (+2% of C-4 epimer)
R¹ = Et, R² = $SiMe_2t$-Bu: 80%

Efforts to realize an intramolecular version of the above reactions met with limited success when monocyclic 4-thio-substituted β-lactams were used. $Cu(acac)_2$-catalyzed decomposition of diazoketone **358** produced the epimeric carbapenams **359a, b** together with the oxapenam derivative **360** [341]; these compounds correspond to the C_4/S insertion products obtained in intermolecular reactions. Oxapenams were obtained exclusively when the acrylate residue in **359** was replaced by an aryl or heteroaryl substituent [275,342]. The different reaction mode of diazoketones **290a, b**, which furnish mainly or exclusively carbonyl ylide rather than sulfur ylide derived products, has already been mentioned (Sect. 5.2).

The C_4/S insertion reaction was suppressed completely upon catalytic decomposition of diazoketones **361**, where the sulfur substituent was alkyl, acyl or thioacyl. It is presumed that sulfonium ylides occur as intermediates which give cepham (or cephem) derivatives in all cases [270,343] rather than products of a Stevens rearrangement.

No S-ylide derived product at all was obtained from the $Rh_2(OAc)_4$-catalyzed decomposition of diazomalonic ester amide **362**; rather, a compound was isolated to which the structure of the Wolff rearrangement product **363** was tentatively assigned [344]. The desired C/S insertion product **364** was accessible, however, by photochemical decomposition of **362**.

The sulfonium ylide derived chemistry of penicillins continues to meet the interest of several research groups. It is well known that intermolecular carbenoid attack at the sulfur atom generates a sulfonium ylide which undergoes spontaneous opening of the thiazolidine ring to furnish a 1,2-*seco*-penicillin [326]. Novel examples of this reaction type were found upon $Rh_2(OAc)_4$-catalyzed decomposition of diazomalonic esters in the presence of various penicillins; this transformation constituted the opening step of a synthetic sequence directed towards 2-alkoxycarbonyl-cephems [345 a)] or modified penicillins [345 b)]. Similar to its reaction with 4-thio-2-azetidinone

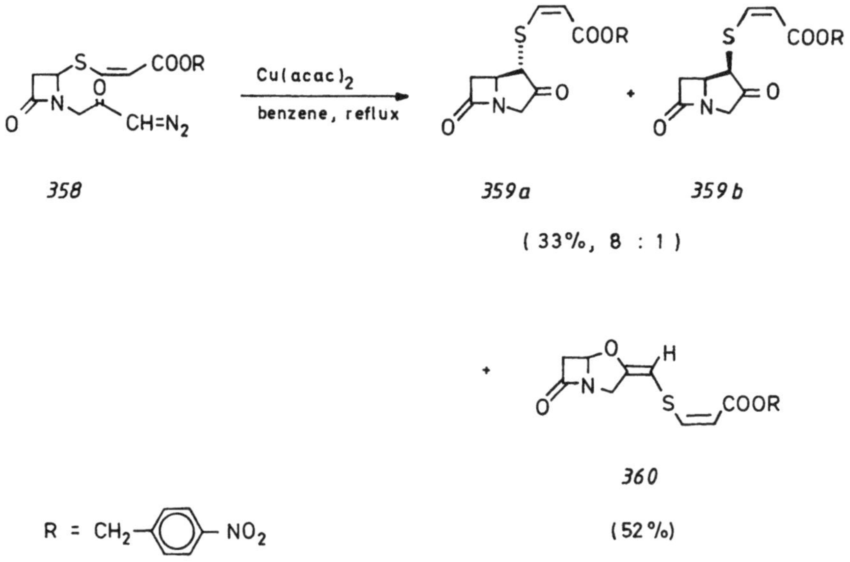

358 359a 359b

(33%, 8 : 1)

360

(52%)

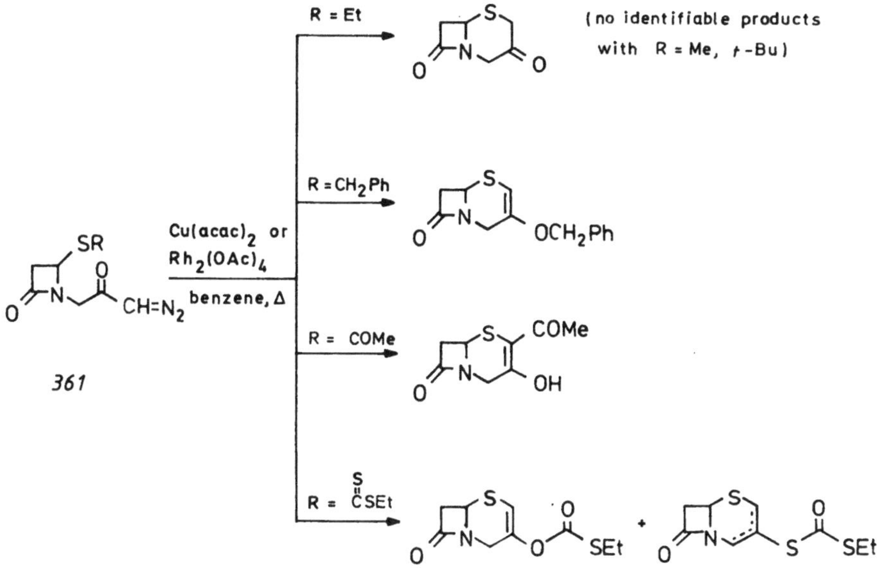

361

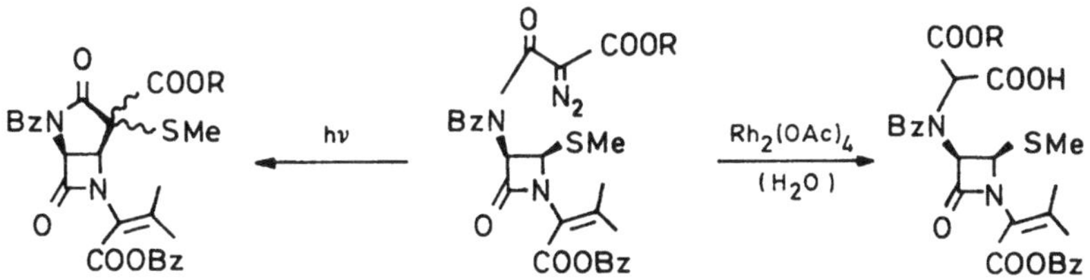

364 362 363

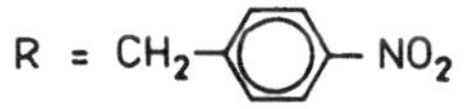

355, *p*-nitrobenzyl α-diazoacetoacetate yields a three-atom insertion product (**366**) by insertion into the C_5—S bond of penicillin G and V esters (**365a, b**). The sulfonium ylide derived from benzyl 6α-phthalimidopenicillanate **365d** gives rise to both the analogous C/S insertion product and the 1,2-*seco*-penicillin **367**, whereas only the latter product results from benzyl 6β-phthalimidopenicillanate **365c** [346]. The insertion products **366a, b, d** are formed with retention of the configuration at C-5 of **365**, irrespective of whether the C-6 amino substituent is in a *cis* or *trans* position. If one attributes this result to intramolecular nucleophilic β-face attack of the oxygen in a sulfonium enolate intermediate, it must also be concluded that β-face attack of the ketocarbenoid at the sulfur atom of **365** is favored due to shielding of the α-face by the $COOR^3$ group at C-3. In **365c**, the R^1 substituent is probably to voluminous to allow the formation of **366c**.

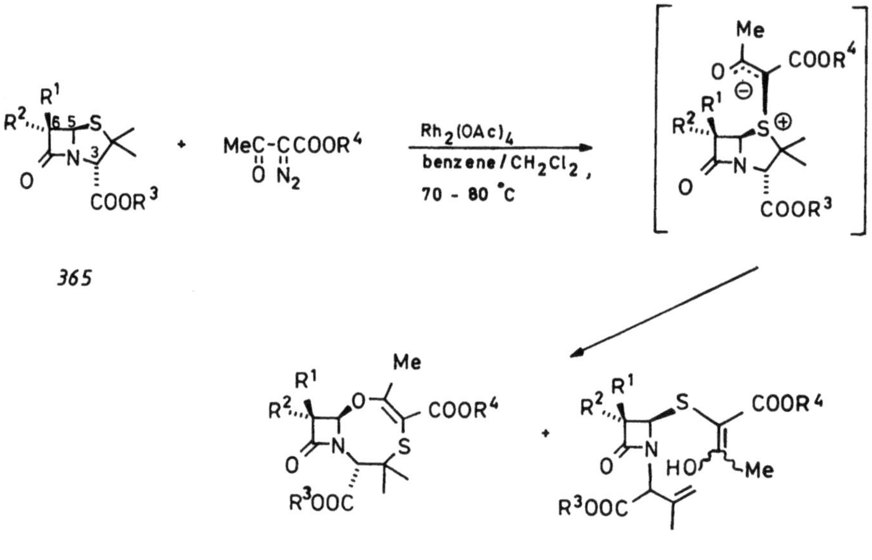

365 *366* *367*

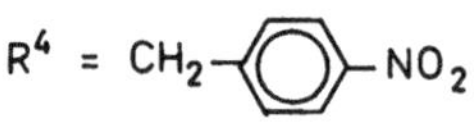

365–367	R^1	R^2	R^3	% *366*	% *367*
a	$NHCOCH_2Ph$	H	Me	27	—
b	$NHCOCH_2OPh$	H	CH_2Ph	26	—
c	phthalimido	H	CH_2Ph	—	81
d	H	phthalimido	CH_2Ph	22	55

The $Cu(acac)_2$-promoted transformation **368** → **369** represents an intramolecular carbenoid insertion into the penicillin C_5—S bond [347]. The original report did not mention the low-yield formation of a second product to which the tricyclic structure **370** was assigned [348,349]. In both **369** and **370**, the original stereochemistry at C-5 of **368** has been inverted; this is seen as a consequence of intramolecular nucleophilic α-face attack in a presumed azetidinium enolate intermediate. Attempts to realize a more flexible intermediate which then would have a chance to undergo β-face attack centered on the chain-extended diazoketone **371**. Its catalytic decomposition led to the tricycle **372** exclusively, however, C_7/N rather than C_5/S insertion having taken place [349].

Diazoamide **373** also failed to give sulfonium ylide derived products upon decomposition with $Cu(acac)_2$ or $Rh_2(OAc)_4$. When the reaction was carried out in benzene, only the cycloheptatriene derived from cyclopropanation of the benzene nucleus was obtained in 25% yield [349]. This result is reminescent of the failure of **362** to give an intramolecular C/S insertion product upon catalytic decomposition.

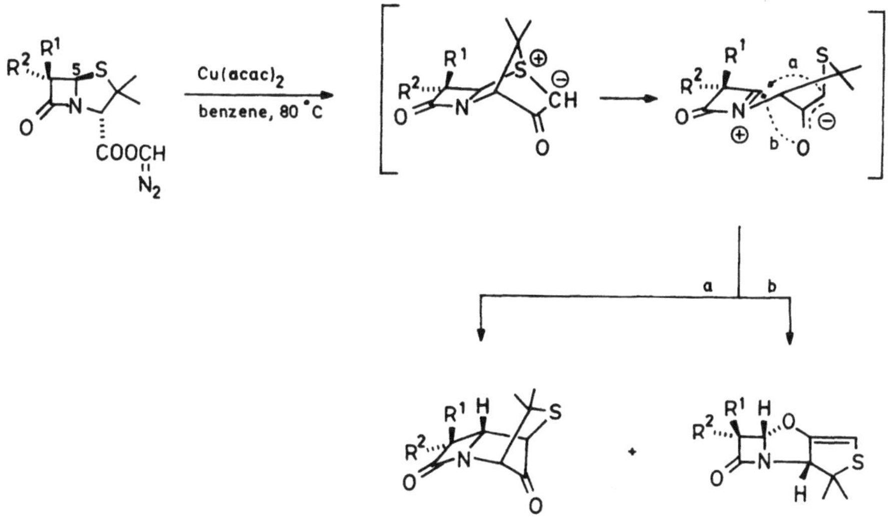

369 *370*

R¹	R²	% *369*	% *370*	Ref.
H	H	60	7	349)
$NHCOCH_2OPh$	H	dec. on workup	6	348)
H	phthalimido	76	7	349)

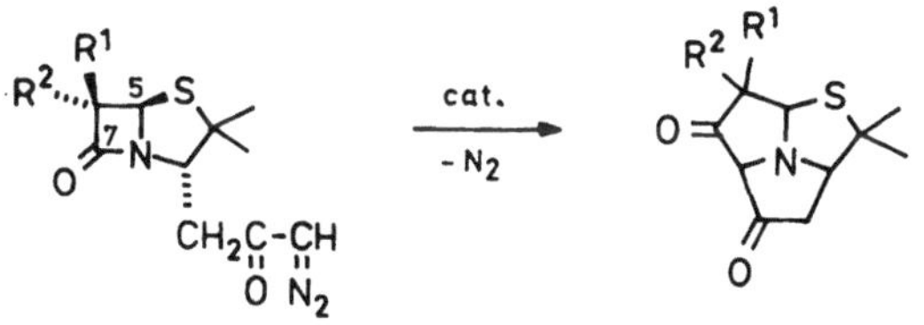

371 *372*

(70%)

R¹ = H, R² = phthalimido
R¹ = phthalimido, R² = H

The reaction of carbenes or carbenoids with compounds containing S—S bonds is likely to begin with sulfonium ylide formation; subsequent [1,2] rearrangement then produces a formal insertion product of the carbene moiety into the S—S bond [152 b)].

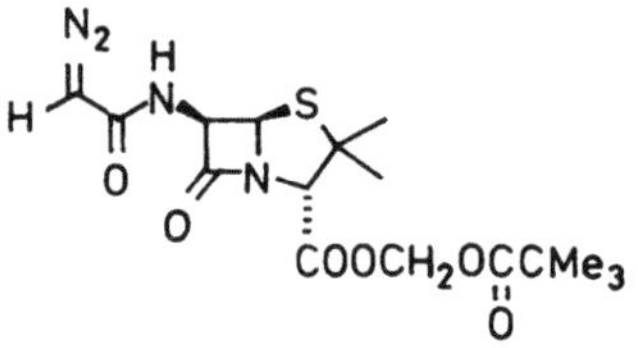

373

Novel example of this reaction type are given by the copper-catalyzed decomposition of ethyl diazoacetate in the presence of bis(dialkoxyphosphoryl)disulfides **374** [350)], where P/S insertion sometimes accompanies the S/S insertion, and of bis(dialkoxythiophosphoryl)trisulfides **375** [351)], where desulfurization to give the disulfide derived product occurs during the reaction. Only P/S insertion product was obtained from bis(dialkoxyphosphoryl)trisulfide or -tetrasulfide **376**; the copper-catalyst is dispensable in this case [351)].

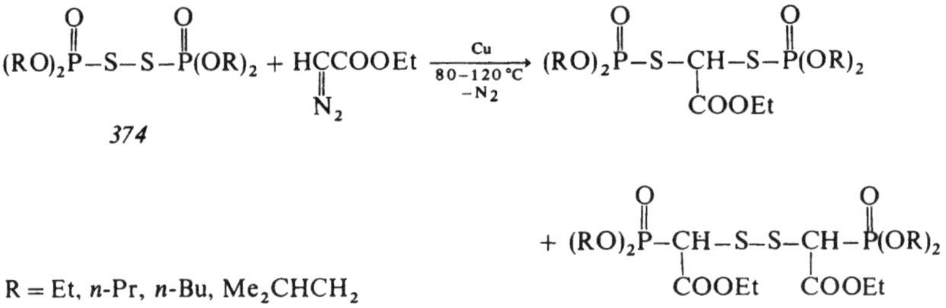

374

R = Et, *n*-Pr, *n*-Bu, Me_2CHCH_2

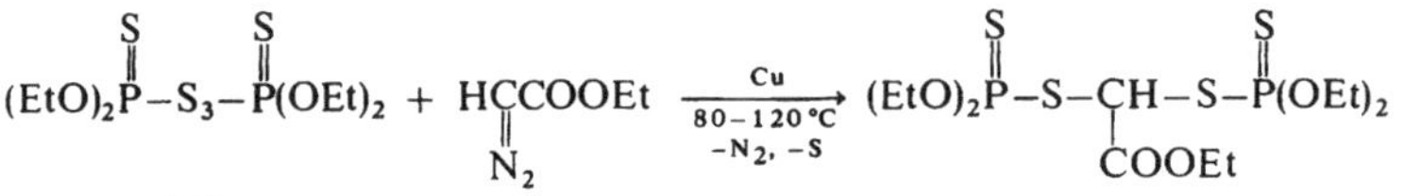

375

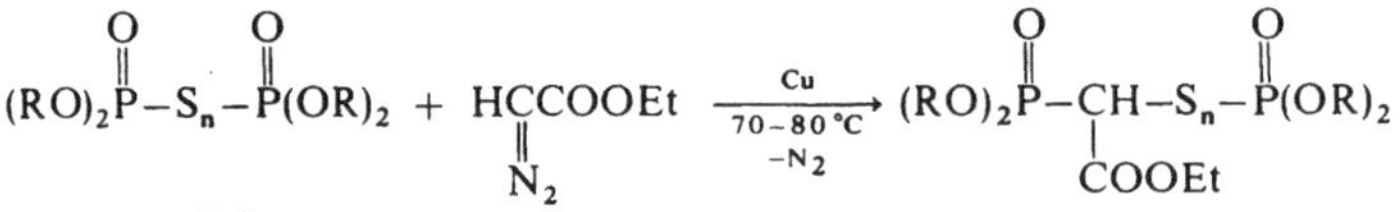

376

R = *i*-Pr, *i*-Bu; n = 3, 4

S/S insertion is also part of the reaction scheme when carbenes (or carbenoids) interact with 1,6,6aλ^4-trithiapentalenes **377** (Scheme 38) [352)]. The origin of the 4-diphenylmethylene-thiopyran **378** resulting from the reaction at higher catalyst concentration has not been elucidated, however.

The carbene moieties of methyl diazoacetate [353)], dimethyl diazomalonate [353)] and diazomethane [354)] have been inserted into the Se—Se bond of diaryl diselenides.

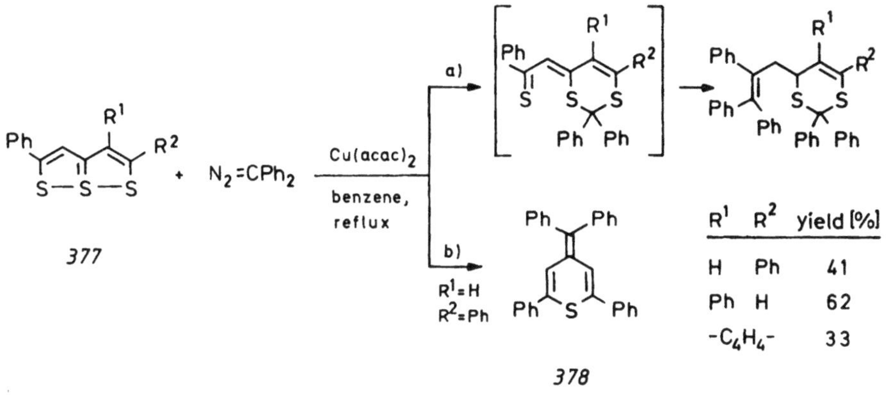

R^1	R^2	yield [%]
H	Ph	41
Ph	H	62
$-C_4H_4-$		33

molar ratio [377] : [Ph_2CN_2] : [cat.] : a) 1 : 2.5 : 0.1

b) 1 : 2.5 : 0.4

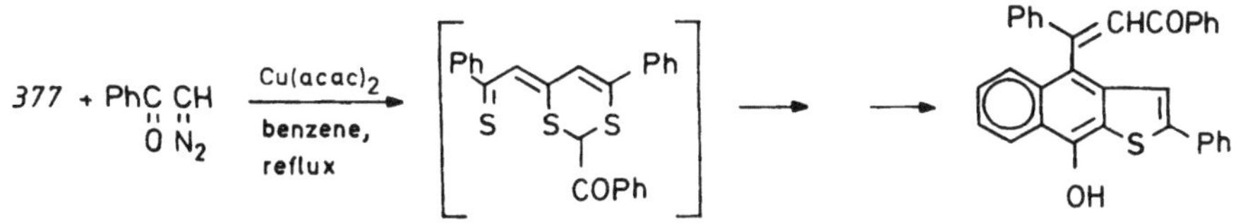

(R^1 = H, R^2 = Ph)

Scheme 38

These reactions, which were promoted either by copper (bronze or powder) or $BF_3 \cdot Et_2O$, are completely analogous to those of disulfides.

The homologation of selenoesters **379** with diazomethane in the presence of Cu or CuI to give α-selenoketones is thought not to involve a carbenoid pathway and an Se-ylide intermediate but rather a tetrahedral species resulting from nucleophilic attack of CH_2N_2 at the carbonyl carbon atom. The role of the catalyst is seen in facilitating nucleophilic attack at C=O by complexation at the selenium atom.

$$\underset{379}{RC(=O)SeR'} + CH_2N_2 \xrightarrow[\text{r.t.}]{\text{Cu or CuI}} \underset{(42-65\,\%)}{RC(=O)CH_2SeR'}$$

R = Ph, Me, $PhCH_2$, 2-thienyl, ⩘$(CH_2)_8$, MeO
R′ = Ph, Me

This would reasonably explain the formation of by-products such as PhSeSePh, $PhSeCH_2SePh$ and $RCOCH_3$ [354]. No C/S insertion was observed with thioester PhCOSPh.

7.3 Miscellaneous

$Rh_2(OAc)_4$-catalyzed decomposition of 2-diazocyclohexane-1,3-dione **380a** or its 5,5-dimethyl derivate **380b** in the presence of an aryl iodide leads to an iodonium ylide **381** [355]. The mild reaction conditions unique to the rhodium catalyst are essential to the successful isolation of the ylide which rearranges to **382** under the more forcing conditions required upon copper catalysis (copper bronze, $Cu(acac)_2$, $CuCl_2$) [355].

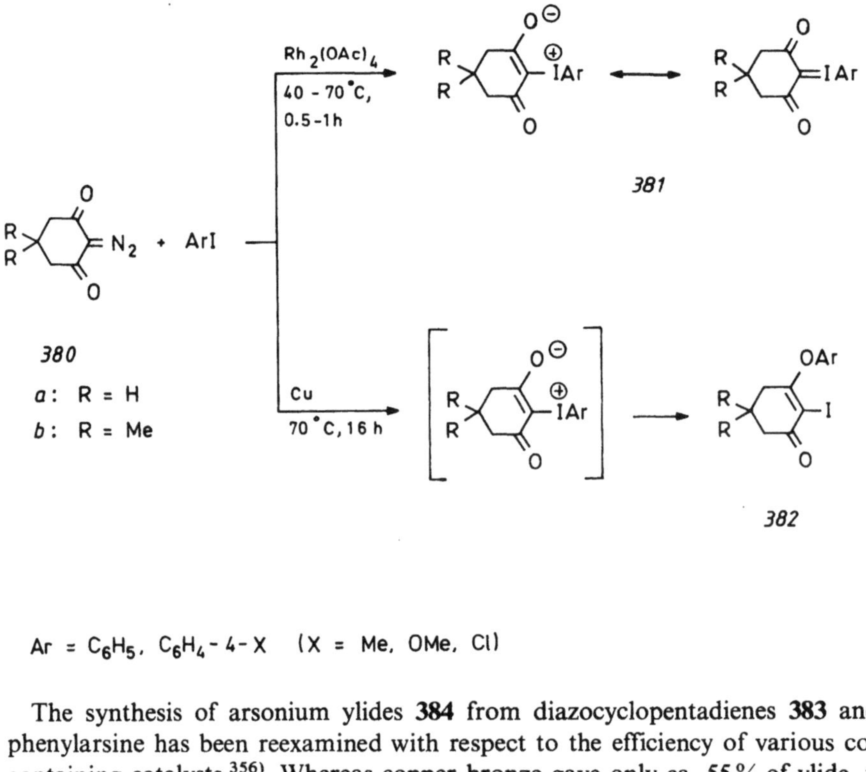

Ar = C_6H_5, C_6H_4-4-X (X = Me, OMe, Cl)

The synthesis of arsonium ylides **384** from diazocyclopentadienes **383** and triphenylarsine has been reexamined with respect to the efficiency of various copper-containing catalysts [356]. Whereas copper bronze gave only ca. 55% of ylide, yields over 80% were provided by the use of Cu(II) complexes of β-diketonates derived from acetylacetone, 3-methylacetylacetone, benzoylacetone or dibenzoylmethane, as well as by bis[4-(phenylimino)-2-pentanonato-N,O-]copper(II) and Cu(II) acetate, all used in boiling benzene. The sterically more demanding complex bis(dipivaloylmethanato)copper(II) as well as dichlorodipyridinecopper(II) proved less efficient. Copper(II) tartrate, the dibenzo-14-crown 6/copper complex and furthermore the acetylacetonate complexes of Co, Ni, Pt and Zn were totally ineffective. When **383a** was decomposed by $Cu(acac)_2$ in the presence of pyridine or thioanisole, pyridinium or methylphenylsulfonium cyclopentadienide was obtained (60 and 54% yield, respectively), but no ylides were obtained from diphenylsulfide or diphenylselenide.

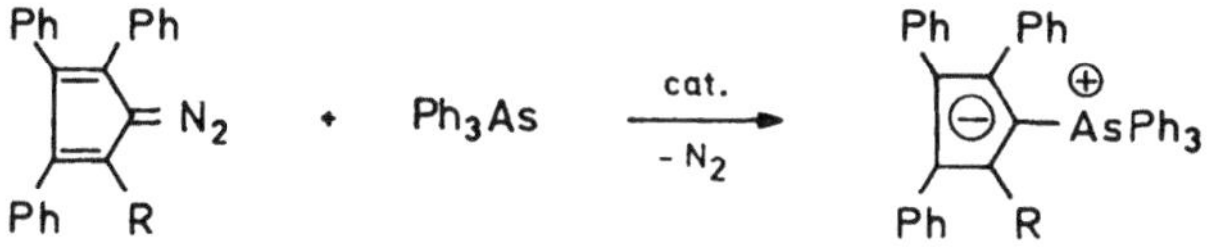

383 384

a: R = H;
b: R = Ph

8 Formation of Carbene Dimers

Occurence of olefins which are, formally speaking carbene dimers, as well as of similar products ($R_2C{=}N{-}N{=}CR_2$, $R_2CH{-}CHR_2$) represents an usually unwanted side-reaction which the chemist endeavors to suppress as far as possible. Nevertheless, conditions for high-yield synthesis of carbene dimers from several diazo compounds have been reported in the past [13,14]. Some novel examples, published since the last review [14] was written, are listed in Table 22.

For the formation of stilbenes from aryldiazomethanes, $Rh_2(OAc)_4$ was shown to be superior to other catalysts such as $Cu(ClO_4)_2$ or $CuBr_2$ [357], LiBr [363] or $Ce(NH_4)_2(NO_3)_6$ [364] in terms of efficiency, *Z*-selectivity and compatibility with substituents on the aromatic ring of the diazoalkane [358]. Even higher *Z*-selectivity was provided by the bulky catalyst iodorhodium(III) *meso*-tetraphenylporphyrin, but reduced yields had to be acknowledged [358]. Contrary to copper catalysts, $Rh_2(OAc)_4$ failed to induce the formation of carbene dimers from secondary aryldiazoalkanes; azines were produced instead [358].

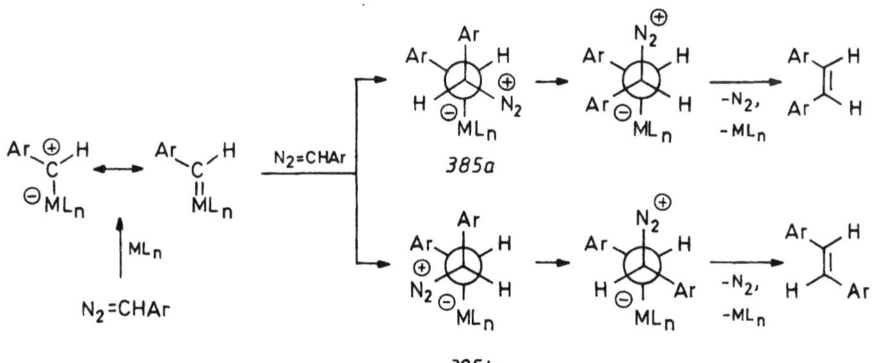

Scheme 39

Table 22. Carbene dimers by catalytic decomposition of diazo compounds

Diazo compound	Catalyst	Conditions	Ratio [diazo comp.]/[catalyst]	Products		Ref.
$ArCH=N_2$	$Cu(ClO_4)_2$	acetonitrile, 20 °C	200:1	$ArCH=CHAr$	Ar = p-Me-C_6H_4: 80%; $Z/E = 1.59$ C_6H_5: 86%; $Z/E = 1.81$ p-Cl-C_6H_4: 93%; $Z/E = 2.47$	357)
	$Rh_2(OAc)_4$	THF, −20 → 0 °C	ca. 90:1 to 160:1	$ArCH=CHAr$	Ar = 4-X-C_6H_4 (X = H, Me, Cl, NO_2); 3-Me-C_6H_4, 2-X-C_6H_4 (X = Me, Cl), 1-naphthyl, 2-naphthyl, 9-phenanthryl; yield: 70–99%; $Z/E = 3.3$–11.9 Ar = 2-thienyl: >38%; $Z/E = 2.7$ Ar = 2,4,6-tri-Me-C_6H_2: 90%; $Z/E = 0.66$	358)
	iodorhodium(III) *meso*-tetraphenylporphyrin	?	?	$ArCH=CHAr$	Ar = C_6H_5, 4-Me-C_6H_4, 4-Cl-C_6H_4: 55–67%; $Z/E = 7.3$–12.6	358)
	$Rh_2(OAc)_4$				Ar = 2,4,6-tri-Me-C_6H_2: 20%; $Z/E = 24.4$	359)
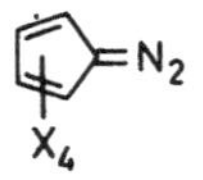	$[(\eta^3\text{-}C_3H_5)PdCl]_2$	hexane, 0 °C		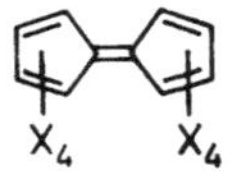	X = Cl: 77% X = Br: 39%	360)
$PhC(=O)-CH=N_2$	$Cu(acac)_2$	benzene, reflux	5:1	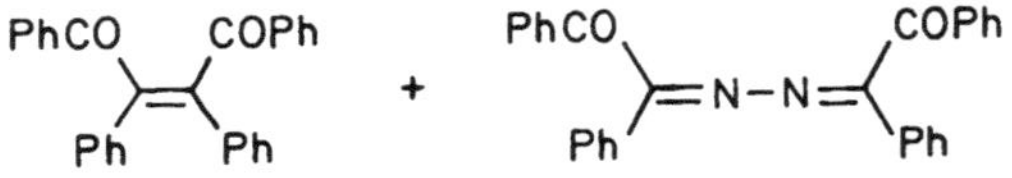 60% + 10%		361)
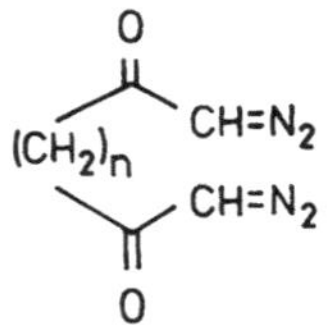	$Cu(acac)_2$	benzene, 60 °C		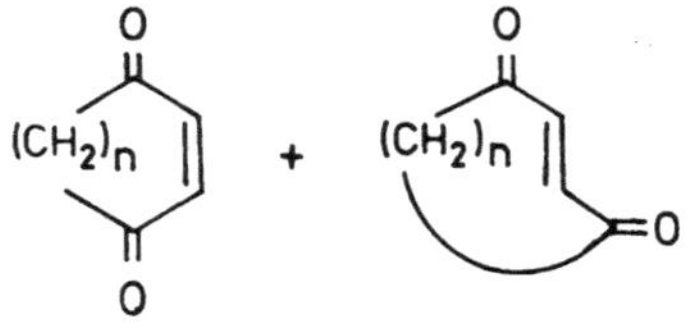		362)

n	4	5	6	7	8	9	10	12	16
yield [%]	20	<10	—	30	70	25	80	44	29
Z/E	>20	—	—	0.33	0.1	0.1	0.11	0.33	0.14

The *E*/*Z* ratio of stilbenes obtained in the $Rh_2(OAc)_4$-catalyzed reaction was independent of catalyst concentration in the range given in Table 22 [357]. This fact differs from the copper-catalyzed decomposition of ethyl diazoacetate, where the ratio diethyl fumarate: diethyl maleate was found to depend on the concentration of the catalyst, requiring two competing mechanistic pathways to be taken into account [365]. The preference for the *Z*-stilbene upon $Cu(ClO_4)_2$-or rhodium-catalyzed decomposition of aryldiazomethanes may be explained by the mechanism given in Scheme 39. Nucleophilic attack of the diazoalkane at the presumed metal carbene leads to two epimeric diazonium intermediates **385**, the sterically less encumbered of which yields the *Z*-stilbene after C/C rotation [357,358]. Thus, steric effects, favoring **385a** over **385b**, ultimately cause the preferred formation of the thermodynamically less stable *cis*-stilbene.

The *Z* selectivity increases with the electron-withdrawing properties of the *p*-substituent of the aryldiazomethane. In the light of the assumed mechanistic scheme, this fact may be rationalized by a more selective formation of **385a**, as the diazo carbon becomes less nucleophilic [357].

Some examples of carbene dimer formation resulting from diazoalkane decomposition on transition-metal surfaces have been reported. Diazomethane is decomposed to give ethylene and N_2 upon passage over a CoO/MoO_3 catalyst as well as on Ni, Pd, Fe, Co, Ru and Cu surfaces [367]. Similarly, 2-diazopropane is readily decomposed on Raney nickel [368]. At room temperature, propene and N_2 were the only detectable products, but above 50 °C, the carbene dimer 2,3-dimethyl-2-butene started to appear which reached its maximum yield at 100 °C, where approximately 40% of the carbene fragments dimerized. It is assumed [367,368], that surface carbenes are formed as intermediates from both diazomethane and 2-diazopropane which either dimerize or desorb by migration of a β-hydrogren, if available (Scheme 40).

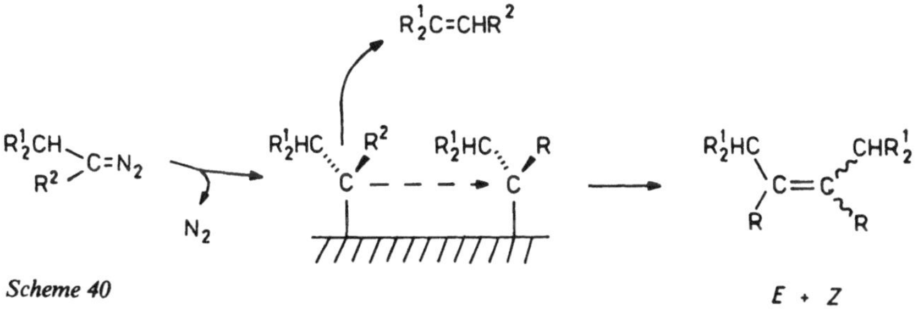

Scheme 40

Methyl diazoacetate is also decomposed on Raney nickel to give quantitatively a mixture of dimethyl fumarate and maleate [369]; N_2 evolution is observed even at room temperature. Most remarkably, dimethyl maleate is formed with high stereoselectivity (at 70 °C: 92% of dimethyl maleate, 7% of dimethyl fumarate [370]). This represents one of the few cases of stereoselective synthesis on metal surfaces which have been found so far.

When a mixture of diazomethane and H_2 was passed over Co, Fe, Ru, Ni or Pd surfaces, a mixture of hydrocarbons was produced (mainly C_1–C_{18}, linear alkanes and monoolefins) whose composition varied with the metal, the temperature and the H_2 partial pressure. The close similarity of this product mixture with that obtained from CO and H_2 over the same metal surfaces points to a common mechanism for both reactions (including surface carbenes as depicted in Scheme 40) and this lends strong support to the carbide/methylene mechanism for the Fischer-Tropsch reaction [371].

A somewhat unusual copper catalyst, namely a zeolite in which at least 25% of its rhodium ions had been exchanged by Cu(II), was active in decomposition of ethyl diazoacetate at room temperature [372]. In the absence of appropriate reaction partners, diethyl maleate and diethyl fumarate were the major products. The selectivity was a function of the zeolite activation temperature, but the maleate prevailed in all cases. Contrary to the copper salt-catalyzed carbene dimer formation [365], the maleate: fumarate ratio was found to be relatively constant at various catalyst concentrations. When Cu(II) was reduced to Cu(I), an improved catalytic activity was observed.

9 Rearrangements

1,2-Hydride and 1,2-alkyl shifts represent the most common rearrangement reactions of carbenes and carbenoids. They may be of minor importance compared to intermolecular or other intramolecular processes, but may also become the preferred reaction modes. Some recent examples for the latter situation are collected in Table 23 (Entries 1–10, 15: 1,2-hydride shifts; Entries 11–15: 1,2-alkyl shifts). Particularly noteworthy is the synthesis of thiepins and oxepins (Entry 11) utilizing such rearrangements, as well as the transformations α-diazo-β-hydroxyester → β-ketoester (Entries 6, 7) and α-diazo-β-hydroxyketone → β-diketone (Entry 8) which all occur under very mild conditions and generally in high yield.

For the synthesis of thiepins and oxepins, $[(\eta^3\text{-}C_3H_5)PdCl]_2$-catalyzed decomposition of 4-diazomethyl-4-methyl-4*H*-thiopyrans [387] or -pyrans [381] is the method of choice. Purely thermal decomposition of the former diazo compounds would require higher temperatures and thus would cause extrusion of sulfur from the primarily formed thiepin, yielding a benzene derivative.

The exclusive and quantitative formation of oxepins upon Pd-catalyzed decomposition of 4-diazomethyl-4-methyl-4*H*-pyrans (Entry 11) contrasts with the results of the CuCl-promoted reaction which affords a 2:1 mixture of oxepin (by 1,2-C migration) and 4-methylene-4*H*-pyran (by 1,2-H migration) under otherwise identical conditions [381]. When the methyl group at C-4 of the diazo precursor is replaced by H, the metal-catalyzed route to thiepins is no longer viable: Pd- or Cu(I)-catalyzed decomposition of 4-diazomethyl-4*H*-thiopyrans invariably leads to 4-methylene-4*H*-thiopyrans [378] (Entry 10). Only the proton-catalyzed decomposition of these diazo compounds affords the desired thiepin, albeit in low yield [378].

The rhodium-catalyzed conversion of α-diazo-β-hydroxy carbonyl into β-dicarbonyl compounds (Table 23, Entries 6–8) in general seems to be preferable to the acid-catalyzed reaction because of higher yields and absence of side-reactions [375,377]. From a screening of 20 metal salts and complexes, $Rh_2(OAc)_4$, $RhCl(PPh_3)_3$, $PdCl_2$ and $CoCl_2$ emerged as the most efficient catalysts for the transformation of α-diazo-β-hydroxy esters into β-ketoesters [376]. This reaction has become part of

Table 23. Rearrangements upon catalytic decomposition of diazo compounds

Entry	Diazo compound	Reaction conditions	Products		Ref.
1	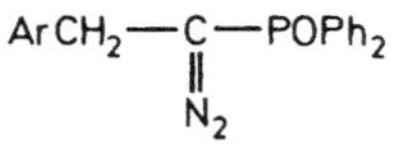	$Cu(acac)_2$, benzene, reflux	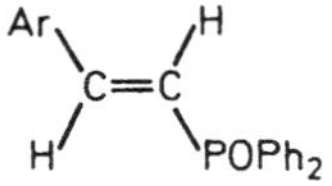	Ar = C_6H_4-4-X (X = H, Me, MeO) 85–100%	373)
2	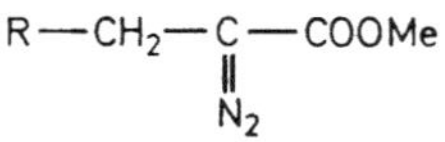	$Rh_2(OAc)_4$ benzene, r.t.	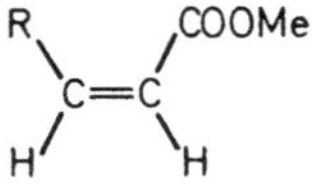	R = *i*-Pr, Ph, $PhCH_2O$ 90–100%	374)
3	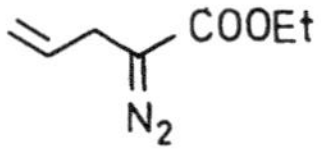	"	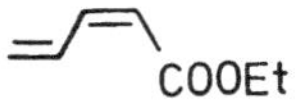 39%	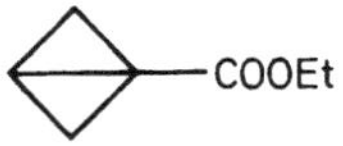 51%	374)
4	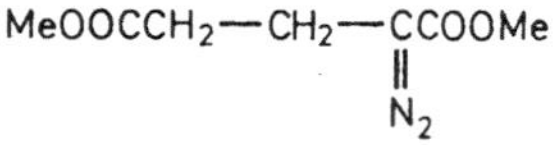	"	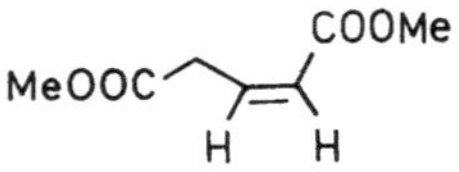 14%	$MeOOCCH_2CH_2$—CH(OH)COOMe or a 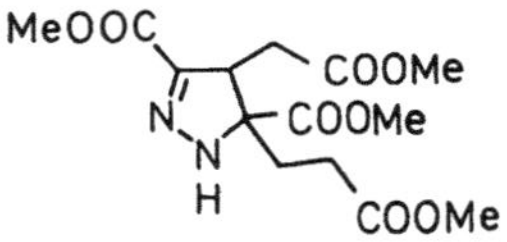	374)
5	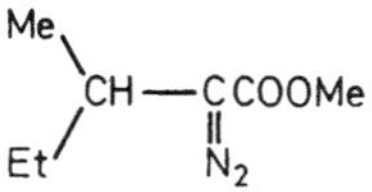	"	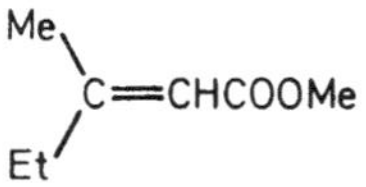	95% *E* : *Z* = 57 : 43	374)
6	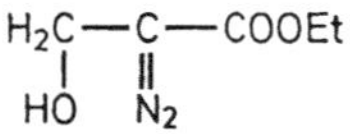	$Rh_2(OAc)_4$, CH_2Cl_2, r.t.	OHC—CH_2COOEt	quantit.	374)
7	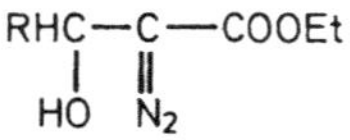	$Rh_2(OAc)_4$, DME, r.t.	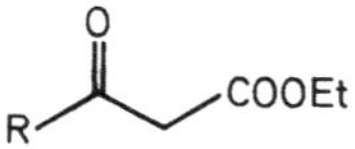	R = *n*-Pr, *i*-Pr, Ph_2CH, CH_2=CH PhCH=CH, Ph: 90–100%	375)
		$RhCl(PPh_3)_3$		R = *trans*-PhCH=CH : 71%	376)
		$PdCl_2$		R = *trans*-PhCH=CH : 58%	376)

Entry	Diazo compound	Reaction conditions	Products		Ref.
8	RHC—C(OH)—COMe, C=N_2	$Rh_2(OAc)_4$, DME, r.t.	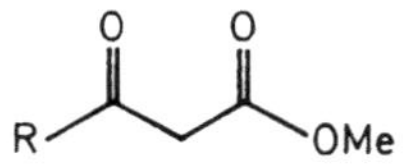	R = *n*-Pr, *i*-Pr, Ph_2CH, CH_2=CH, PhCH=CH, Ph: 68–81% R = 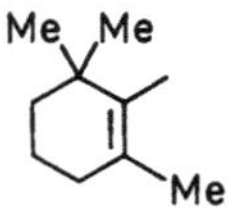 93%	377)
9	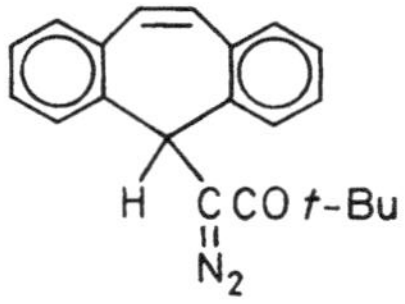	$Cu(acac)_2$, toluene, reflux	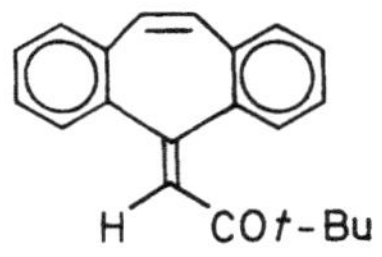	78%	180)
10	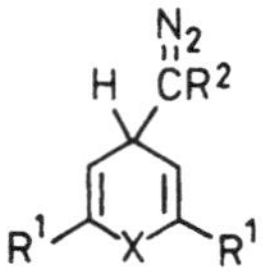	$[(\eta^3\text{-}C_3H_5)PdCl]_2$, r.t.	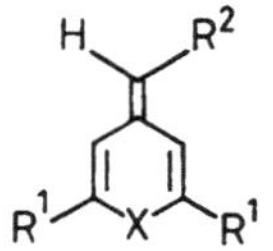	X = S; R^1 = *t*-Bu; R^2 = COOEt: 98% R^1 = *t*-Bu; R^2 = $PO(OMe)_2$, $POPh_2$: 70–76% X = O; R^1 = *t*-Bu, Ar, Me; R^2 = $POPh_2$, $PO(OMe)_2$, P(OMe)Ph, $PO(t\text{-}Bu)_2$, COOEt: 60–100%	328) 380, 381)
11	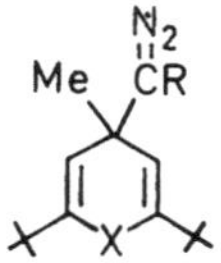	$[(\eta^3\text{-}C_3H_5)PdCl]_2$, $CHCl_3$ or benzene, r.t.	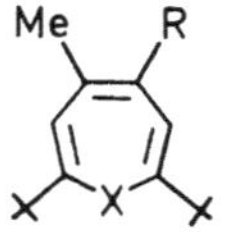	X = S, R = COOEt: 99% X = O, R = $POPh_2$, $PO(OMe)_2$, P(OMe)Ph, COOEt: 92–99%	382, 383) 381)
12	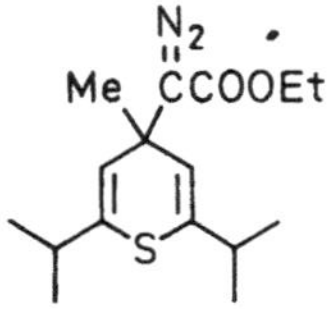	$[\eta^3\text{-}C_3H_5)PdCl]_2$, $CD_2Cl_2/CDCl_3/CCl_4$, −70 → −50°C	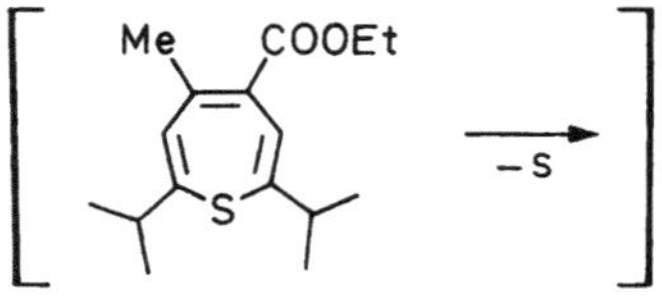	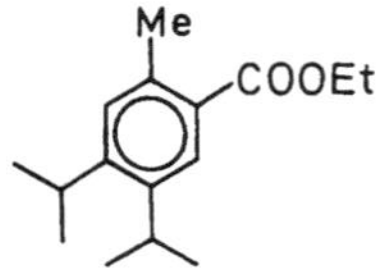 70%	384)

13 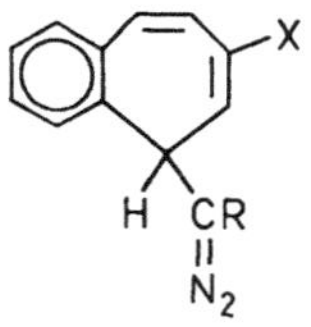	Cu(acac)$_2$, benzene, reflux	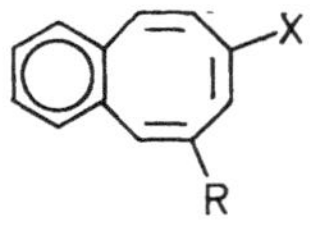	X = Br, R = POPh$_2$, PO(OMe)$_2$, P(OMe)Ph, COOMe, COO*t*-Bu, COPh: 38–66%	385)
			X = Cl, R as before	386)
			X = OCH$_2$C≡CH; R = POPh$_2$: 60%	180)
			X = SMe, SCH$_2$Ph; R = POPh$_2$: 18–71%	180)
14[b] 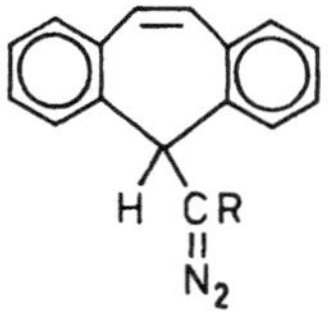	Cu(acac)$_2$, toluene, reflux	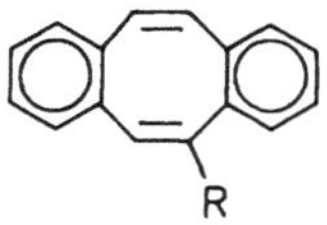 + 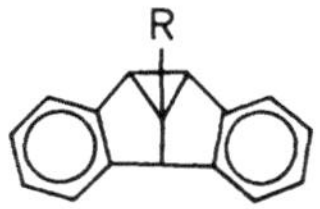 R = , COOMe, COPh		180)
15 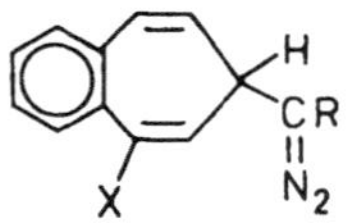 R = POPh$_2$, PO(OMe)$_2$ X = Cl, Br	Cu(acac)$_2$, benzene, reflux	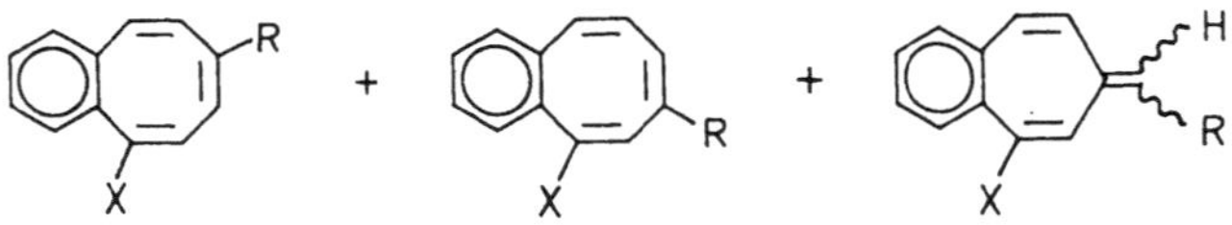 41–69% (isomers not separated) + 10–13%		180)

[a] Formation of these two products depends on the concentration of diazoester.
[b] For yields, see p. 149.

a three-step sequence for the one-carbon homologation of ketones (**386** → **389** + **390**) [376]. When unsymmetrical alicyclic ketones or acyclic ones having residues with very different steric demand are used, the smaller group migrates with very high selectivity in the carbenoid rearrangement step. In contrast, catalyst-dependent ratios of both **387** and **388** resulted from other acyclic ketones [376].

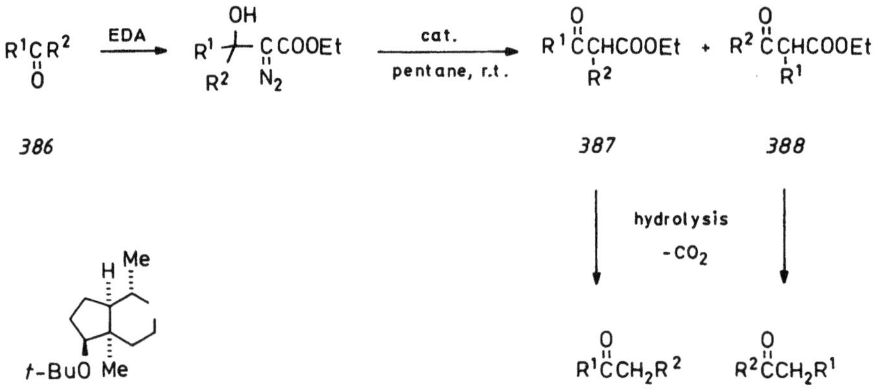

386 *387* *388*

(for table)

389 *390*

A remarkable difference exists between (1-diazo-2-arylethyl)diphenylphosphine oxides and methyl (diazo)phenylacetate, since the former yield exclusively *E*-olefins after N_2 loss and 1,2-H shift, whereas the *Z*-olefin results from the diazoester (Table 23, Entries 1 and 2). It has been speculated that an intermediate carbenoid adopts conformation **391** where the hydrogen atom perpendicular to the metal carbene bond undergoes 1,2-migration. The tendency of the aryl group to avoid the environment of the bulky metal residue would then determine the formation of the *cis*-cinnamic ester [374]. Along these lines of reasoning, structure **392** would have to be assumed for the carbenoid generated from a diazomethylphosphine oxide. Considering the bulkiness of the phosphoryl substituent, the positional change of the β-aryl group, as compared to **391**, would make sence. (It has not been established, however, whether the different catalysts have an influence on the stereoselectivity of the rearrangement.)

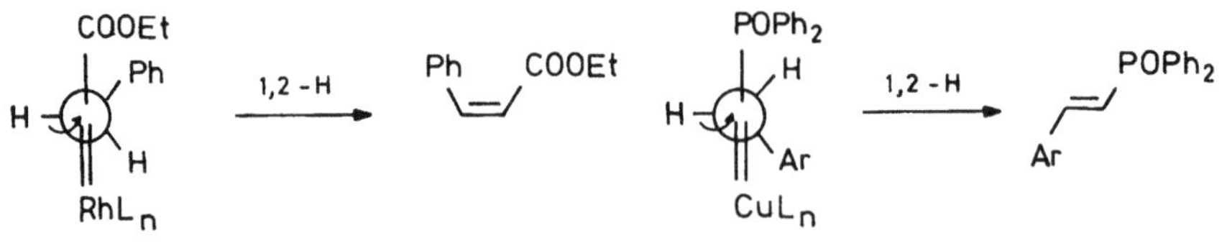

391 *392*

Ring contraction by (Si → C) vinyl migration took place when the diazosilacyclohexadiene **393** was heated in the presence of anhydrous $CuSO_4$ [388]. The silafulvene **394** was not isolated, but it could be trapped by *t*-butanol, benzophenone or benzaldehyde to give **395**, **396a** and **396b**, respectively. Silane **395** was also obtained as the sole product in the absence of the catalyst, but at more vigorous reaction conditions. Contrary to **393**, diazo-silanaphthalene **397** showed no tendency to undergo ring contraction or (Si → C) phenyl migration. Thermolysis in the presence of $CuSO_4$ and methanol gave only the O/H insertion product [388].

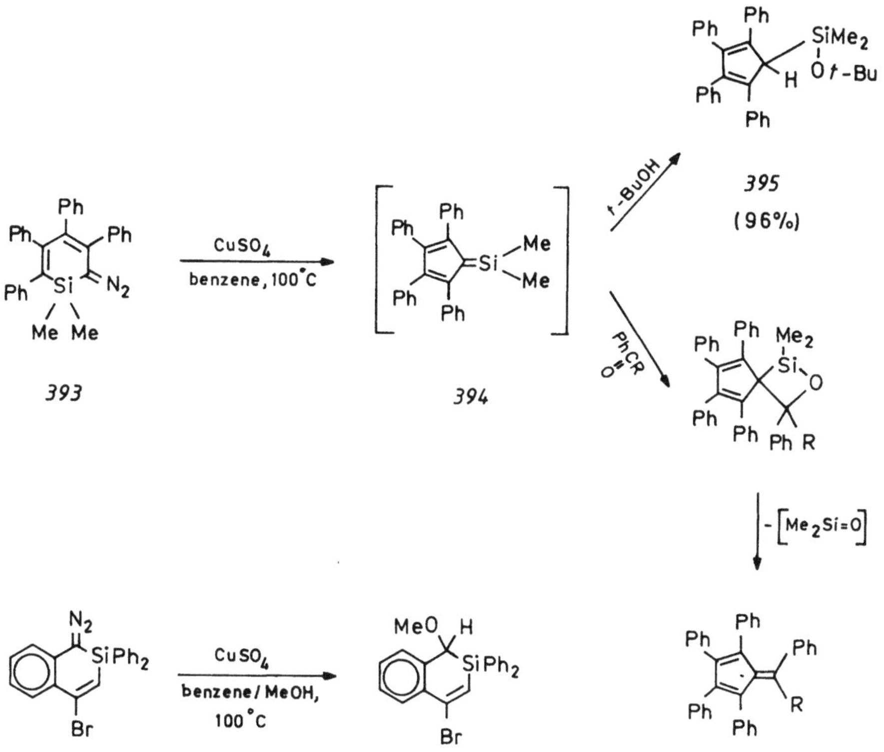

393 394 395 (96%) 396

397 (96%)

a: R = Ph (66%)
b: R = H (44%)

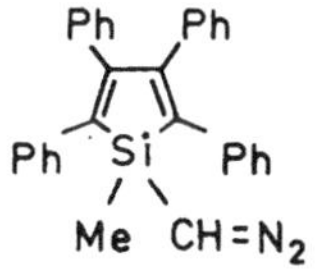

398

$CuSO_4$-catalyzed decomposition of the (1-sila-cyclopentadienyl)diazomethane **398** did not furnish defined products. The desired rearrangement reactions to a silabenzene and a 1-ethylidene-1-sila-2,4-cyclopentadiene, both trapped by *t*-butanol, were brought about by irradiation of **398**, however [388].

A rearrangement reaction involving migration of a benzoyloxy group explains the rhodium-catalyzed transformation **399** → **400** [374]; olefins from a 1,2-H shift could not be detected. Similarly, acetoxy migration accounts for the major product (**403**) from the copper-catalyzed decomposition of 5-*endo*-acetoxy-3-diazo-6,6-dimethyl-2-norbornanone **401** [389]. In accordance with a isotope labelling study, carbonyl ylide **402** may be formed intramolecularly, which then undergoes bond reorganization as shown. This pathway would account for ca. 80% of the tricyclic product. An analogous reaction occurs with Br instead of OCOMe, but other "carbene reactions" take over in this case [389].

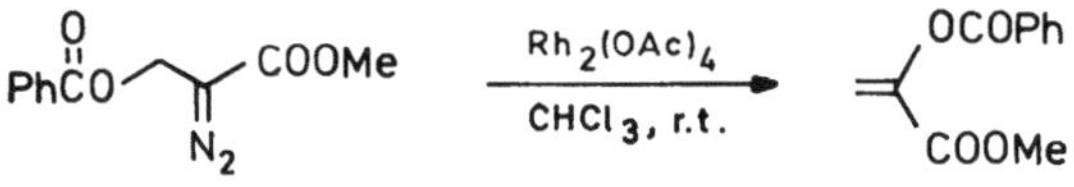

399 *400* (94%)

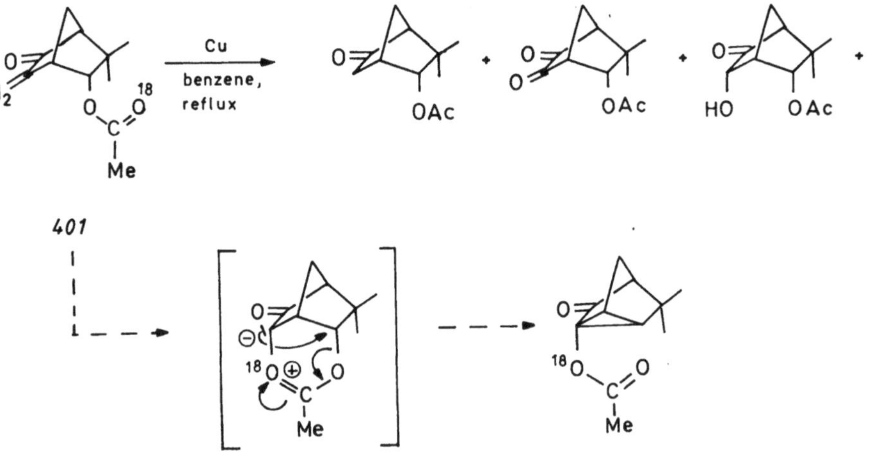

401 *402* *403*

Wolff rearrangement of α-diazoketones to give ketenes or subsequent products is an often used synthetic procedure the scope and limitations of which are well established [13,390], so that only a few new features of this reaction need to be considered here. Concerning its catalytic version, one knows that copper, rhodium and palladium catalysts tend to suppress the rearrangement [390]. A recent case to the contrary is provided by the $Rh_2(OAc)_4$-catalyzed decomposition of ethyl *E*-2-diazo-3-oxopent-4-enoates **404** from which the β,γ-unsaturated esters **405** are ultimately obtained via a Wolff rearrangement [236]. The *Z*-5-aryl-2-diazo-3-oxopent-4-enoates undergo intramolecular insertion into an aromatic C—H bond instead (see Sect. 4.1).

Metal-catalyzed decomposition of β,γ-unsaturated α′-diazoketones **406** in the presence of an alcohol affords rearranged γ,δ-unsaturated esters **407**; this process has been termed the vinylogous Wolff rearrangement [193]. Full accounts dealing with the scope and limitations [193] as well as the mechanism [391] of this transformation have appeared recently.

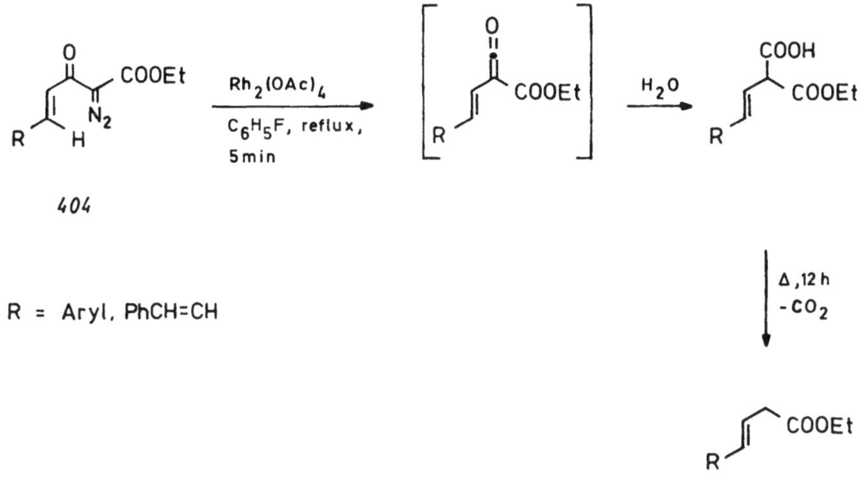

404

405

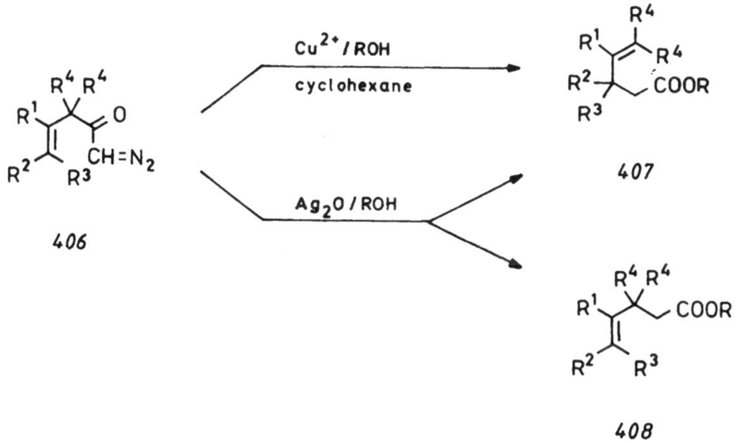

406

407

408

Whereas under the usual conditions for Wolff rearrangement (Ag_2O, MeOH, Δ) both the normal and the vinylogous rearrangement product (**408** and **407**) were obtained from **406** (R^1–R^2 = —$(CH_2)_n$—, n = 3–5); R^3 = H, R^4 = Me). Application of Cu(II) catalysts such as $CuSO_4$, $Cu(acac)_2$ and $Cu(OTf)_2$ resulted in the exclusive formation of **407**. Copper(II)triflate was the most efficient catalyst, especially in combination with benzyl alcohol. Under these conditions, the vinylogous Wolff rearrangement is a general reaction of unsaturated diazoketones **406**, be the olefinic bond acyclic or incorporated into a ring, monosubstituted or tetrasubstituted. Given this versatility and the ready availability of the diazoketones **406**, the vinylogous Wolff rearrangement offers a synthetic alternative to the orthoester version of the

Claisen rearrangement. As for the mechanism, the reaction begins with intramolecular cyclopropanation; the resulting bicyclo[2.1.0]pentan-2-one then undergoes fragmentation to a β,γ-unsaturated ketene which finally is trapped by the added alcohol to afford a β,γ-unsaturated ester (Scheme 41). The intermediates could be observed in selected cases.

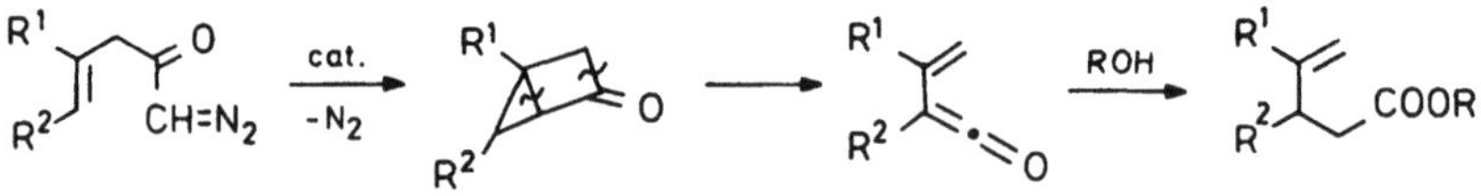

Scheme 41

10 Metalloporphyrins

A number of diazo compounds are known to be decomposed by Zn(II), Co(II), Co(III) and Rh(III) complexes of porphyrins (**409**) to give 1:1 and 1:2 adducts between the porphyrin and the formal carbene unit. Depending on the metal ion, different products may result (Scheme 42): Zinc octaethylporphyrin or *meso*-tetraphenylporphyrin yield N-alkylated porphyrins **410** with ethyl diazoacetate [392] and ethyl 2-diazopropionate [393]. In the latter case, a homoporphyrin **411** is obtained additionally. Cu(I)-catalyzed decomposition of diazomethane or alkyl diazoacetates in the presence of zinc *meso*-tetraphenylporphyrin leads to cyclopropanation of a pyrrolic ββ double bond, besides an N-alkylated product of type **410** [394,395]. The stoichiometric reaction between cobalt porphyrins and diazoacetates furnishes products of formal carbene insertion into a metal-nitrogen bond (**412**) [396]. Whereas such products are thermally moderately stable, the corresponding adducts with R^3 = >CH and R^4 = COR or $PO(OMe)_2$ were never isolated. They rapidly go on to alkylcobalt(III) porphyrins (**413**) after rearrangement and HX elimination [393,397]. Haloalkylcobalt(III) porphyrins result from diazomethane. 9-Diazofluorene is catalytically decomposed by bromocobalt(III) porphyrins to give the azine in high yield [397].

Iodorhodium(III) porphyrins generally lead to alkylrhodium(III) porphyrins (Scheme 42) [398]. This is also true for the reaction with ethyl diazoacetate in the presence of HOAc or an alcohol, and the insertion product **412** (M = Rh) could not be detected, in contrast to the corresponding cobalt porphyrin. A mechanistic scheme, which includes the diverse reaction modes of metalloporphyrins towards diazo compounds, has been proposed by Callot [393,398].

Another remarkable property of iodorhodium(III) porphyrins is their ability to decompose excess diazo compound, thereby initiating carbene transfer reactions [398]. This observation led to the use of iodorhodium(III) *meso*-tetraarylporphyrins as cyclopropanation catalysts with enhanced *syn*:*anti* selectivity (see Sect. 2.2.3) [87,100] as well as catalysts for carbenoid insertion into aliphatic C—H bonds, whereby an unusually high affinity for primary C—H bonds was achieved (see Sect. 6.1) [287]. These selectivities, unapproached by any other transition metal catalyst, have been credited to the steric encumbrance in the alleged rhodium carbene intermediate, as depicted in Formulae 48, 314 and 315. In line with this picture is the finding that catalysis by iodorhodium(III) *meso*-tetraphenylporphyrin completely alters the regioselectivity of cyclopropanation of zinc *meso*-tetraphenylporphyrin. With ethyl diazoacetate, reaction takes place at one of the phenyl groups at the almost complete expense of cyclopropanation of a pyrrolic ββ bond, which is the exclusive reaction upon catalysis by CuCl [87].

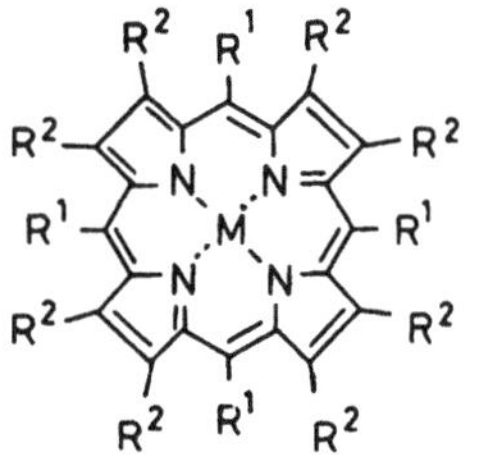

R^1 = Ph, R^2 = H
R^1 = H, R^2 = Et

409

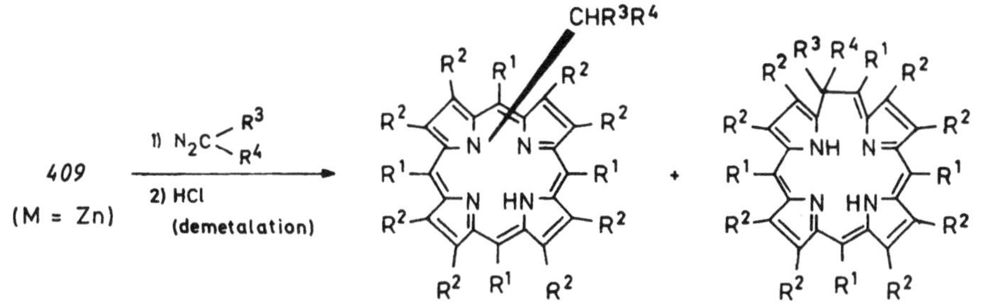

410 *411*

R^3 = H, R^4 = COOEt
R^3 = Me, R^4 = COOEt

R^3 = Me, R^4 = COOEt

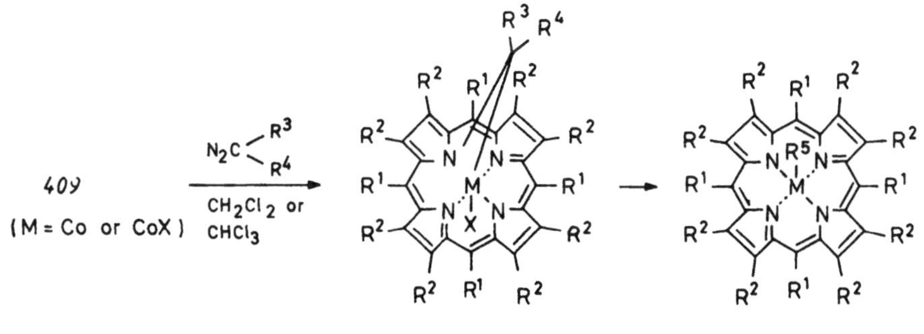

R^3 = H, R^4 = COOEt *412* M = Co *413*

R^3	R^4	R^5
H	H	CH_2X
Me	COOR	$H_2C{=}\overset{|}{C}COOR$
Me	COMe	$H_2C{=}\overset{|}{C}COMe$
$MeOOCCH_2$	COOMe	$MeOOCCH{=}\overset{|}{C}COOMe$
Me	$PO(OMe)_2$	$H_2C{=}\overset{|}{C}PO(OMe)_2$
$PhCH_2$	COOMe	$PhCH{=}\overset{|}{C}COOMe$
$Me_2CH{=}CH$	$PO(OMe)_2$	$Me_2C{=}C{=}\overset{|}{C}PO(OMe)_2$
$Me_2\underset{OH}{\underset{|}{C}}-$	COOEt	$Me(OMe)C{=}\overset{|}{C}COOEt$
	and others more	

409
(M = RhI, R^1 = Ph,
R^2 = H)

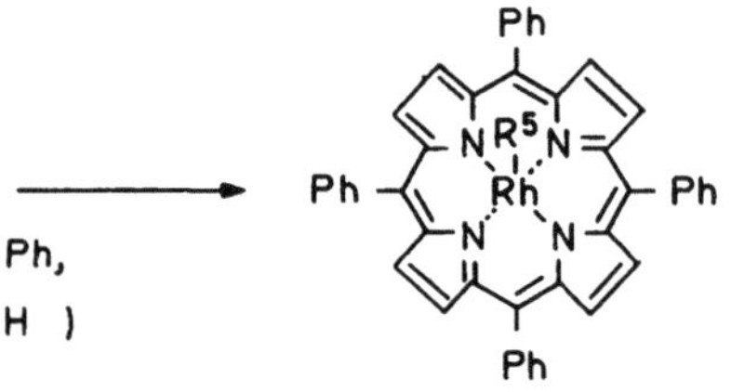

414

$N_2CHCOOEt/ROH: R^5 = \overset{OR}{\overset{|}{C}}H{-}COOEt$

R = Me, $PhCH_2$, Ac

$CH_2N_2/CH_2Cl_2: R^5 = CH_2I$

$CH_2N_2/MeOH: R^5 = CH_2I$ and CH_2OMe

$Me\underset{N_2}{\underset{\|}{C}}COOEt$: $\left[R^5 = \overset{COOEt}{\overset{|}{C}}{=}CH_2\right] \xrightarrow{*)} R^5 = COOEt$

*) with an excess of diazo compound

$MeOOC\underset{N_2}{\underset{\|}{C}}{-}CH_2COOMe: R^5 = \underset{COOMe}{\underset{|}{C}}{=}CHCOOMe$

Scheme 42

11 The Role of the Catalyst in Diazoalkane Decomposition

Whilst the active involvement of a transition-metal catalyst along the whole reaction coordinate (or at least part of it) of diazoalkane decomposition in the absence or presence of additional substrates is nowadays undisputed, the actual metal-containing intermediates continue to be subjects of conjecture, working hypotheses and conclusions based on analogy considerations [14,399]. The fundamental difficulty to detect reaction intermediates is accompanied by the diversity of processes to be expected. The outcome of a reaction may or may not depend on the nature, valence state and ligands of a metal center, furthermore on the particular diazo compound as well as on the redox potentials of the metal ion and diazo compound. Besides, a substrate rather than the diazo compound may be involved in the primary interaction with the catalyst. Wulfman [14] has given a very informative and detailed theoretical analysis of possible mechanistic schemes for copper-ion induced reactions. Some new evidence which sheds light on the primary stages of diazoalkane decomposition will be reported in the following.

The interaction between catalyst and diazo compound may be initialized by electrophilic attack of the catalyst metal at the diazo carbon, with simultaneous or subsequent loss of N_2, whereupon a metal-carbene complex (**415**) or the product of carbene insertion into a metal/ligand bond (**416**) or its ionic equivalent (**417**) are formed. This is outlined in a simplified manner in Scheme 43, which does not speculate on the kinetics of such a sequence, nor on the possible interconversion of **415** and **416**/**417** or the primarily formed Lewis acid — Lewis base adducts.

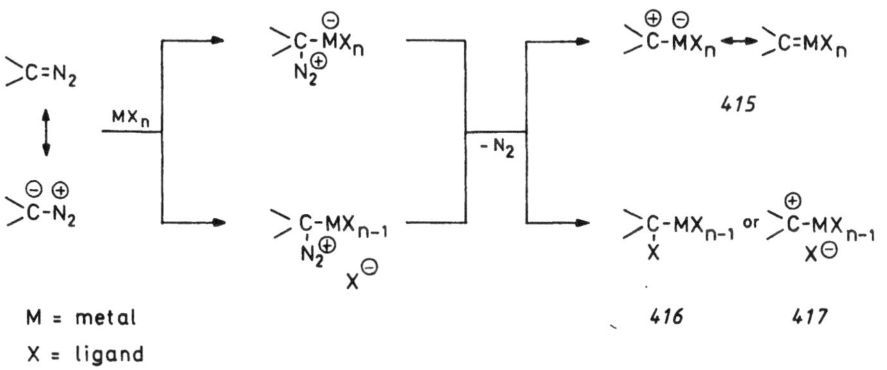

Scheme 43

Based on this scheme, the kinetics of decomposition of diazodiphenylmethane and diazofluorene in acetonitrile by $ZnCl_2$, $ZnBr_2$ or ZnI_2 has been explained [400]. Copper complexes of type **416** or **417** have been invoked to account for kinetics and products of the decomposition of diazodiphenylmethane by $CuBr_2$ in acetonitrile [401]. Copper carbenoids of unspecified structure were assumed in analogous investigations with CuBr and $CuClO_4$ [402] [but not with $Cu(ClO_4)_2$ which initiates a radical cation pathway, vide infra]. Carbenoids **416** (X = Hal) have been isolated from reactions between electrophilic metal halides and diazo compounds in a number of cases [4,5,14,19,397,398]. Some stoichiometric reactions of stable carbenoids support the assumption that similar species also occur in metal halide-catalyzed reactions of diazo compounds [48,400].

The Lewis acid-Lewis base interaction outlined in Scheme 43 also explains the formation of alkylrhodium complexes **414** from iodorhodium(III) *meso*-tetraphenylporphyrin **409** and various diazo compounds (Scheme 42) [398]. It seems reasonable to assume that intermediates **418** or **419** (corresponding to **415** and **417** in Scheme 43) are trapped by an added nucleophile in the reaction with ethyl diazoacetate, and that similar intermediates, by proton loss, give rise to vinylrhodium complexes from ethyl 2-diazopropionate or dimethyl diazosuccinate. As the rhodium porphyrin **409** is also an efficient catalyst for cyclopropanation of olefins with ethyl diazoacetate [87,100], stilbene formation from aryl diazomethanes [358] and carbene insertion into aliphatic C—H bonds [287], intermediates **418** or **419** are likely to be part of the mechanistic scheme of these reactions, too.

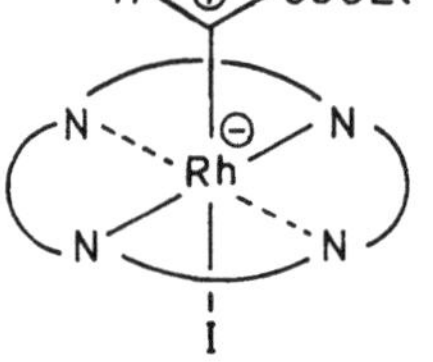

418

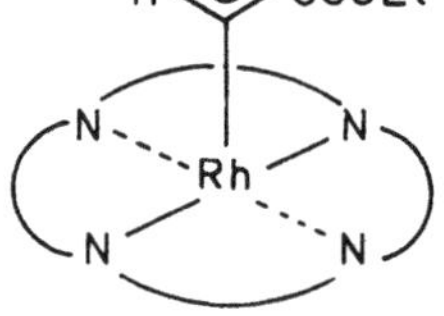

419

Metal carbenes **415** (X ≠ Hal) are widely assumed to participate in transformations of diazo compounds catalyzed by complexes of Cu, Rh, Ru, Pd, Co and Ni (for examples, see the preceding chapters). Formally speaking, they result from coordination of a carbene moiety to a metal derivative MX_n if the metal can accomodate an increase in coordination number or from replacing a neutral two-electron ligand at MX_n with the carbene unit. All evidence concerning the occurrence and properties of such a metal carbene is indirect. In assigning a metal carbene structure to the elusive intermediates of transition-metal catalyzed decomposition of diazo compounds, one refers to the existence and properties of stable metal carbene complexes, especially those with a group 6–8 transition metal, which have an electrophilic carbene ligand and undergo some typical carbene transfer reactions such as cyclopropanation and carbene dimer formation [403,404] as well as C/H insertion [405]. Interaction between diazo compounds and low-valent transition-metal complexes gave acces to some relatively stable diazoalkane complexes of Mo, W, Mn, Ru, Rh, Ir, Ni and Pd [19,23,41,43,51,52,406–408] as well as carbene complexes with some of these metals [19–22], and the rare investigations on their chemical behavior revealed a few instances of carbenoid chemistry. From all those diazoalkane complexes in which the diazo group acts as an end-on or side-on coordinated two-electron donor ligand, the intact diazo compound can be displaced by appropriate ligands. In contrast, their thermal stability varies within wide limits. For instance, 9-diazofluorene forms a complex with $Ni(C_2H_4)(PPh_3)_2$ which is stable at room temperature and above, whereas it is decomposed by $Pt(C_2H_4)(PPh_3)_2$ to give the fluorenone azine at low temperatures (—78 → 0 °C) [51]. Focussing on the diazoalkane ligand, one finds that 9-diazofluorene [43,51], dicyanodiazomethane [409] and diazocyclopentadiene [52] form isolable Ni(0) and Pd(0) complexes such as **9**, whereas diazomethane is rapidly decomposed with the same zerovalent metal complexes even at —78 °C, allowing carbene transfer to methyl acrylate [43]. By analogy with the stable Ni(0)- or Pd(0)-

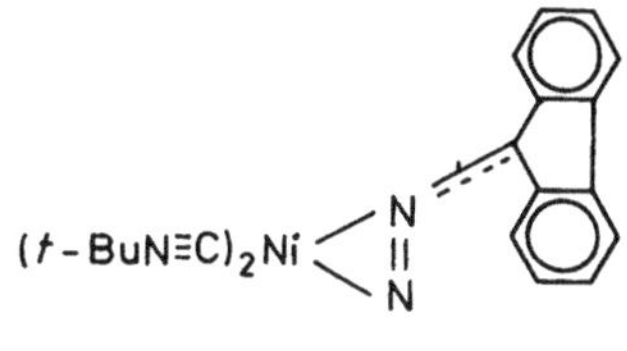

9

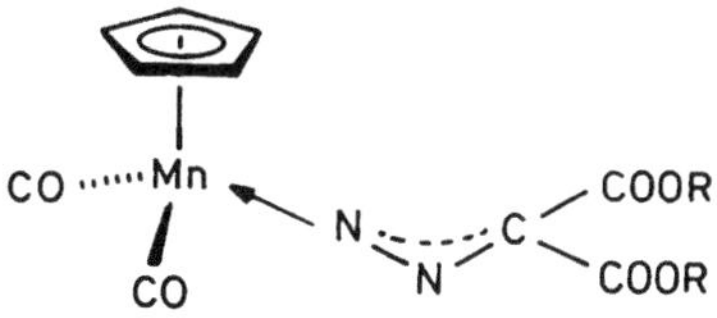

420

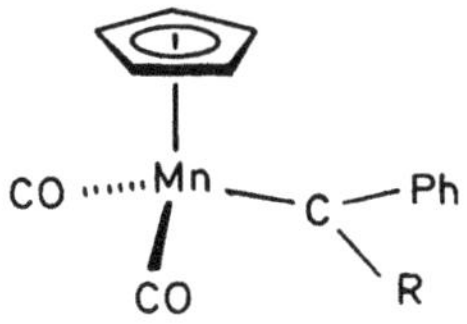

421

R = Ph, COPh

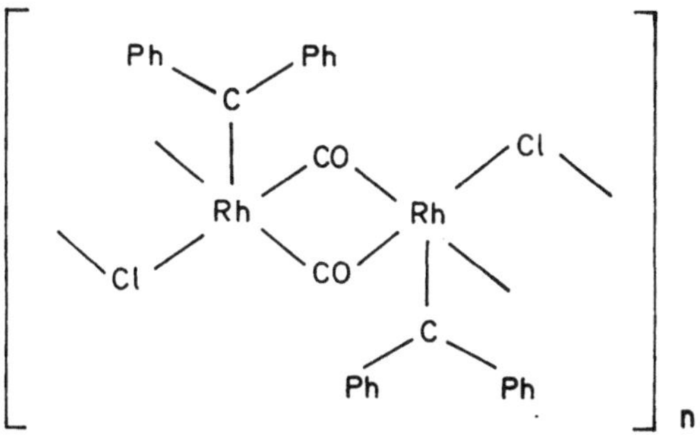

422

diazoalkane complexes, it has been assumed that such a complex is formed initially with diazomethane as well, but rearranges to a metal-carbene complex capable of subsequent carbenoid reactions. It has been suggested that such a sequence could be valid generally for diazoalkane decomposition involving a catalyst with no marked Lewis acid character [43]. Some reactions of *isolable* Ni(0)-diazoalkane complexes fit into the scenario of carbenoid reactivity [51,52], but cyclopropanation of olefinic substrates [43] or complex ligands [e.g. (1,5-cyclooctadiene)diazofluorene nickel(0) [51]] seems not to occur (for the origin of cyclopropanes on heating **9** in the presence of diethyl maleate, see discussion in Sect. 2.1).

It is not known whether there is any carbenoid chemistry of end-on coordinated manganese-diazoalkane complexes **420** [410]. However, it is known that diaryldiazomethanes [408,411] and azibenzil [411] yield carbene rather than diazoalkane complexes with $(\eta^5\text{-}C_5H_5)Mn(CO)_2THF$.

The η^1-coordinated carbene complexes **421** (R = Ph) [411] and **422** [412] are rather stable thermally. As metal-free product of thermal decomposition [**421** (R = Ph): 110 °C, **422**: PPh_3, 105 °C], one finds the formal carbene dimer, tetraphenylethylene, in both cases. Carbene transfer from **422** onto 1,1-diphenylethylene does not occur, however. Among all isolated carbene complexes, **422** may be considered the only connecting link between stoichiometric diazoalkane reactions and catalytic decomposition [except for the somewhat different results with rhodium(III) porphyrins, see above]: **422** is obtained from diazodiphenylmethane and $[Rh(CO)_2Cl]_2$, which is also known to be an efficient catalyst for cyclopropanation and S-ylide formation with diazoesters [66].

New evidence as to the nature of the intermediates in catalytic diazoalkane decomposition comes from a comparison of olefin cyclopropanation with the electrophilic metal carbene complex $(CO)_5W{=}CHPh$ on one hand and $Rh_2(OAc)_4/N_2CHCOOEt$ or $Rh_2(OAc)_4/N_2CHPh$ on the other [45]. For the same set of monosubstituted alkenes, a linear log-log relationship between the relative reactivities for the stoichiometric reaction with $(CO)_5W{=}CHPh$ and the catalytic reaction with $Rh_2(OAc)_4$ was found (reactivity difference of $2.2 \cdot 10^4$ in the former case and 14 in the latter). No such correlation holds for di- and trisubstituted olefins, which has been attributed to "steric and/or electronic differences in olefin interaction with the reactive electrophile" [45]. A linear relationship was also found between the relative reactivities of $(CO)_5W{=}CHPh$ and $Rh_2(OAc)_4/N_2CHPh$. These results lead to the conclusion that the intermediates in the Rh(II)-catalyzed reaction are very similar to stable electrophilic carbenes in terms of electron demand. As far as *cis/trans* stereoselectivity of cyclopropanation is concerned, no obvious relationship between $Rh_2(OAc)_4/N_2CHCOOEt$ and $Rh_2(OAc)_4/N_2CHPh$ was found, but the log-log plot displays an excellent linear relationship between $(CO)_5W{=}CHPh$ and $Rh_2(OAc)_4/N_2CHPh$, including mono-, 1,1-di-, 1,2-di- and trisubstituted alkenes [45]. In the phenylcarbene transfer reactions, *cis*- (*syn*-) cyclopropanes are formed preferentially, whereas *trans*- (*anti*-) cyclopropanes dominate when the diazoester is involved.

Taking together the results of reactivity and stereoselectivity comparisons, one may conclude that the cyclopropanation mechanism as such is quite similar in all cases and involves a metal carbene, but that the stereoselectivity is determined by the nature of the diazoalkane substituent. Doyle has developed a mechanistic scheme which accounts for these observations (Scheme 44).

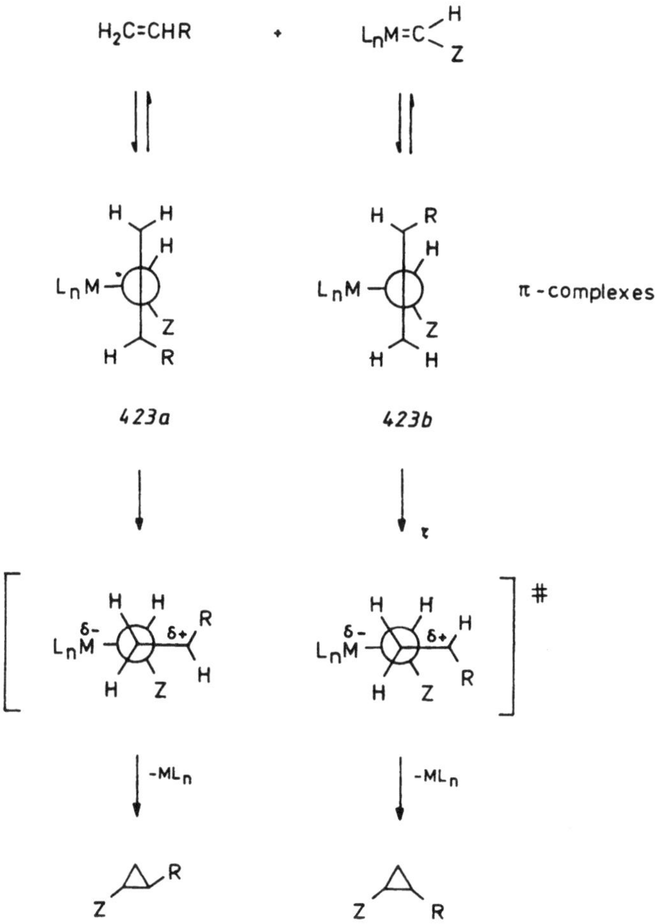

Scheme 44

Initial interaction between alkene and metal carbene is envisaged to be that of a π complex. For a monosubstituted alkene, the two sterically least hindered arrangements are given by **423a** and **423b**, but for steric reasons, **423b** is favored over **423a**. As shown in Scheme 44, **423b** leads to the *cis*-cyclopropane, and this is indeed found when Z = Ph [$Rh_2(OAc)_4/N_2CHPh$ or $(CO)_5W$ = CHPh]. When the metal carbene is derived from a diazoester or a diazoketone (Z = COR), stabilizing electronic interaction between the incipient positive charge on the alkene and the lone pairs of the carbonyl group comes into play as the π complex **423a** moves towards the transition state of cyclopropane formation. This situation does not hold for the reaction pathway of **423b**. Thus, *trans*-cyclopropane formation is expected and indeed observed for α-diazocarbonyl compounds. Scheme 44 can be expanded to understand the deviations in relative reactivity correlations for higher substituted olefins as well as the formation of dihydrofurans from olefins which allow a higher degree of positive charge development and diazocarbonyl compounds having a more nucleophilic carbonyl oxygen (see Sect. 2.3.1).

When the *cis/trans* stereoselectivity of cyclopropanation with ethyl diazoacetate in the presence of $CuCl \cdot P(O\text{-}i\text{-}Pr)_3$, $Rh_6(CO)_{16}$ or $PdCl_2 \cdot 2\,PhCN$ was plotted against that obtained with $Rh_2(OAc)_4$, a linear correlation was observed in every case, with slopes of 1.74, 1.04 and 0.59, respectively (based on 22 olefins, T = 298 K) [59]. These relationships as well as the results of regioselectivity studies carried out with 1,3-dienes point to the similar nature of the intermediates involved in Cu-, Rh- and Pd-catalyzed cyclopropanation. Furthermore, obvious parallels in reactivity in the transformations of Scheme 45 for a variety of catalysts based on Cu, Rh, Fe, Ru, Re and Mo suggest the conclusion that electrophilic metal carbenes are not only involved in cyclopropanation but also in ylide-forming reactions [66].

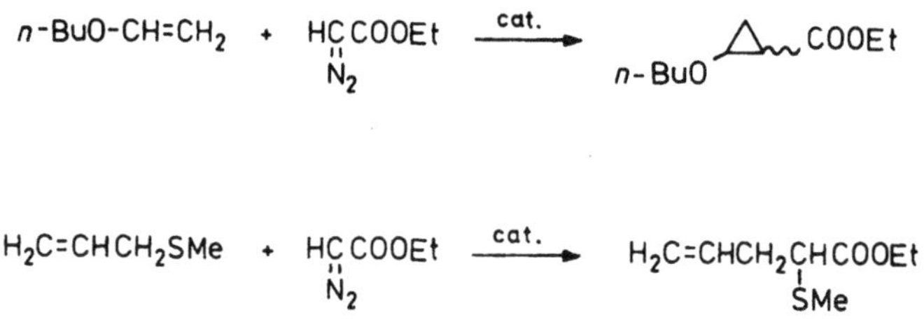

Scheme 45

Strong evidence exists for the intermediacy of a tungsten ethoxycarbonyl carbene **425** in cyclopropanation of various enol ethers, 1,3-dienes and cyclohexene with ethyl diazoacetate in the presence of catalytic amounts of $(CO)_5W = C(OMe)Ph$ [413]. The following equations could account for the obtained products:

$$(CO)_5W{=}C(OMe)Ph + N_2CHCOOEt \rightarrow \underset{\mathbf{424}}{(CO)_5W[Ph(OMe)C{=}CHCOOEt]} + N_2$$

$$\mathbf{424} + \text{alkene} \rightleftharpoons (CO)_5W(\text{alkene}) + Ph(OMe)C{=}CHCOOEt$$

$$(CO)_5W(\text{alkene}) \rightleftharpoons (CO)_5W + \text{alkene}$$

$$(CO)_5W + N_2CHCOOEt \rightarrow \underset{\mathbf{425}}{(CO)_5W{=}CHCOOEt} + N_2$$

$$\mathbf{425} + \text{alkene} \rightarrow (CO)_5W + \text{cyclopropane}$$

$$\mathbf{425} + N_2CHCOOEt \rightarrow (CO)_5W + EtOOCCH{=}CHCOOEt + N_2$$

Inhibition of diazoester decomposition by a large excess of olefin speaks in favor of intermediarily liberated $W(CO)_5$ as direct metal precursor of **425**. Stereoselectivities in the cyclopropanation reaction are very similar to those observed in the $Rh_2(OAc)_4$ catalyzed version, which underlines once more the close relationship of tungsten and rhodium carbene complexes.

In the stoichiometric reaction between $(CO)_5W = C(OMe)Ph$ and ethyl diazoacetate, the coupling product $Ph(OMe)C{=}CHCOOEt$ is obtained in high yield. Pyridine is needed to liberate this alkene from **424** [413]. The analogous reaction sequence of diazomethane has also been reported [414].

In order to rationalize the catalyst-dependent selectivity of cyclopropanation reaction with respect to the alkene, the ability of a transition metal for olefin coordination has been considered to be a key factor (see Sect. 2.2.1 and 2.2.2). It was proposed that palladium and certain copper catalysts promote cyclopropanation through intramolecular carbene transfer from a metal carbene to an alkene molecule coordinated to the same metal atom [25,64]. The preferential cyclopropanation of terminal olefins and the less hindered double bond in dienes spoke in favor of metal-olefin coordination. Furthermore, stable and metastable metal-carbene-olefin complexes are known, some of which undergo intramolecular cyclopropane formation, e.g. **426** → **427** [415].

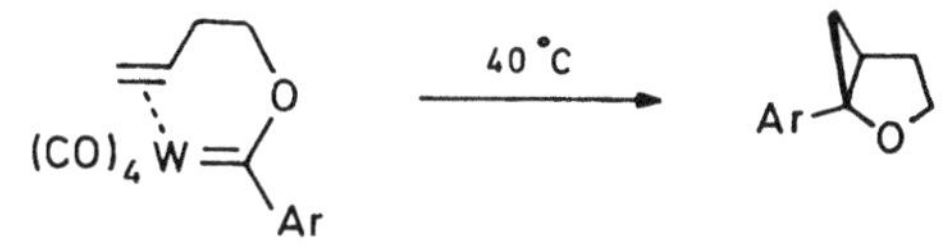

426 427

Ar = p-tolyl

Doyle has put forward arguments against the intermediacy of such complexes in catalytic cyclopropanation [47]. Firstly, metal coordination activates the alkene to nucleophilic attack. Hence, an electrophilic metal carbene would add only reluctantly or not at all. Secondly, the stable $PdCl_2$ complexes of dienes **8** and **428** do not react with ethyl diazoacetate, even if $Rh_2(OAc)_4$ or $PdCl_2(PhCN)_2$ is added. The diazoester is decomposed only when it is added to a mixture of the Pd complex and excess diene. These results exclude the metal-carbene-olefin intermediate, but they leave open the possibility of metal carbene interaction with an uncomplexed olefin molecule. The preferred formation of *exo*-cyclopropanes in the $PdCl_2(PhCN)_2$-catalyzed reactions between **8** and $N_2CHCOOEt$ or N_2CPh_2, with *exo*:*endo* ratios virtually identical to those observed upon cyclopropanation of monoolefin **429**, also rule out coordination of a palladium carbene to the exocyclic double bond of **8** prior to cyclopropanation of the endocyclic double bond.

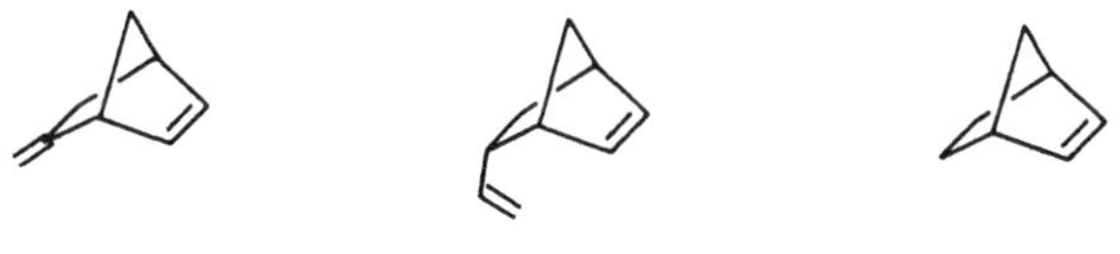

8 428 429

For the copper-induced decomposition of diazodiphenylmethane in acetonitrile, a fundamental difference in the catalytic action of $Cu^{I}ClO_4$ and $Cu^{II}(ClO_4)_2$ was detected. Whilst with $CuClO_4$, intermediary copper carbenoids are believed to be responsible for the mainly formed benzophenone azine [402], $Cu(ClO_4)_2$ initiates a chain reaction, promoted by radical cations and yielding mainly tetraphenylethene and some benzophenone azine[416]. The second-order rate constant (k_{obs}/[catalyst]) of the $CuClO_4$ reaction is some 300 times smaller than k_{obs} for the decomposition by $CuClO_4$. Similar kinetics and product ratios were found for both the $Cu(ClO_4)_2$ reaction and the decomposition of the same diazo compound in the presence of the stable radical cation salt tris(*p*-bromophenyl)-ammoniumyl perchlorate, and both reactions obey a first-order rate law. According to these findings, a mechanistic picture can be drawn with the catalyst involved in the first step only, namely one-electron oxidation of the diazoalkane (Scheme 46). Reaction of the radical cation $N_2CPh_2^{\cdot +}$ with traces of water is a side-reaction which ultimately leads to diphenylmethanol, benzophenone and benzpinacol. The liberation of a proton in this sequence causes the acid-catalyzed decomposition of the diazo compound to take over at larger water concentration.

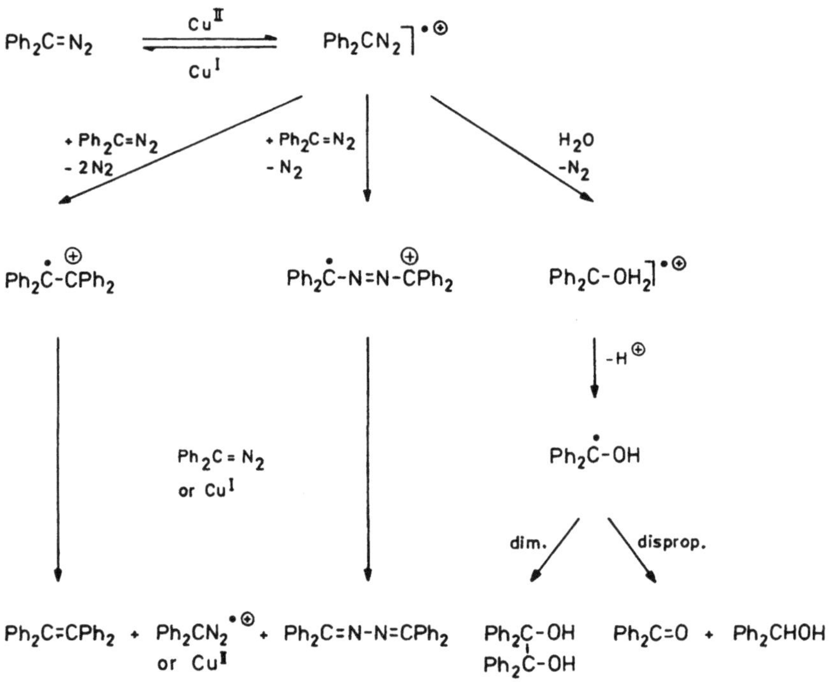

Scheme 46

It is known that the oxidation potentials of diazodiphenylmethane and Cu(I) in acetonitrile are very similar. With $CuBr_2$ however, no radical-chain reaction takes place. Contrary to the copper perchlorates, $CuBr_2$ and CuBr initiate identical reaction pathways involving copper carbenoids. No definite answer to this discrepancy is available[402].

Initial one-electron oxidation of the diazo compound by Ag(I), Hg(II) or Cu(II) acetates may also be responsible for the formation of $Ph_2C(OAc)$—$C(OAc)Ph_2$ from diazodiphenylmethane and of EtOOCCH(OAc)—CH(OAc)COOEt from ethyl diazoacetate in DMF/H_2O[417]. Direct evidence for reduction of Cu(II) triflate to the Cu(I) salt by alkyl diazoacetates has been furnished by the disappearance of the

Cu(II) EPR signal in nitriles as solvent as well as by polarographic measurements [144]. Similarly, the EPR signal disappeared when $Cu(OTf)_2$ was used for catalytic cyclopropanation of olefins with diazoesters [64]. In these cases, no evidence for radical-chain reactions has been reported, however. The $Cu(acac)_2$- or $Cu(hfacac)_2$-catalyzed decomposition of $N_2CHCOOEt$, $N_2C(COOEt)_2$, $MeCOC(N_2)COOEt$ and $N_2CHCOCOOEt$ in the presence of cyclopropyl-substituted ethylenes did not furnish any products derived from a cyclopropylcarbinyl → butenyl rearrangement [128]. These results rule out the possible participation of electron-transfer processes and radical intermediates which would arise from interaction between the olefin and a radical species derived from the diazocarbonyl compound.

12 Acknowledgement

This work was supported by the Fonds der Chemischen Industrie. Reprint permission has been granted by the American Chemical Society for Tables 2, 3, 5 and 7 and by Synthesis for Table 6.

13 References

1. Loose, A.: J. prakt. Chem. *79*, 507 (1909)
2. Wolff, L.: Liebigs Ann. Chem. *394*, 23 (1912)
3. Eistert, B.; Regitz, M. Heck, G.; Schwall, H.: in Houben-Weyl (Müller, E. (ed.)), Vol. X/4, p. 810, Thieme, Stuttgart 1981
4. Lappert, M. F.; Poland, J. S.: Adv. Organomet. Chem. *9*, 397 (1970)
5. Cowell, G. W.; Ledwith, A.: Quart. Reviews *24*, 119 (1970)
6. Dave, V.; Warnhoff, E.: Org. Reactions *18*, 217 (1970)
7. Kirmse, W.: Carbene Chemistry, 2nd edition, p. 85, Academic Press, New York 1971
8. Jones, M.; Moss, R. A. (eds.): Carbenes, Vol. I, pp. 1–151, Wiley, New York 1973
9. Peace, B. W.; Wulfman, D. S.: Synthesis *1973*, 137
10. Marchand, A. R.; Brockway, N. M.: Chem. Reviews *74*, 431 (1974)
11. Mandel'shtam, T. V.: Sovrem Probl. Org. Khim. *5*, 87 (1976)
12. Burke, S. T.; Grieco, P. A.: Org. Reactions *26*, 361 (1979)
13. Wulfman, D. S.; Linstrumelle G.; Cooper, C. F.: in The Chemistry of Diazonium and Diazo Groups (Patai, S. (ed.)); Vol. 2, Chapter 18, Wiley, New York 1978
14. Wulfman, D. S.; Poling, B.: in Reactive Intermediates (Abramovitch, R. A. (ed.)); Vol. 1, p. 321, Plenum, New York 1980
15. Taylor, K. G.: Tetrahedron *38*, 2751 (1982)
15a. Kruglaya, O. A.; Vyazankin, N. S.: Russian Chem. Reviews *49*, 357 (1980)
16. Wulfman, D. S.: Tetrahedron *32*, 1231 (1976)
17. Regitz, M.: Diazoalkane, Thieme, Stuttgart 1977
18. Regitz, M.; Maas, G.: Aliphatic Diazo Compounds — Properties and Synthesis; Academic Press, New York, in press
19. Herrmann, W. A.: Angew. Chem. *90*, 855 (1978); Angew. Chem. Int. Ed. Engl. *17*, 800 (1978)
20. Herrmann, W. A.: Adv. Organomet. Chem. *20*, 159 (1982)
21. Herrmann, W. A.: Pure Appl. Chem. *54*, 65 (1982)
22. Herrmann, W. A.: J. Organomet. Chem. *250*, 319 (1983)
23. Albini, A.; Kisch, H.: Topics Curr. Chem. *65*, 105 (1976)
24. Müller, E.; Kessler, H.; Zech, B.: Fortschr. chem. Forsch. *7*, 128 (1966)

25. Salomon, R. G.; Kochi, J. K.: J. Am. Chem. Soc. *95*, 3300 (1973)
26. Salomon, R. G.; Salomon, M. F.; Heyne, T. R.: J. Org. Chem. *40*, 756 (1975)
27. Salomon, R. G.; Salomon, M. F.; Kachinski, J. L. C.: J. Am. Chem. Soc. *99*, 1043 (1977)
28. Mueller, L. G.; Lawton, R. G.: J. Org. Chem. *44*, 4741 (1979)
29. Paulissen, R.; Hubert, A. J.; Teyssié, P.: Tetrahedron Lett. *1972*, 1465
30. Suda, M.: Synthesis *1981*, 714
31. Kottwitz, J.; Vorbrüggen, H.: Synthesis *1975*, 636
32. Tomilov, Y. V.; Bordakov, V. G.; Dolgii, I. E.; Nefedov, O. M.: Izv. Akad. Nauk SSSR, Ser. Khim. *1984*, 582
33. Rehse, K.; Behncke, S.; Siemann, U.; Kehr, W.: Arch. Pharm. (Weinheim) *313*, 221 (1980)
34. Mende, U.; Radüchel, B.; Skuballa, W.; Vorbrüggen, H.: Tetrahedron Lett. *1975*, 629
35. Radüchel, B.; Mende, U.; Cleve, G.; Hoyer, G.-A.; Vorbrüggen, H.: Tetrahedron Lett. *1975*, 633
36. Spur, B.; Crea, A.; Peters, W.: Z. Naturforsch. *39b*, 125 (1984)
37. Abdallah, H.; Grée, R.; Carrié, R.: Tetrahedron Lett. *23*, 503 (1982)
38. Nefedov, O. M.; Dolgii, I. E.; Tomilov, Y. V.; Bordakov, V. G.: Izv. Akad. Nauk SSSR, Ser. Khim. *1984*, 119
39. Dinulescu, I.; Enescu, L. N.; Ghenciulescu, A.; Avram, M.: J. Chem. Research (S) *1978*, 456
40. Rüchardt, C.; Schrauzer, G.: Chem. Ber. *93*, 1840 (1960)
41. Bogdanovic, B.; Kroner, M.; Wilke, G.: Liebigs Ann. Chem. *699*, 1 (1966)
42. Werner, H.; Richards, J. H.: J. Am. Chem. Soc. *90*, 4976 (1968)
43. Nakamura, A.; Yoshida, T.; Cowie, M.; Otsuka, S.; Ibers, J. A.: J. Am. Chem. Soc. *99*, 2108 (1977).
44. Takaya, H.; Suzuki, T.; Kumagai, Y.; Hosoya, M.; Kawauchi, H.; Noyori, R.: J. Org. Chem. *46*, 2854 (1981)
45. Doyle, M. P.; Griffin, J. H.; Bagheri, V.; Dorow, R. L.: Organometallics *3*, 53 (1984)
46. Goh, S. H.; Closs, L. E.; Closs, G. L.: J. Org. Chem. *34*, 25 (1966)
47. Doyle, M. P.; Wang, L. C.; Loh, K.-L.: Tetrahedron Lett. *25*, 4087 (1984)
48. Crumrine, D. S.; Haberkamp, T. J.; Suther, D. J.: J. Org. Chem. *40*, 2274 (1975)
49. Crumrine, D. S.; Yen, H.-H. B.: J. Am. Chem. Soc. *98*, 297 (1976)
50. Nozaki, H.; Takaya, H.; Morluti, S.; Noyori, R.: Tetrahedron *24*, 3655 (1968)
51. Otsuka, S.; Nakamura, A.; Koyama, T.; Tatsuno, Y.: Liebigs Ann. Chem. *1975*, 626
52. Schramm, K. D.; Ibers, J. A.: Inorg. Chem. *19*, 2441 (1980)
53. House, H. O.; Blankley, C. I.: J. Org. Chem. *33*, 53 (1968)
54. House, H. O.; Fischer Jr., W. F.; Gall, M.; McLaughlin, T. E.; Peet, N. P.: J. Org. Chem. *36*, 3429 (1971)
55. Wulfman, D. S.; McDaniel Jr. R. S.; Peace, B. W.: Tetrahedron *32*, 1241 (1976)
56. Doering, W. v. E.; Mole, T.: Tetrahedron *10*, 65 (1960)
57. Wulfman, D. S.; McGibboney, B. G.; Steffen, E. K.; Thinh, N. V.; McDaniel Jr., R. S.; Peace, B. W.: Tetrahedron *32*, 1257 (1976)
58. Moser, W. R.: J. Am. Chem. Soc. *91*, 1135, 1141 (1969)
59. Doyle, M. P.; Dorow, R. L.; Buhro, W. E.; Griffin, J. H.; Tamblyn, W. H.; Trudell, M. L.: Organometallics *3*, 44 (1984)
60. Kunkel, E.; Reichelt, I.; Reißig, H.-U.: Liebigs Ann. Chem. *1984*, 512
61. Wulfman, D. S.; Peace, B. W.; Steffen, E. K.: J. Chem. Soc. Chem. Commun. *1971*, 1360
62. Paulissen, R.; Hubert, A. J.; Teyssié, P.: Tetrahedron Lett. *1972*, 1465
63. Hubert, A. J.; Noels, A. F.; Anciaux, A. J.; Teyssié, P.: Synthesis *1976*, 600
64. Anciaux, A. J.; Hubert, A. J.; Noels, A. F.; Petiniot, N.; Teyssié, P.: J. Org. Chem. *45*, 695 (1980)
65. Doyle, M. P.; Tamblyn, W. H.; Buhro, W. E.; Dorow, R. L.: Tetrahedron Lett. *22*, 1783 (1981)
66. Tamblyn, W. H.; Hoffmann, S. R.; Doyle, M. P.: J. Organomet. Chem. *216*, C64 (1981)
67. (a) Porai-Koshits, M. A.; Antsyshkina, A. S.: Dokl. Akad. Nauk SSSR *146*, 1102 (1962). — (b) Johnson, S. A.; Hunt, H. R.; Neumann, H. M.: Inorg. Chem. *2*, 960 (1963). — (c) Richman, R. M.; Kuechler, T. C.; Tanner, S. P.; Drago, R. S.: J. Am. Chem. Soc. *99*, 1055 (1977).
68. Maitlis, P. M.: The Organic Chemistry of Palladium, Vol. 1, p. 106, Academic Press, New York 1971
69. Salomon, R. G.; Kochi, J. K.: J. Chem. Soc. Chem. Commun. *1972*, 559

70. Kirmse, W.; Hellwig, G.: Chem. Ber. *115*, 2744 (1982)
71. Doyle, M. P.; Dorow, R. L.; Tamblyn, W. H.; Buhro, W. E.: Tetrahedron Lett. *23*, 2261 (1982)
72. Anciaux, A. J.; Demonceau, A.; Noels, A. F.; Warin, R.; Hubert, A. J.; Teyssié, P.: Tetrahedron *39*, 2169 (1983)
73. Doyle, M. P.; van Leusen, D.; Tamblyn, W. H.: Synthesis *1981*, 787
74. Elzinga, J.; Hogeveen, H.; Schudde, E. P.: J. Org. Chem. *45*, 4337 (1980)
75. Ryang, H.-S.; Foote, C. S.: J. Am. Chem. Soc. *102*, 2129 (1980)
76. Doyle, M. P.; Colsman, M. R.; Chinn, M. S.: Inorg. Chem. *23*, 3684 (1984)
77. Demonceau, A.; Noels, A. F.; Anciaux, A. J.; Hubert, A. J.; Teyssié, P.: Bull. Soc. Chim. Belges *93*, 949 (1984)
78. Milner, D. J.: J. Organomet. Chem. *262*, 85 (1984)
79. Wenkert, E.; Goodwin, T. E.; Ranu, B. C.: J. Org. Chem. *42*, 2137 (1977)
80. Bohlmann, F.; Rotard, W.: Liebigs Ann. Chem. *1982*, 1211
81. Decock—Le Reverend, B.; Durand, M.; Merenyi, R.: Bull. Soc. Chim. France *78* (2), 369 (1978)
82. De Smet, A.; Anteunis, M.; Tavernier, D.: Bull. Soc. Chim. Belges *84*, 67 (1975)
83. Neidlein, R.; Wesch, K. F.: Helv. Chim. Acta *66*, 891 (1983)
84. Paquette, L. A.; Thompson, G. L.; Heyd, W. E.: J. Am. Chem. Soc. *96*, 3177 (1974)
85. Komendantov, M. I.; Bespalov, V. Y.; Bezrukova, O. A.; Bekmukhametov, R. R.: Zh. Org. Khim. *11*, 27 (1975)
86. Nakamura, A.; Koyama, T.; Otsuka, S.: Bull. Soc. Chim. Japan *51*, 593 (1978)
87. Callot, H. J.; Metz, F.; Piechocki: Tetrahedron *38*, 2365 (1982)
88. Nakamura, A.; Konishi, A.; Tsujitani, R.; Kudo, M.; Otsuka, S.: J. Am. Chem. Soc. *100*, 3449 (1978)
89. Cowan, D. O.; Couch, M. M.; Kopecky, K. R.; Hammond, G. S.: J. Org. Chem. *29*, 1922 (1964)
90. Knorr, R.: Chem. Ber. *113*, 2441 (1980)
91. Aratani, T.; Yoneyoshi, Y.; Nagase, T.: Tetrahedron Lett. *23*, 685 (1982)
92. Aratani, T.; Yoneyoshi, Y.; Nagase, T.: Tetrahedron Lett. *1977*, 2599
93. Laidler, D. A.; Milner, D. J.: J. Organomet. Chem. *270*, 121 (1984)
94. Vasil'vitskii, A. E.; Shostakovskii, V. M.; Nefedov, O. M.: Izv. Akad. Nauk SSSR, Ser. Khim. *1979*, 690
95. Nakamura, A.; Konishi, A.; Tatsuno, Y.; Otsuka, S.: J. Am. Chem. Soc. *100*, 3443 (1978)
96. Williams, J. L.; Rettig, M. F.: Tetrahedron Lett. *22*, 385 (1981)
97. Doyle, M. P.; van Leusen, D.: J. Org. Chem. *47*, 5326 (1982)
98. Holland, D.; Milner, D. J.: J. Chem. Research (S) *1979*, 317; (M) *1979*, 3734
99. Milner, D. J.; Holland, D.: Ger. Offen. DE 2810098 (1978); Chem. Abstr. *90*, 38578r (1979)
100. Callot, H. J.; Piechocki, C.: Tetrahedron Lett. *21*, 3489 (1980)
101. Pazynina, G. V.; Luk'yanets, E. A.; Bolesov, I. G.: Zh. Org. Khim. *20*, 802 (1984)
102. Wenkert, E.; Mueller, R. A.; Reardon Jr., E. J.; Sathe, S. S.; Scharf, D. J.; Tosi, G.: J. Am. Chem. Soc. *92*, 7428 (1970)
103. (a) Wenkert, E.; Acc. Chem. Res. *13*, 27 (1980). — (b) Wenkert, E.: Heterocycles *14*, 1703 (1980)
104. House, H. O.; Blankley, C. J.: J. Org. Chem. *33*, 47 (1968)
105. Wenkert, E.; McPherson, C. A.; Sanchez, E. L.; Webb, R. A.: Synth. Commun. *3*, 255 (1973)
106. Beckmann, K.; Bruza, K. J.: Tetrahedron *37*, 3997 (1981)
107. Jendralla, H.; Pflaumbaum, W.: Chem. Ber. *115*, 210, 229 (1982)
108. Shatzmiller, S.; Neidlein, R.: Liebigs Ann. Chem. *1977*, 910
109. Morrison, N. J.: J. Chem. Soc. Perkin Trans. I *1982*, 3027
110. Ceccherelli, P.; Coccia, R.; Curini, M.; Pellicciari, R.: Gazz. Chim. Ital. *113*, 453 (1983)
111. Kunz, H.; Lindig, M.: Chem. Ber. *116*, 220 (1983)
112. Reichelt, I.; Reißig, H.-U.: Chem. Ber. *116*, 3895 (1983)
113. Wenkert, E.; Alonso, M. E.; Buckwalter, B. L.; Sanchez, E. L.: J. Am. Chem. Soc. *105*, 2021 (1983)

114. Wenkert, E.; Alonso, M. E.; Buckwalter, B. L.; Chou, K. J.: J. Am. Chem. Soc. *99*, 4778 (1977)
115. Böhm, I.; Hirsch, E.; Reißig, H.-U.: Angew. Chem. *93*, 593 (1981); Angew. Chem. Int. Ed. Engl. *20*, 574 (1981)
116. Saigo, K.; Okagawa, S.; Nohira, H.: Bull. Chem. Soc. Japan *54*, 3603 (1981)
117. Saigo, K.; Kurihara, H.; Miura, H.; Hongu, A.; Kubota, N.; Nohira, H.; Hasegawa, M.: Synth. Commun. *14*, 787 (1984)
118. LeGoaller, R.; Pierre, J.-L.: C. R. Acad. Sci. Paris *C276*, 193 (1973)
119. Marino, J. P.; Laborde, E.: J. Am. Chem. Soc. *107*, 734 (1985)
120. Doyle, M. P.; van Leusen, D.: J. Am. Chem. Soc. *103*, 5917 (1981)
121. Graziano, M. L.; Scarpati, R.: J. Chem. Soc. Perkin 1 *1985*, 289
122. Alonso, M. E.; Jano, P.; Hernandez, M. I.; Greenberg, R. S.; Wenkert, E.: J. Org. Chem. *48*, 3047 (1983)
123. Dull, M. F.; Abend, P. G.: J. Am. Chem. Soc. *81*, 2588 (1959)
124. Dowd, P.; Kaufman, C.; Paik, Y. H.: Tetrahedron Lett. *26*, 2283 (1985)
125. Alonso, M. E.; Jano, P.; Hernández, M. I.: J. Org. Chem. *45*, 5299 (1980)
126. Bien, S.; Segal, Y.: J. Org. Chem. *42*, 1685 (1977)
127. Gallucci, R. R.; Jones Jr., M.: J. Am. Chem. Soc. *98*, 7704 (1976)
128. Alonso, M. E.; Hernández, M. I.; Gómez, M.; Jano, P.; Pekerar, S.: Tetrahedron *41*, 2347 (1985)
129. Wenkert, E.; Halls, T. D. J.; Kwart, L. D.; Magnusson, G.; Showalter, H. D. H.: Tetrahedron *37*, 4017 (1981)
130. Alonso, M. E.; Morales, A.; Chitty, A. W.: J. Org. Chem. *47*, 3747 (1982)
131. Alonso, M. E.; Morales, A.: J. Org. Chem. *45*, 4530 (1980)
132. Huisgen, R.: in Cycloaddition Chemistry (Padwa, A. (ed.)); Vol. 1, Chapter 1, Wiley, New York 1984
133. Alonso, M. E.; García, M. C.: J. Org. Chem. *50*, 988 (1985)
134. Scarpati, R.; Cioffi, M.; Scherillo, G.; Nicolaus, R. A.: Gazz. Chim. Ital. *66*, 1164 (1966)
135. Wenkert, E.; Buckwalter, B. L.; Craveiro, A. A.; Sanchez, E. L.; Sathe, S. S.: J. Am. Chem. Soc. *100*, 1267 (1978)
136. Wenkert, E.; Bakuzis, M. L. F.: Synth. Commun. *11*, 533 (1981)
137. Campbell, M. M.; Harcus, R. G.; Ray, S. J.: Tetrahedron Lett. *1979*, 1441
138. Coates, R. M.; Sandefur, L. O.; Smillie, R. D.: J. Am. Chem. Soc. *97*, 1619 (1975)
139. Regitz, M.; Heydt, H.: in 1,3-Dipolar Cycloaddition Chemistry (Padwa, A. (ed.)); Volume 2, p. 393; Wiley, New York 1984
140. Kaiser, C.; Zirkle, C. L.: US 3010971 (1960); Chem. Abstr. *56*, 15484 (1962)
141. Majchrzak, M. W.; Kotelko, A.; Lambert, J. B.: Synthesis *1983*, 469
142. Chang, S.-J.; Shankar, B. K. R.; Shechter, H.: J. Org. Chem. *47*, 4226 (1982)
143. Paulissen, R.; Moniotte, P.; Hubert, A. J.; Teyssié, P.: Tetrahedron Lett. *1974*, 3311
144. Moniotte, P. G.; Hubert, A. J.; Teyssié, P.: J. Organomet. Chem. *88*, 115 (1975)
145. Doyle, M. P.; Dorow, R. L.; Tamblyn, W. H.: J. Org. Chem. *47*, 4059 (1982)
146. Doyle, M. P.; Davidson, J. G.: J. Org. Chem. *45*, 1538 (1980)
147. Thiellier, H. P. M.; Koomen, G. J.; Pandit, U. K.: Tetrahedron *33*, 1493 (1977)
148. Elguero, J.; Ochoa, C.; Stud, M.: Heterocycles *17*, 401 (1982)
149. Ando, W.: Tetrahedron Lett. *1969*, 929
150. Alonso, M. E.; Gómez, M.: Tetrahedron Lett. *1979*, 2763
151. Dzhemilev, U. M.; Fakhretdinov, R. N.; Marvanov, R. M.; Nefedov, O. M.: Izv. Akad. Nauk SSSR, Ser. Khim. *1984*, 588
152. (a) Ando, W.; Yagihara, T.; Kondo, S.; Nakayama, K.; Yamato, H.; Nakaido, S.; Migita, T.: J. Org. Chem. *36*, 1732 (1971). — (b) Ando, W.: Acc. Chem. Res. *10*, 179 (1977). — (c) Ando, W.: in The Chemistry of Diazonium or Diazo Groups (Patai, S. (ed.)); Vol. 1, p. 341; Wiley, New York 1978
153. Doyle, M. P.; Tamblyn, W. H.; Bagheri, V.: J. Org. Chem. *46*, 5094 (1981)
154. Doyle, M. P.; Griffin, J. H.; Chinn, M. S.; van Leusen, D.: J. Org. Chem. *49*, 1917 (1984)
155. Giddings, P. J.; John, D. I.; Thomas, E. J.: Tetrahedron Lett. *21*, 395 (1980)
156. Vedejs, E.; Hagen, J. P.: J. Am. Chem. Soc. *97*, 6878 (1975)
157. Vedejs, E.; Hagen, J. P.; Roach, B. L.; Spear, K. L.: J. Org. Chem. *43*, 1185 (1978)

158. De March, P.; Moreno-Mañas, M.; Ripoll, I.: Synth. Commun. *14*, 521 (1984)
159. Giddings, P. J.; John, D. I.; Thomas, E. J.; Williams, D. J.: J. Chem. Soc. Perkin Trans. I *1982*, 2757
160. Doyle, M. P.; Trudell, M. L.: J. Org. Chem. *49*, 1196 (1984)
161. Nickon, A.; Rodriguez, A. D.; Ganguly, R.; Shirhatti, V.: J. Org. Chem. *50*, 2767 (1985)
162. Noels, A. F.; Demonceau, A.; Petiniot, N.; Hubert, A. J.; Teyssié, P.: Tetrahedron *38*, 2733 (1982)
163. a) Ganem, B.; Ikota, N.; Muralidharan, V. B.; Wade, W. S.; Young, S. D.; Yukimoto, Y.: J. Am. Chem. Soc. *104*, 6787 (1982). — b) Teng, C.-Y. P.; Ganem, B.; Doktor, S. Z.; Nichols, B. P.; Bhatnagar, R. G.; Vining, L. C.: J. Am. Chem. Soc. *107*, 5008 (1985). — c) Hoare, J. H.; Policastro, P. P.; Berchthold, G. A.: J. Am. Chem. Soc. *105*, 6264 (1983)
164. Beecham Group Ltd.; Brit. 572259 (1980); Chem. Abstr. *94*, 103343j (1981)
165. Maier, G.; Hoppe, M.; Reisenauer, H. P.: Angew. Chem. *95*, 1009 (1983); Angew. Chem. Int. Ed. Engl. *22*, 990 (1983)
166. Dowd, P.; Garner, P.; Schappert, R.; Irngartinger, H.; Goldman, A.: J. Org. Chem. *47*, 4240 (1982)
167. Dowd, P.; Schappert, R.; Garner, P.; Go, C. L.: J. Org. Chem. *50*, 44 (1985)
168. Miyashi, T.; Nakajo, T.; Kawamoto, H.; Akiyama, K.; Mukai, T.: Tetrahedron Lett. *1979*, 151
169. Kreiser, W.; Below, P.: Liebigs Ann. Chem. *1985*, 203
170. Ghatak, U. R.; Roy, S. C.: J. Chem. Research (S) *1981*, 5; (M) *1981*, 159
171. Chakraborti, A. K.; Saha, B.; Ghatak, U. R.: Indian J. Chem. *20B*, 911 (1981)
172. Ghatak, U. R.; Chakraborti, P. C.: J. Org. Chem. *44*, 4562 (1979)
173. Chen, E. Y.: J. Org. Chem. *49*, 3245 (1984)
174. Taber, D. F.; Krewson, K. R.; Maman, K.; Rheingold, A.: Tetrahedron Lett. *25*, 5283 (1984)
175. Callant, P.; D'Haenens, L.; Vandewalle, M.: Synth. Commun. *14*, 155 (1984)
176. Danishefsky, S.; McKee, R.; Singh, R. K.: J. Am. Chem. Soc. *99*, 7712 (1977)
177. Danishefsky, S.; Regan, J.; Doehner, R.: J. Org. Chem. *46*, 5255 (1981)
178. Mandai, T.; Hara, K.; Kawada, M.; Nokami, J.: Tetrahedron Lett. *24*, 1517 (1983)
179. Danishefsky, S.: Acc. Chem. Res. *12*, 66 (1979)
180. Böhshar, M.: PhD Thesis, Univ. of Kaiserslautern (1985)
181. Hudlicky, T.; Sheth, J. P.; Gee, V.; Barnvos, D.: Tetrahedron Lett. *1979*, 4889
182. Hudlicky, T.; Koszyk, F. J.; Kutchan, T. M.; Sheth, J. P.: J. Org. Chem. *45*, 5020 (1980)
183. Hudlicky, T.; Kutchan, T. M.; Wilson, S. R.; Mao, D. T.: J. Am. Chem. Soc. *102*, 6351 (1980)
184. Govindan, S. V.; Hudlicky, T.; Koszyk, F. J.: J. Org. Chem. *48*, 3581 (1983)
185. Hudlicky, T.; Ranu, B. C.; Naqvi, S. M.; Srnak, A.: J. Org. Chem. *50*, 123 (1985)
186. Hudlicky, T.; Short, R. P.: J. Org. Chem. *47*, 1522 (1982)
187. Short, R. P.; Revol, J.-M.; Ranu, B. C.; Hudlicky, T.: J. Org. Chem. *48*, 4453 (1983)
188. Piers, E.; Jung, G. L.; Moss, N.: Tetrahedron Lett. *25*, 3959 (1984)
189. Kondo, K.; Umemoto, T.; Yako, K.; Tunemoto, D.: Tetrahedron Lett. *1978*, 3927
190. Tunemoto, D.; Takahatake, Y.; Kondo, K.: Chem. Letters *1978*, 189
191. Hudlicky, T.; Reddy, D. B.; Govindan, S. V.; Kulp, T.; Still, B.; Sheth, J. P.: J. Org. Chem. *48*, 3422 (1983)
192. Monti, S. A.; Bucheck, D. J.; Shepard, J. C.: J. Org. Chem. *34*, 3080 (1969)
193. Smith III, A. B.; Toder, B. H.; Branca, S. J.: J. Am. Chem. Soc. *106*, 3995 (1984)
194. Bien, S.; Gillon, A.; Kohen, S.: J. Chem. Soc. Perkin Trans. I, 489 (1976)
195. Gillon, A.; Ovadia, D.; Kapon, M.; Bien, S.: Tetrahedron *38*, 1477 (1982)
196. Bien, S.; Ovadia, D.: J. Chem. Soc. Perkin Trans I, 333 (1974)
197. (a) Baldwin, J. E.; Carter, C. G.: J. Am. Chem. Soc. *104*, 1362 (1982). — (b) Baldwin, J. E.; Patapoff, T. W.; Barden, T. C.: J. Am. Chem. Soc. *106*, 1421 (1984). — (c) Baldwin, J. E.: Barden, T. C.: J. Am. Chem. Soc. *106*, 6364 (1984)
198. Brunner, H.; Miehling, W.: Monatsh. Chem. *115*, 1237 (1984)
199. Daniewski, A. R.; Kowalczyk-Przewloka, Tetrahedron Lett. *23*, 2411 (1982); J. Org. Chem. *50*. 2976 (1985)

200. Matlin, S. A.; Lough, W. J.; Chan, L.; Abram, D. M. H.; Zhou, Z.: J. Chem. Soc. Chem. Commun. *1984*, 1038
201. Nakamura, A.: Pure Appl. Chem. *50*, 37 (1978)
202. Elliott, M.; Janoo, N. F.: Chem. Soc. Reviews *7*, 473 (1978)
203. Arlt, D.; Jautelat, M.; Lantzsch, R.: Angew. Chem. *93*, 719 (1981); Angew. Chem. Int. Ed. Engl. *20*, 719 (1981)
204. Aratani, T.; Yoneyoshi, Y.; Fujita, F.; Nagase, T.: USSR 719491 (1980) (Chem. Abstr. *93*, 132129d (1980)); Ger. Offen. 2634663 Al (1977) (Chem. Abstr. 87, 68506w (1977)); Jpn. Kokai 77/17448 (1977) (Chem. Abstr. *87*, 102480v (1977))
205. Aratani, T.; Yoneyoshi, Y.; Nagase, T.: Tetrahedron Lett. *1975*, 1707
206. Laidler, D. A.; Milner, D. J.: J. Organomet. Chem. *270*, 121 (1984)
207. Imperial Chemical Industries Ltd.; Jpn. Kokai Tokkyo Koho JP 56/18956 (1981); Chem. Abstr. *95*, 115793q)
208. Holland, D.; Laidler, D. A.; Milner, D. J.: J. Mol. Catal. *11*, 119 (1981)
209. Laidler, D. A.; Milner, D. J.: Eur. Pat. Appl. EP 23075 (1981); Chem. Abstr. *95*, 132349s (1981). — (b) Imperial Chemical Industries Ltd.; Jpn. Kokai Tokkyo Koho JP 56/16498 (1981); Chem. Abstr. *95*, 97173f (1981)
210. Holland, D.; Laidler, D. A.; Milner, D. J.: Inorg. Chim. Acta *54*, L21 (1981)
211. Crosby, J.; Holland, D.; Laidler, D. A.; Milner, D.J.: Eur. Pat. Appl. EP 22608 (1981); Chem. Abstr. *95*, 114886k (1981)
212. Imperial Chemical Industries Ltd.: Jpn. Kokai Tokkyo Koho JP 56/16468 (1981); Chem. Abstr. *95*, 97174g (1981)
213. Hirai, H.; Matsui, M.: Agr. Biol. Chem. *40*, 169 (1974)
214. Krieger, P. E.; Landgrebe, J. A.: J. Org. Chem. *43*, 4447 (1978)
215. Doyle, M. P.; Dorow, R. L.; Terpstra, J. W.; Rodenhouse, R. A.: J. Org. Chem. *50*, 1663 (1985)
216. Izumi, Y.; Tai, A.: Stereo-differentiating Reactions; p. 137; Academic Press, New York (1977)
217. Domnin, I. N.; Zhuravleva, E. F.; Pronina, N. V.: Zh. Org. Khim. *14*, 2323 (1978)
218. Petiniot, N.; Anciaux, A. J.; Noels, A. F.; Hubert, A. J.; Teyssié, P.: Tetrahedron Lett. *1978*, 1239
219. Armstrong, R. K.: J. Org. Chem. *31*, 618 (1966)
220. Dolgii, I. E.; Okonnishnikova, G. P.; Nefedov, O. M.: Izv. Akad. Nauk SSSR, Ser. Khim. *1979*, 822
221. Shapiro, E. A.; Lun'kova, G. V.; Nefedov, A. O.; Dolgii, I. E.; Nefedov, O. M.: Izv. Akad. Nauk SSSR, Ser. Khim. *1981*, 2535
222. Maier, G.; Hoppe, M.; Reisenauer, H. P.; Krüger, C.: Angew. Chem. *94*, 445 (1982); Angew. Chem. Int. Ed. Engl. *21*, 437 (1982); Angew. Chem. Suppl. *1982*, 1061
223. Donaldson, W. A.; Hughes, R. P.: J. Am. Chem. Soc. *104*, 4846 (1982)
224. Zefirov, N. S.; Averina, N. V.; Boganov, A. M.; Koz'min, A. S.; Anufriev, V. S.; Tatevskii, V. M.; Yarovoi, S. S.; Shchelokov, R. N.; Baranovskii, I. B.: U.S.S.R. SU 825502 A1 (1981); Chem. Abstr. *96*, 6244m (1982)
225. Shapiro, E. A.; Romanova, T. N.; Dolgii, I. E.; Nefedov, O. M.: Izv. Akad. Nauk SSSR, Ser. Khim. *1984*, 2535
226. Zefirov, N. S.; Koz'min, A. S.; Lapteva, V. L.; Livantsov, M. V.; Lutsenko, I. F.; Proskurnina, M. V.; Tatevskii, V. M.; Fel'dblym, V. Sh.; Stepanov, G. A.: U.S.S.S.R. SU 852849 (1981); Chem. Abstr. *95*, 203413w (1981)
227. Mandel'shtam, T. V.; Ivanova, T. V.; Kharicheva, E. M.: Zh. Org. Khim. *12*, 761 (1976)
228. Anciaux, A. J.; Demonceau, A.; Noels, A. F.; Hubert, A. J.; Warin, R.; Teyssié, P.: J. Org. Chem. *46*, 873 (1981)
229. McKervey, M. A.; Russell, D. N.; Twohig, M. F.: J. Chem. Soc. Chem. Commun. *1985*, 491
230. Chan, L.; Matlin, S. A.: Tetrahedron Lett. *22*, 4025 (1981)
231. Constantino, A.; Linstrumelle, G.; Julia, S.: Bull. Soc. Chim. Fr. *1970*, 907, 912
232. Scott, L. T.; Minton, M. A.; Kirms, M. A.: J. Am. Chem. Soc. *102*, 6311 (1980)
233. McKervey, M. A.; Tuladhar, S. M.; Twohig, M. F.: J. Chem. Soc. Chem. Commun. *1984*. 129

234. Garst, M. E.; Roberts, V. A.: J. Org. Chem. *47*, 2188 (1982)
235. Iwata, C.; Miyashita, K.; Imao, T.; Masuda, K.; Kondo, N.; Uchida, S.: Chem. Pharm. Bull. *33*, 853 (1985)
236. Taylor, E. C.; Davies, H. M. L.: Tetrahedron Lett. *24*, 5453 (1983)
237. Saba, A.: Synthesis *1984*, 268
238. McNeil Laboratories, Inc.: Jpn. Kokai Tokkyo Koho 78/124263 (1978); Chem. Abstr. *90*, 137669c (1979)
239. Maryanoff, B. E.: J. Org. Chem. *44*, 4410 (1979)
240. Maryanoff, B. E.: J. Org. Chem. *47*, 3000 (1982)
241. Skvortsov, I. M.; Kolesnikov, S. A.; Samitov, Y. Y.; Shcherbakova, G. D.: Khim. Geterotsikl. Soedin. *1977*, 1090
242. Jefford, C. W.; Johncock, W.: Helv. Chim. Acta *66*, 2666 (1983)
243. Galeazzi, E.; Guzman, A.; Pinedo, A.; Saldana, A.; Torre, D.; Muchowski, J. M.: Can. J. Chem. *61*, 454 (1983)
244. Pellicciari, R.; Curini, M.; Spagnoli, N.; Ceccherelli, P.: Synthesis *1981*, 629
245. (a) Pellicciari, R.; Fringuelli, R.; Natalini, B.; Brucato, L.; Contessa, A. R.: Arch. Pharm. (Weinheim) *318*, 393 (1985). — (b) Pellicciari, R.; Curini, M.; Spagnoli, N.: Arch. Pharm. (Weinheim) *317*, 38 (1984)
246. Gillespie, R. J.; Porter, A. E. A. : J. Chem. Soc. Perkin Trans. 1, *1979*, 2624
247. Gillespie, R. J.; Porter, A. E. A.; Willmott, W. E.: Brit. UK Pat. Appl. 2013652; Chem. Abstr. *92*, 198259g (1980)
248. Porter, A. E. A.; Gillespie, R. J.; Willmott, W. E.: Brit. UK Pat. Appl. 2008570 (1979); Chem. Abstr. *92*, 76264n (1980)
249. Shostakovskii, V. M.; Zlatkina, V. L.; Vasil'vitskii, A. E.; Nefedov, G. M.: Izv. Akad. Nauk SSSR, Ser. Khim. *1982*, 2126
250. Gillespie, R. J.; Porter, A. E. A.; Willmott, W. E.: J. Chem. Soc. Chem. Commun. *1978*, 85
251. Cuffe, J.; Gillespie, R. J.; Porter, A. E. A.: J. Chem. Soc. Chem. Commun. *1978*, 641
252. Gillespie, R. J.; Porter, A. E. A.: J. Chem. Soc. Chem. Commun. *1979*, 50
253. Storflor, H.; Skramstad, J.; Nordenson, S.: J. Chem. Soc. Chem. Commun. *1984*, 208
254. Murray-Rost, P.; McManus, J.; Lennon, S. P.; Porter, A. E. A.; Rechka, J. A.: J. Chem. Soc. Perkin Trans. I *1984*, 713
255. Nefedov, O. M.; Shostakovsky, V. M.; Vasilvizky, A. E.: Angew. Chem. *89*, 674 (1977); Angew. Chem. Int. Ed. Engl. *16*, 646 (1977)
256. Nefedov, O. M.; Shostakovskii, V. M.; Vasil'vitskii, A. E.; Kravchenko, M. I.: Izv. Akad. Nauk SSSR, Ser. Khim. *1980*, 607
257. Rokach, J.; Adams, J.; Perry, R.: Tetrahedron Lett. *24*, 5185 (1983)
258. Adams, J.; Rokach, J.: Tetrahedron Lett. *25*, 35 (1984)
259. Adams, J.; Leblanc, Y.; Rokach, J.: Tetrahedron Lett. *25*, 1227 (1984)
260. Marchand, A. P.; in The chemistry of double-bonded functional groups (Patai, S. (ed.)); Suppl. A, Part 1, Chapter 7; Wiley, London (1977)
261. Bartnik, R.; Mlostón, G.: Synthesis *1983*, 924
262. Mara, A. M.; Singh, O.; Thomas, E. J.; Williams, D. J.: J. Chem. Soc. Perkin I, *1982*, 2169
263. Mehrotra, K. N.; Prasad, G.: Tetrahedron Lett. *1978*, 4179
264. Mehrotra, K. N.; Prasad, G.: J. Org. Chem. *17*, 2806 (1982)
265. Mehrotra, K. N.; Singh, S. B.; Singh, K. N.: Indian J. Chem. *21B*, 146 (1982)
266. Drapier, J.; Feron, A.; Warin, R.; Hubert, A. J.; Teyssié, P.: Tetrahedron Lett. *1979*, 559
267. Hubert, A. J.; Feron, A.; Warin, R.; Tessié, P.: Tetrahedron Lett. *1976*. 1317
268. Ibata, T.; Toyoda, J.: Chem. Lett. *1983*, 1453
269. (a) Hamaguchi, M.; Nagai, T.: J. Chem. Soc. Chem. Commun. *1985*, 190, 1319. — (b) Toyoda, J.; Ibata, T.; Tamura, H.; Ogawa, K.; Nishino, T.; Takebayashi, M.: Bull. Chem. Soc. Jpn. *58*, 2212 (1985)
270. Oida, S.; Yoshida, A.; Ohki, E.: Heterocycles *14*, 1999 (1980)
271. Landgrebe, J. A.; Iranmanesh, H.: J. Org. Chem. *43*, 1244 (1978)
272. Kharash, M. S.; Rudy, T.; Nudenberg, W.; Büchi, G.: J. Org. Chem. *18*, 1030 (1953)
273. De March, P.; Huisgen, R.: J. Am. Chem. Soc. *104*, 4952 (1982)
274. Husigen, R.; De March, P.: J. Am. Chem. Soc. *104*, 4953 (1982)

275. Alonso, M. E.; Chitty, A. W.: Tetrahedron Lett. *22*, 4181 (1981)
276. Alonso, M. E.; Jano, P.: J. Heterocycl. Chem. *17*, 721 (1980)
277. Wenkert, E.; Davis, L. L.; Mylari, B. L.; Solomon, M. F.; da Silva, R. R.; Shulman, S.; Warnet, R. J.; Ceccherelli, P.; Curini, M.; Pellicciari, R.: J. Org. Chem. *47*, 3242 (1982)
278. Chakraborti, A. K.; Ray, J. K.; Kundu, K. K.; Chakrabarty, S.; Mukherjee, D.; Ghatak, U. R.: J. Chem. Soc. Perkin I *1984*, 261
279. Agosta, W. C.; Wolff, S.: J. Org. Chem. *40*, 1027 (1975)
280. Taber, D. F.; Petty, E. H.: J. Org. Chem. *47*, 4808 (1982)
280a. Cane, D.; Thomas, P.: J. Am. Chem. Soc. *106*, 5295 (1984)
281. Sakai, K.; Fujimoto, T.; Yamashita, M.; Kondo, K.: Tetrahedron Lett. *26*, 2089 (1985)
281a. Taber, D. F.; Ruckle, R. E. Jr.: Tetrahedron Lett. *26*, 3059 (1985)
282. Ledon, H.; Linstrumelle, G.; Julia, S.: Tetrahedron Lett. *1973*, 25
283. Taber, D. F.; Petty, E. H.; Raman, K.: J. Am. Chem. Soc. *107*, 196 (1985)
284. Taber, D. F.; Raman, K.: J. Am. Chem. Soc. *105*, 5935 (1983)
285. Bateson, J. H.; Baxter, A. J. G.; Dickinson, K. H.; Hickling, R. I.; Ponsford, R. J.; Roberts, P. M.; Smale, T. C.; Southgate, R.: Spec. Publ. Royal Chem. Soc., Vol 38 (1981). — Ponsford, R. J.; Southgate, R.: J. Chem. Soc. Chem. Commun. *1979*, 846
286. Demonceau, A.; Noels, A. F.; Hubert, A.; Teyssié, P.: J. Chem. Soc. Chem. Commun. *1981*, 688
287. (a) Callot, H. J.; Metz, F.: Tetrahedron Lett. *23*, 4321 (1982); (b) Nouv. J. Chim. *9*, 167 (1985)
288. Demonceau, A.; Noels, A. F.; Hubert, A. J.; Teyssié, P.: Bull. Soc. Chim. Belges *93*, 945 (1984)
289. Scott, L. T.; DeCicco, G. J.: J. Am. Chem. Soc. *96*, 322 (1974)
290. Tomioka, H.; Itoh, M.; Yamakawa, S.; Izawa, Y.: J. Chem. Soc. Perkin Trans. II *1980*, 603
291. Mehrotra, K. N.; Singh, G. S.: Indian J. Chem. *20B*, 672 (1982)
292. Singh, S. B.; Mehrotra, K. N.: Can. J. Chem. *59*, 2475 (1981)
293. Singh, G. S.; Singh, S. B.; Mehrotra, K. N.: Bull. Chem. Soc. Jpn. *57*, 1667 (1984)
294. (a) Shapiro, E. A.; Romanova, T. N.; Dolgii, I. E.; Nefedov, O. M.: Izv. Akad. Nauk SSSR, Ser. Khim. *1983*, 1933. — (b) Moody, C. J.; Pearson, C. J.; Lawton, G.: Tetrahedron Lett. *26*, 3171 (1985)
295. Young, J.-J.; Sha, C.-K.: Heterocycles *22*, 2571 (1984)
296. Mattingly, P. G.; Miller, M. J.: J. Org. Chem. *46*, 1557 (1981)
297. Kametani, T.; Kanaya, N.; Mochizuki, T.; Honda, T.: Heterocycles *19*, 1023 (1982)
298. Brooks, G.; Howarth, T. T.; Hunt, E.: J. Chem. Soc. Chem. Commun. *1981*, 642
299. Cuffe, J.; Porter, A. E. A.: J. Chem. Soc. Chem. Commun. *1980*, 1257
300. Ratcliffe, R. W.; Salzmann, T. N.; Christensen, B. G.: Tetrahedron Lett. *21*, 31 (1980)
301. Kametani, T.; Honda, T.; Sasaki, J.; Terasawa, H.; Fukumoto, K.: J. Chem. Soc. Perkin Trans. I *1981*, 1884
302. Ueda, Y.; Damas, C. E.; Belleau, B.: Can. J. Chem. *61*, 1996 (1983)
303. Berges, D. A.; Spines, E. R.; Chan, G. W.; Kingsbury, W. D.; Kinzig, C. M.: Tetrahedron Lett. *22*, 3557 (1981)
304. Salzman, T. N.; Ratcliffe, R. W.; Christensen, B. G.; Bouffard, F. A.: J. Am. Chem. Soc. *102*, 6161 (1980)
305. Hirai, H.; Sawada, K.; Aratani, M.; Hashimoto, M.: Tetrahedron Lett. *25*, 5075 (1984)
306. Aratani, M.; Hirai, H.; Sawada, K.; Yamada, A.; Hashimoto, M.: Tetrahedron Lett. *26*, 223 (1985)
307. Shih, D. H.; Cama, L. D.; Christensen, B. G.: Tetrahedron Lett. *26*, 587 (1985)
308. Ona, H.; Uyeo, S.: Tetrahedron Lett. *25*, 2237 (1984)
309. Kametani, T.; Honda, T.; Nakayama, A.; Sakai, Y.; Mochizuki, T.; Fukumoto, K.: J. Chem. Soc. Perkin Trans. I *1981*, 2228
310. Mak, C.-P.; Mayerl, C.; Fliri, H.: Tetrahedron Lett. *24*, 347 (1983)
311. a) Umezawa, H.; Oono, M.; Ishihama, H.; Kyotani, Y.; Takahashi, Y.: E. P. 91239 (1983). — b) Okano, K.; Kyotani, Y.; Ishihama, H.; Kobayashi, S.; Oono, M.: J. Am. Chem. Soc. *105*, 7186 (1983)
312. a) Buynak, J. D.; Pajouhesh, H.; Lively, D. L.; Ramalakshmi, Y.: J. Chem. Soc. Chem.

Commun. *1984*, 948. — b) Buynak, J. D.; Rao, M. N.; Pajouhesh, H.; Chandrasekaran, R. Y.: Finn, K.; de Meester, P.; Chu, S. C.: J. Org. Chem. *50*, 4245 (1985)
313. Heck, J. V.; Szymonifka, M. J.; Christensen, B. G.: Tetrahedron Lett. *23*, 1519 (1982)
314. Heck, J. V.; Christensen, B. G.: Tetrahedron Lett. *22*, 5027 (1981)
315. Brennan, J.; Pinto, I. L.: Tetrahedron Lett. *24*, 4731 (1983)
316. Cama, L. D.; Christensen, B. G.: Tetrahedron Lett. *1978*, 4233
317. Salzman, T. N.; Ratcliffe, R. W.; Christensen, B. G.: Tetrahedron Lett. *21*, 1193 (1980). — (b) Ratcliffe, R. W.; Salzmann, T. N.: U.S. 4275207 (1981); Chem. Abstr. *95*, 150461b (1981)
318. Evans, D. A.; Sjogren, E. B.: Tetrahedron Lett. *26*, 3787 (1985)
319. Häbich, D.; Hartwig, W.: Tetrahedron *40*, 3667 (1984)
320. Yamamoto, S.; Itani, H.; Takahashi. H.; Tsuji, T.; Nagata, W.: Tetrahedron Lett. *25*, 4545 (1984)
321. Taylor, E. C.; Davies, H. M. L.: J. Org. Chem. *49*, 113 (1984)
322. Bartels, G.; Hinze, R. P.; Wullbrandt, D.: Liebigs Ann. Chem. *1980*, 168
323. Kulkowit, S.; McKervey, M. A.: J. Chem. Soc. Chem. Commun. *1981*, 616
324. Matlin, S. A.; Chan, L.: J. Chem. Soc. Chem. Commun. *1980*, 798
325. Ramsay, B. G.; Stoodley, R. J.: J. Chem. Soc. Chem. Commun. *1971*, 450
326. Sammes, P. G.: Chem. Reviews *76*, 113 (1976)
327. McKervey, M. A.; Ratananukul, P.: Tetrahedron Lett. *23*, 2509 (1982)
328. Ando, W.; Kondo, S.; Nakayama, K.; Ichibori, K.; Kohoola, H.; Yamato, H.; Imai, I.; Nakaido, S.; Migata, T.: J. Am. Chem. Soc. *94*, 3870 (1972)
329. Kirmse, W.; Kapps, M.: Chem. Ber. *101*, 994 (1968)
330. Kirmse, W.; Chiem, P. V.; Schurig, V.: Tetrahedron Lett. *26*, 197 (1985)
331. Martin, M. G.; Ganem, B.: Tetrahedron Lett. *25*, 251 (1984)
332. Nozaki, H.; Takaya, H.; Noyori, R.: Tetrahedron *22*, 3393 (1966)
333. Thijs, L.; Zwanenburg, B.: Tetrahedron *36*, 2145 (1980)
334. Smeets, F. L. M.; Thijs, L.; Zwanenburg, B.: Tetrahedron *36*, 3269 (1980)
335. Tamura, Y.; Takebe, Y.; Mukai, C.; Ikeda, M.: Heterocycles *15*, 875 (1981)
336. Benati, L.; Montevecchi, P. C.; Spagnolo, P.: J. Chem. Soc. Perkin Trans. I *1982*, 917
337. Crow, W. D.; Gosney, I.; Ormiston, R. A.: J. Chem. Soc. Chem. Commun. *1983*, 643
338. Pellicciari, R.; Curini, M.; Ceccherelli, P.: J. Chem. Soc. Perkin Trans. 1 *1977*, 1155
339. Tamura, Y.; Mukai, C.; Nakajima, N.; Ikeda, M.; Kido, M.: J. Chem. Soc. Perkin Trans. I *1981*, 212
339a. Kametani, T.; Kawamura, K.; Tsubuki, M.; Honda, T.: J. Chem. Soc. Chem. Commun. *1985*, 1324
340. Prassad, K.; Kneussel, P.; Schulz, G.; Stütz, P.: Tetrahedron Lett. *23*, 1247 (1982)
341. Prassad, K.; Schulz, G.; Mak, C.-P.; Hamberger, H.; Stütz, P.: Heterocycles *16*, 1305 (1981)
342. Prasad, K.; Stütz, P.: Heterocycles *19*, 1597 (1982)
343. Fliri, H.; Mak, C.-P.; Prasad, K.; Schulz, G.; Stütz, P.: Heterocycles *20*, 205 (1983)
344. Kametani, T.; Nakayama, A.; Itoh, A.; Honda, T.: Heterocycles *20*, 2355 (1983)
345. (a) Kametani, T.; Kanaya, N.; Mochizuki, T.; Honda, N.: Heterocycles *20*, 435 (1983). — (b) ibid. 455 (1983)
346. Kametani, T.; Kanaya, N.; Mochizuki, T.; Honda, T.: Tetrahedron Lett. *24*, 221 (1983)
347. Ernst, I.: Tetrahedron *33*, 547 (1977)
348. Ponsford, R. J.: Tetrahedron Lett. *21*, 2451 (1980)
349. Mak, C.-P.; Baumann, K.; Mayerl, F.; Mayerl, C.; Fliri, H.: Heterocycles *19*, 1647 (1982)
350. Khaskin, B. A.; Sheluchenko, O. D.; Torgasheva, N. A.: Zh. Obshch. Khim. *53*, 780 (1983)
351. Khaskin, B. A.; Tolmacheva, N. A.; Koroleva, T. I.: Zh. Obshch. Khim. *53*, 1219 (1983)
352. McKinnon, D. M.: Can. J. Chem. *61*, 1161 (1983)
353. Pellicciari, R.; Carini, M.; Ceccherelli, P.; Fringuelli, R.: J. Chem. Soc. Chem. Commun. *1979*, 440
354. Back, T. G.; Kerr, R. G.: Tetrahedron Lett. *23*, 3241 (1982); Tetrahedron *41*, 4759 (1985)
355. Moriarty, R. M.; Bailey III, B. R.; Prakash, O.; Prakash, I.: J. Am. Chem. Soc. *107*, 1375 (1985)
356. Hood, J. N. C.; Lloyd, D.; MacDonald, W. A.; Shepherd, T. M.: Tetrahedron *38*, 3355 (1982)

357. Oshima, T.; Nagai, T.: Tetrahedron Lett. *21*, 1251 (1980)
358. Shankar, B. K. R.; Shechter, H.: Tetrahedron Lett. *23*, 2277 (1982)
359. Shankar, B. K. R.: PhD Thesis, Ohio State Univ. (1981); Diss. Abstr. Int. *B 42*, 639 (1981)
360. Abel, E. W.; Pring, G. M.: Inorg. Chim. Acta *44*, L161 (1980)
361. Singh, S. B.: Indian J. Chem. *20B*, 810 (1981)
362. Kulkowit, S.; McKervey, M. A.: J. Chem. Soc. Chem. Commun. *1978*, 1069
363. Nakajima, M.; Anselme, J.-P.: J. Chem. Soc. Chem. Commun. *1980*, 796
364. Trahanovsky, W. S.; Robins, M. D.; Smick, D.: J. Am. Chem. Soc. *93*, 2086 (1971)
365. Wulfman, D. S.; Peace, B. W.; McDaniel Jr. R. S.: Tetrahedron *32*, 1251 (1976)
366. O'Neill, P. P.; Rooney, J. J.: J. Chem. Soc. Chem. Commun. *1972*, 104
367. Brady III, R. C.; Pettit, R.: J. Am. Chem. Soc. *102*, 6181 (1980)
368. Bock, H.; Tschmutova, G.; Wolf, H. P.: J. Chem. Soc. Chem. Commun. *1986*, 1068
369. Bock, H.; Wolf, H. P.: Angew. Chem. *97*, 411 (1985); Angew. Chem. Int. Ed. Engl. *23*, 418 (1985)
370. Bock, H.: Personal communication
371. Brady III, R. C.; Pettit, R.: J. Am. Chem. Soc. *103*, 1287 (1981)
372. Oudejans, J. C.; Kaminska, J.; van Bekkum, H.: Rec. Trav. Chim. Pays-Bas *102*, 537 (1983)
373. Regitz, M.; Weber, B.; Eckstein, U.: Liebigs Ann. Chem. *1979*, 1002
374. Ikota, N.; Takamura, N.; Young, S. D.; Ganem, B.: Tetrahedron Lett. *22*, 4163 (1981)
375. Pellicciari, R.; Fringuelli, R.; Ceccherelli, P.; Sisani, E.: J. Chem. Soc. Chem. Commun. *1979*, 959
376. Nagao, K.; Chiba, M.; Kim, S.-W.: Synthesis *1983*, 197
377. Pellicciari, R.; Fringuelli, R.; Sisani, E.; Curini, M.: J. Chem. Soc. Perkin Trans I *1981*, 2566
378. Yamamoto, U.; Yamazaki, S.; Kohashi, Y.; Matsukawa, A.; Murata, I.: Chem. Letters *1982*, 1843
379. Khbeis, S. G.: PhD Thesis, Univ. of Kaiserslautern (1982)
380. Regitz, M.; Khbeis, S. G.: Chem. Ber. *117*, 2233 (1984)
381. (a) Hoffmann, K.-L.; Regitz, M.: Tetrahedron Lett. *24*, 5355 (1983). — (b) Hoffmann, K.-L.; Maas, G.; Regitz, M.: Chem. Ber. *118*, 3700 (1985)
382. Nishino, K.; Yano, S.; Kohashi, Y.; Yamamoto, K.; Murata, I.: J. Am. Chem. Soc. *101*, 5059 (1979)
383. Murata, I.; Nishino, K.; Yano, S.; Kohashi, Y.; Yamamoto, K.: Croat. Chem. Acta *53*, 615 (1980)
384. Yano, S.; Nishino, K.; Nakasuji, K.; Murata, I.: Chem. Lett. *1978*, 723
385. Böhshar, M.; Heydt, H.; Regitz, M.: Chem. Ber. *117*, 3093 (1984)
386. Böhshar, M.; Maas, G.; Heydt, H.; Regitz, M.: Tetrahedron *40*, 5171 (1984)
387. Murata, I.; Nakasuji, K.: Topics Curr. Chem. *97*, 33 (1981)
388. Sekiguchi, A.; Tanikawa, H.; Ando, W.: Organometallics *4*, 584 (1985)
389. Yates, P.; Hambly, G. F.: Can. J. Chem. *57*, 1668 (1979)
390. Meier, H.; Zeller, K.-P.: Angew. Chem. *87*, 52 (1975); Angew. Chem. Int. Ed. Engl. *14*, 32 (1975)
391. Smith III, A. B.; Toder, B. H.; Richmond, R. E.; Branca, S. J.: J. Am. Chem. Soc. *106*, 4001 (1984)
392. Callot, H. J.; Tschamber, T.: J. Am. Chem. Soc. *97*, 6175 (1975)
393. Callot, H. J.; Schaeffer, E.: Nouv. J. Chim. *4*, 307 (1980)
394. Callot, H. J.: Tetrahedron Lett. *1972*, 1011; Bull. Soc. Chim. Fr. *1978*, 4387
395. Callot, H J.; Tschamber, T.: Bull. Soc. Chim. Fr. *1973*, 3192
396. (a) Johnson, A. W.; Ward, D.; Batten, P.; Hamilton, A. L.; Shelton, G.; Elson, C. M.: J. Chem. Soc. Perkin Trans. I *1975*, 2076. — (b) Johnson, A. W.; Ward, D.: ibid. *1977*, 720. — (c) Batten, P.; Hamilton, A. L.; Johnson, A. W.; Mahendran, M.; Ward, D.; King, T. J.: ibid. *1977*, 1623
397. Callot, H. J.; Schaeffer, E.: Tetrahedron Lett. *1977*, 239; J. Organomet. Chem. *145*, 91 (1978)
398. Callot, H. J.; Schaeffer, E.: Nouv. J. Chim. *4*, 311 (1980)
399. Cardin, D. J.; Cetinkaya, B.; Doyle, M. J.; Lappert, M. F.: Chem. Soc. Reviews *2*, 99 (1973)

400. Bethell, D.; Brown, K. C.: J. Chem. Soc. Perkin Trans. II *1972*, 895
401. Bethell, D.; Eeles, M. F.: J. Chem. Soc. Perkin Trans. II *1974*, 704
402. Bethell, D.; Eeles, M. F.; Handoo, K. L.: J. Chem. Soc. Perkin Trans. II *1979*, 714
403. Casey, C. P., in: Reactive Intermediates (Jones, M. Jr.; Moss, R. A., eds.), Vol. II, p. 135, Wiley, New York 1981
404. Dötz, K. H.: Angew. Chem. *96*, 573 (1984); Angew. Chem. Int. Ed. Engl. *23*, 587 (1984)
405. Fischer, H.; Schmid, J.; Märkl, R.: J. Chem. Soc. Chem. Commun. *1985*, 572
406. Schramm, K. D.; Ibers, J. A.: Inorg. Chem. *19*, 1231 (1980)
407. (a) Messerle, L.; Curtis, M. D.: J. Am. Chem. Soc. *104*, 889 (1982). — (b) Hillhouse, G. L.; Haymore, B. L.: J. Am. Chem. Soc. *104*, 1537 (1982)
408. Herrmann, W. A. et al.: Z. Naturforsch. *38b*, 1392 (1983)
409. Yarrow, D. J.; Ibers, J. A.; Tatsuno, Y.; Otsuka, S.: J. Am. Chem. Soc. *95*, 8590 (1973)
410. (a) Herrmann, W. A.: J. Organomet. Chem. *84*, C25 (1975). — (b) Herrmann, W. A.; Kriechbaum, G.; Ziegler, M. L.; Wülknitz, P.: Chem. Ber. *114*, 276 (1981)
411. Herrmann, W. A.: Angew. Chem. *86*, 556 (1974); Angew. Chem. Int. Ed. Engl. *13*, 599 (1974); Chem. Ber. *108*, 486 (1975)
412. Hong, P.; Nishi, N.; Sonogashira, K.; Hagihara, N.: J. Chem. Soc. Chem. Commun. *1972*, 993
413. Doyle, M. P.; Griffin, J. H.; da Conceição, J.: J. Chem. Soc. Chem. Commun. *1985*, 328
414. Casey, C. P.; Bertz, S. A.; Burkhardt, T. J.: Tetrahedron Lett. *1973*, 1421
415. Casey, C. P.; Shusterman, A. J.: Organometallics *4*, 736 (1985) and references cited therein
416. Bethell, D.; Handoo, K. L.; Fairhurst, S. A.; Sutcliffe, L. H.: J. Chem. Soc. Perkin Trans. II *1979*, 707
417. (a) D'yakonov, I. A.; Vitenberg, A. B.: Zh. Org. Khim. *3*, 1153 (1967). — (b) Shirafuji, T.; Yamamoto, Y.; Nozaki, H.: Tetrahedron *27*, 5353 (1971)

Author Index Volumes 101–137

Contents of Vols. 50–100 see Vol. 100
Author and Subject Index Vols. 26–50 see Vol. 50

The volume numbers are printed in italics

Bei Fragen zur Produktsicherheit wenden Sie sich bitte an:
If you have any questions regarding product safety,
please contact:

Walter de Gruyter GmbH
Genthiner Straße 13
10785 Berlin
productsafety@degruyterbrill.com